RENEWABLE ENERGY
FROM THE OCEAN

THE JOHNS HOPKINS UNIVERSITY
Applied Physics Laboratory Series
In Science and Engineering

SERIES EDITOR
JOHN R. APEL

WILLIAM H. AVERY AND CHIH WU.
Renewable Energy from the Ocean. A Guide to OTEC

BRUCE I. BLUM.
Software Engineering: A Holistic View

ROBERT M. FRISTROM.
Flame Structure and Processes

RICHARD A. HENLE AND BORIS W. KUVSHINOFF.
Desktop Computers: In Perspective

VINCENT L. PISACANE AND ROBERT C. MOORE (EDS.).
Fundamentals of Space Systems

RENEWABLE ENERGY
FROM THE OCEAN
A Guide to OTEC

William H. Avery

DIRECTOR (RETIRED)
Ocean Energy Programs
The Johns Hopkins University
Applied Physics Laboratory

and

Chih Wu

PROFESSOR
Department of Mechanical Engineering
United States Naval Academy

New York   Oxford
OXFORD UNIVERSITY PRESS
1994

Oxford University Press

Oxford   New York   Toronto
Delhi   Bombay   Calcutta   Madras   Karachi
Kuala Lumpur   Singapore   Hong Kong   Tokyo
Nairobi   Dar es Salaam   Cape Town
Melbourne   Auckland   Madrid

and associated companies in
Berlin   Ibadan

Published by Oxford University Press, Inc.
200 Madison Avenue, New York, New York 10016

Oxford is a registered trademark of Oxford University Press

Library of Congress Cataloging-in-Publication Data
Avery, William H.
Renewable energy from the ocean : a guide to OTEC /
William H. Avery and Chih Wu.
p.   cm. — (Johns Hopkins University/
Applied Laboratory series in science and engineering)
Includes bibliographical references and index.
ISBN 0-19-507199-9
1. Ocean thermal power plants   I. Wu, Chih, 1936–
II. Title.   III. Series.
TK1073.A94   1993
621.31'244 — dc20   93-20885

9 8 7 6 5 4 3 2 1

Printed in the United States of America
on acid-free paper

To the Pioneers

Foreword

All too often when societies face crises involving population, energy, resources, or environment they procrastinate until the problem can no longer be ignored. This phase is followed by a period in which the society turns to panaceas or palliatives. These cosmetic solutions mask and defer the problem until a new crisis erupts in a new mode with heightened potential for catastrophe. Happily, some real solutions to past problems have emerged. These solutions resulted from the work of problem solvers who recognized that there were no short-term answers.

The energy–environmental crisis with which the world is wrestling, and has wrestled for more than 2 decades, when coupled with the population explosion, will be the most chronic and crucial of these world crises. We are now aware that the panaceas of direct solar conversion, nuclear energy, and natural gas, even with the palliatives of energy conservation, are ineffective solutions to the onset of global warming. World competition surrounds the depleting pools of oil like a crowd of thirsting dinosaurs.

Once again the establishment machinery is in motion to pump new life into short-term solutions. But the readers of this book will recognize that long-term solutions are an absolute necessity. One such solution is the development of ocean thermal energy and the transportable nonpolluting fuels that can be obtained from this large renewable reservoir.

Dr. William Avery and his colleagues were among the few teams who, in 1973, asked themselves the scientific question—how best to meet the energy crisis— without any preconceived notion or prejudice about what the solution might be. Their long experience in the development of advanced technology made them familiar with the scientific approach to innovation that strips away the conventional wisdom of the past and reduces the problem to one of fundamental physics and geophysics. This approach led to the identification of ocean thermal energy conversion as a plentiful and practical way of alleviating world energy needs by utilizing solar energy stored in the vast tropical oceans that is available night and day, throughout the year.

Once identified, there was the analytical and experimental documentation of practical ways of transferring this thermal energy into electrical energy and from electrical energy to chemical energy in the form of hydrogen, ammonia, or methanol. Each step required consideration of economics and environment as well as technology with recognition that solutions must be economically viable and environmentally sustainable.

As early as 1975 Avery and his colleagues had identified the sea-based manufacture of ammonia as an economically and environmentally feasible energy option, but the implementation of this option was held back by the availability of "low-cost fuel." The economic and environmental impacts of dependence on Middle East fuel supplies were disregarded. As a consequence the team was reduced in size and then all but abandoned. Avery and Wu have pooled the knowledge gained and in this book have carefully and meticulously documented the studies and experimental results of the small band of investigators who persisted in the exploration of this vital resource.

Who would deny that we are once again in the middle of an energy/environmental/population crisis? Now, however, a critical mass of investigators employing the waters of Hawaii as their laboratory have developed the proof that ocean thermal energy and its by-products are an important element in a rational and environmentally sustainable solution. This important work is being recognized. Relevant pilot projects now exist in Britain and Hawaii, and developments are under serious consideration in the Cook Islands, the Marshall Islands, and the Cape Verde Islands. The entrepreneurs who have independently entered the development process will soon be joined by others. This book will be their bible.

John P. Craven
Former Dean of Marine Engineering
University of Hawaii

Preface

The upper layers of the tropical oceans are a vast reservoir of warm water that is held at a temperature near 27°C (80°F) by a balance between the absorption of heat from the sun and the loss of heat by evaporation, convection, and long-wavelength radiation. On an average day, the water near the surface absorbs more heat from the sun in one square mile of ocean area than could be produced by burning 7000 barrels of oil. For the whole tropical ocean area, the solar energy absorbed per day by the surface waters is more than 10,000 times the heat content of the daily oil consumption of the United States. A practical and economical technology for converting even 0.01% of the absorbed solar energy into electricity or fuel in a form suitable for delivery to consumers on land could have a profound impact on world energy availability and economics.

The technology called ocean thermal energy conversion (OTEC) fulfills the technical and economic requirements for productive use of the solar energy continually absorbed by the oceans. However, the unfamiliarity of the OTEC concept and misconceptions about the difficulty of implementing it have slowed the development of interest by industry and government.

This book presents the scientific and engineering fundamentals of OTEC technology in a context that will make clear how this technology can be applied in a practical way to a future world economy based on renewable, nonpolluting energy sources. The status of OTEC development programs is presented, and technical and cost information are provided from which valid estimates can be made of OTEC plant investment requirements and delivered costs of OTEC fuels, chemicals, and electric power. The book shows that the technology base for OTEC is sufficiently well established for large-scale demonstration plants to be built as forerunners to commercial plants and plantships that will be economically attractive and environmentally benign. The manufacture of OTEC plants and plantships can draw on the talents and facilities available in the U.S. chemical and power production industries and in shipyards, steel companies, and concrete construction firms. OTEC commercialization offers many new employment opportunities, including large requirements for low-skilled laborers.

OTEC-produced fuels can be an inexhaustible substitute for petroleum-based fuels and could ultimately be produced in quantities and at costs that would make it practical for this technology to supply a large part of the energy needs of the world.

After an introductory overview, the book reviews the history of OTEC development and presents a discussion of the scientific and engineering fundamentals of the technology, including the technical options for electrical and chemical energy delivery. The engineering status, projected costs (which are favorable in comparison with other renewable energy alternatives), manufacturing capabilities, and potential markets are then discussed. The book concludes with a critical examination of the economic implications and prospects of OTEC, including the environmental and social impacts of its large-scale commercial development.

Acknowledgments

A few farseeing and dedicated individuals created programs in the early 1970s that expanded rapidly to establish OTEC technology as a renewable energy resource with the potential to make major contributions to world requirements for safe, benign, and inexhaustible energy sources. The senior author (WHA) has been privileged to know and have the support of the early American investigators, and to play a part in the efforts of the large teams of scientists and engineers who worked to bring OTEC into operation in the early 1980s. Unfortunately only a small number of those whose contributions were vital to the success of the technical program can be given credit for their contributions. Since team efforts were the rule, it is appropriate to recognize individuals in conjunction with the organizations that supported their work. To all of them we are grateful.

J. Hilbert Anderson, Sea Solar Power, Inc.: His innovative engineering concepts and tireless enthusiasm brought OTEC to the attention of the engineering community and inspired future efforts worldwide.

Clarence Zener, Carnegie–Mellon University: His physical insight into energy options led to his interest in OTEC, and his international prestige created significant support for OTEC within the National Science Foundation, ERDA, and, finally, DOE.

Robert Cohen, NSF, ERDA, and DOE: He was able to maintain funding for OTEC despite unenthusiastic support from the federal bureaucracy and the energy establishment.

Marvin Pitkin, MARAD: His recognition of the commercial promise of OTEC brought funding from the U.S. Maritime Administration (MARAD) for vital studies of the shipbuilding and commercialization requirements.

Eugene Schorsch: His interest in OTEC gained the support of the management of the Sun Shipbuilding and Drydock Company, who, at their expense, assigned a team to work with JHU/APL to define the marine engineering requirements and conceptual design of a concrete barge that would minimize OTEC shipbuilding costs.

John Craven, University of Hawaii: His deep understanding of ocean science and engineering combined with political prowess induced the government of Hawaii to recognize the unique capabilities of Hawaii to become a world center for renewable energy development, with particular emphasis on OTEC. This led to the establishment of the Natural Energy Laboratory in Hawaii.

The Lockheed team: Under the leadership of James Wentzel and Lloyd Trimble, Lockheed, with support from the State of Hawaii and industrial partners, built and operated the Mini-OTEC vessel with private funding, demonstrating for the first time net power production from solar thermal energy stored in the oceans.

The TRW team: Under the leadership of Arthur Butler and Robert Douglass, the TRW team conducted the program that developed and performed the at-sea demonstration of the OTEC-1 heat exchanger test vehicle.

ABAM Engineers: Robert Mast and his associates generously provided engineering and cost data acquired in the design and construction of the concrete ARCO LPG vessel, information that was vital to the design of the 40-MWe OTEC baseline demonstration vessels.

SOLARAMCO: J. F. Babbitt graciously furnished detailed information for the design of an ammonia plant suitable for installation on an OTEC plantship, based on his wide background in ammonia plant construction and operation.

Brown and Root Development, Inc. (BARDI), EBASCO Engineering, and Rocketdyne International Corp.: Teams from these organizations, led by J. D. Shoemaker, W. G. Niemeyer, and Farouk Mian of BARDI, Malcolm Jones of EBASCO, and Arthur Kohl of Rocketdyne, cooperated in the conceptual design of the 200-MWe OTEC methanol plantship, generously making available much of their relevant in-house development and construction experience.

The DOE Ocean Energy team: William Richards, William Sherwood, Carmine Castellano, and Peter Ritzcoven of DOE, and Abraham Lavi, on leave from Carnegie–Mellon University, recognized the potential of OTEC to make large contributions to world energy needs and were able temporarily to manage a program in DOE that provided solutions to the perceived technological problems. As a consequence, OTEC is ready to proceed to large-scale demonstration when financing becomes available.

The pioneering work of the French and Japanese teams is reviewed in Chapter 3.

We are deeply indebted to our associates at JHU/APL and the Naval Academy whose engineering understanding and program management capabilities provided insights essential to the success of the program and the writing of this book. Particular credit is due to Gordon Dugger (deceased), Dennis Richards, E. J. Francis, R. W. Blevins, James George, W. B. Shippen, and Peter Pandolfini of JHU/APL. We are also deeply grateful to Michael McCormick of the U.S. Naval Academy for his support and counsel.

The vital help of those concerned with the production of the book is acknowledged with deep gratitude: Jeffrey Robbins and Anita Lekhwani of the Oxford University Press; JHU/APL management under the direction of Carl Bostrom; John Apel, Laurie Fetcher, Carol West, Terry Gordon, Murrie W. Burgan, and Ruth Novick of JHU/APL; P. R. Wells and C. A. Mayr of the United States Naval Academy; and G. A. Bare of the United States Air Force.

Finally, both authors are deeply aware that nothing of value would have been accomplished without the enthusiastic, devoted, untiring, and patient support of their wives, for whom no words of gratitude can be sufficient.

Chapter 5 and Sections 9.2 and 9.3 of Chapter 9 were written by C. Wu. The remaining text was written by W. H. Avery.

Of all . . . inorganic substances . . . water is the most wonderful . . . if we think of it as the source of all the changefulness and beauty which we have seen in clouds; then as the instrument by which the earth we have contemplated was modelled into symmetry, and its crags chiselled into grace; then as, in the form of snow, it robes the mountains it has made, with that transcendent light which we could not have conceived if we had not seen; then as it exists in the form of the torrent . . . in the iris which spans it, in the morning mist which rises from it in the deep crystalline pools which mirror its hanging shore, in the broad lake and glancing river; finally, in that which is to all human minds the best emblem of unwearied, unconquerable power, the wild, various, fantastic, tameless unity of the sea; what shall we compare to this mighty, this universal element, for glory and for beauty? or how shall we follow its eternal changefulness of feeling? It is like trying to paint a soul.

John Ruskin

Water is the noblest of the elements.

Pindar

CONTENTS

FIGURES

TABLES

RENEWABLE ENERGY
FROM THE OCEAN

1

INTRODUCTION AND OVERVIEW

1.1 THE OCEAN THERMAL ENERGY RESOURCE

The sunlight that falls on the oceans is so strongly absorbed by the water that effectively all of its energy is captured within a shallow "mixed layer" at the surface, 35 to 100 m (100 to 300 ft) thick, where wind and wave actions cause the temperature and salinity to be nearly uniform. In the regions of the tropical oceans between approximately 15° north and 15° south latitude, the heat absorbed from the sun warms the water in the mixed layer to a value near 28°C (82°F) that is nearly constant day and night and from month to month. The annual average temperature of the mixed layer throughout the region varies from about 27°C to about 29°C (80 to 85°F).

Beneath the mixed layer, the water becomes colder as depth increases until at 800 to 1000 m (2500 to 3300 ft), a temperature of 4.4°C (40°F) is reached. Below this depth, the temperature drops only a few degrees further to the ocean bottom at an average depth of 3650 m (12,000 ft). Thus, a huge reservoir of cold water exists below a depth of 3000 ft. This cold water is the accumulation of ice-cold water that has melted from the polar regions. Because of its higher density and minimal mixing with the warmer water above, the cold water flows along the ocean bottom from the poles toward the equator, displacing the lower-density water above. The result of the two physical processes is to create an oceanic structure with a large reservoir of warm water at the surface and a large reservoir of cold water at the bottom, with a temperature difference between them of 22 to 25 degrees Celsius (40 to 45 degrees Fahrenheit); this structure is found throughout the entire area of the tropical oceans where the depth exceeds 1000 m (3300 ft). The temperature difference is maintained throughout the year, with variations of a few degrees Fahrenheit due to the seasonal effects and weather, and day-to-night changes on the order of one degree.

The ocean thermal energy conversion (OTEC) process uses this temperature difference to operate a heat engine, which produces electric power. Calculations show that OTEC plants sited in the tropical oceans can be operated continuously, without significant environmental effects, if the power generated is limited to approximately 0.5 MWe (net) per square mile of ocean surface (0.19 MWe/km^2). This amount of power corresponds to the conversion of 0.07% of the average absorbed solar energy to electricity.

A map prepared by Wolff for the U.S. Department of Energy showing the temperature difference in the tropical oceans between the surface and a depth of 1000 m (3287 ft) is presented in Fig. 1-1. The regions most suitable for OTEC operation, in which the change in temperature (ΔT) exceeds 22 degrees Celsius (40 degrees Fahrenheit), have a total area of approximately 60 million km^2 (23 million miles^2). Thus, if floating OTEC power plants were uniformly spaced throughout the

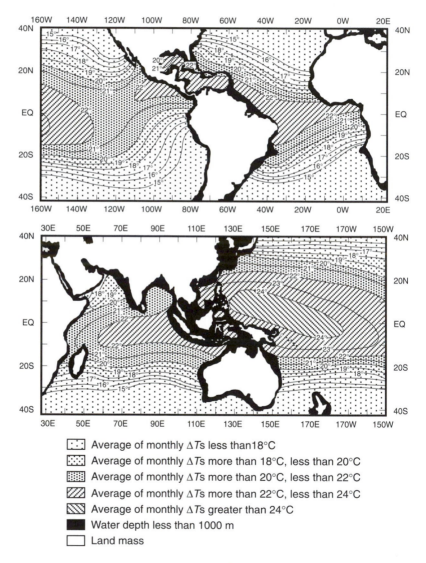

	Average of monthly ΔTs less than18°C
	Average of monthly ΔTs more than 18°C, less than 20°C
	Average of monthly ΔTs more than 20°C, less than 22°C
	Average of monthly ΔTs more than 22°C, less than 24°C
	Average of monthly ΔTs greater than 24°C
	Water depth less than 1000 m
	Land mass

Fig. 1-1. Ocean temperature resource for OTEC (U.S. Department of Energy, Assistant Secretary Energy Technology, Division of Solar Technology).

useful tropical ocean area, the total power generated on board would exceed 10 million MWe; if each plant generated 200 MWe of net power, the plants would be spaced 32 km (20 miles) apart. For comparison, the total U.S. electricity-generating capacity in 1987 was 165 thousand MWe.

1.2 DESIGN REQUIREMENTS FOR OTEC SYSTEMS

The OTEC power plant uses the heat in the surface water of the tropical oceans to generate electricity for on-land facilities or for ship-mounted plants that produce fuels or other products.

The major subsystems of an OTEC system, shown schematically in Fig. 1-2, are:

1. A heat engine or power plant, including heat exchangers, turbines, electric generator, water and working-fluid pumps, and associated piping and controls;
2. A water ducting system, which includes a cold-water pipe (CWP) through which water is drawn from a depth of about 900 to 1000 m (3000 ft) and warm-water inlet and exhaust flow pipes;
3. An energy transfer system to carry the energy produced on board to on-land users as either electricity or fuel;
4. A position-control system, including propulsion or mooring equipment, controls, and standby power systems; and
5. A platform to support the power plant, ducting systems, auxiliary ship equipment, and accommodations for operating personnel, along with safety equipment and other habitability requirements (on-land buildings may serve some of these functions for near-shore or shore-based systems).

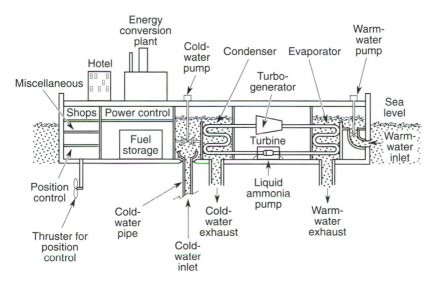

FIG. 1-2. Diagram of closed-cycle OTEC plantship.

Where deep water exists near the shore, for example, at tropical islands and coral atolls and at some continental sites, OTEC plants may be on shore or shelf mounted.

The following sections in this chapter present brief discussions of the general design features of the major OTEC subsystems to show the basic relationships among engineering and thermodynamic requirements, component design, and system construction and performance that are necessary for a preliminary assessment of OTEC technical feasibility and cost. Specific engineering information on the design and construction of systems and components, and detailed cost and performance estimates are presented in succeeding chapters.

1.2.1 OTEC Power Systems

OTEC power systems may be divided into two categories: closed cycle and open cycle. In closed-cycle operation, the working fluid is conserved (i.e., pumped back to the evaporator after condensation), as shown in Fig. 1-2.

In the open-cycle system, the working fluid is vented after use, as shown in Fig. 1-3. In this case, the working fluid is water vapor. Warm seawater is pumped into a chamber in which the pressure is reduced by a vacuum pump to a value low enough to cause the water to boil. The low-pressure steam, after passing through a turbine, is condensed by cold water in a similar chamber and is then discharged into the ocean. Instead of being condensed by direct contact with cold water, the vapor may be directed to a heat exchanger cooled by the cold seawater. In this case, the condensed vapor becomes a source of fresh water.

A complete closed-cycle OTEC system (Mini-OTEC) was tested by Lockheed at sea in Hawaii in 1979 at a 50-kWe gross power output. A cold-water pipe (CWP) of 0.71 m diameter (28 in.) and 670 m length (2200 ft) was successfully deployed and operated in this experiment. A diagram of the at-sea deployment is shown in

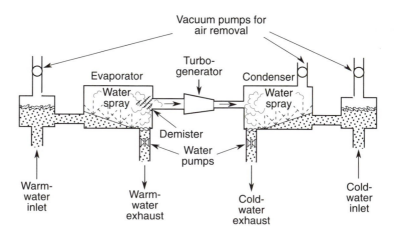

FIG. 1-3.   Diagram of open-cycle OTEC power system.

Fig. 1-4. In early 1981, an ocean-based OTEC power system (OTEC-1), including a 670-m (2200-ft) CWP of 2.55 m diameter (8.4 ft), and cold- and warm-water ducting equipment, was deployed and operated at 1-MWe scale, also in Hawaii (Castellano, 1981). A photograph of the ship and a layout diagram are shown in Fig. 1-5. These successful tests demonstrate that closed-cycle OTEC is ready for the next major step: scale-up to a size large enough to provide detailed engineering data on performance and cost for the design of commercial OTEC systems.

Open-cycle OTEC systems are still in the research and development phase but offer promise of competitive performance, particularly if fresh water, and/or mariculture products, produced along with or instead of electric power, are marketable products at the plant site.

Since the purpose of this overview chapter is to give the reader an understanding of the general features of OTEC design and operation, along with cost information that will allow evaluation of the appropriate role of OTEC in future U.S. and world energy development, discussion of open-cycle OTEC is postponed until Chapter 5.

1.2.1.1 Closed-Cycle OTEC Process

The discussion in this section is confined to OTEC power systems that use ammonia as a working fluid. The characteristics of heat-engine cycles based on alternative fluids, which could be preferred in some situations, are discussed in Chapter 4.

A schematic of a closed-cycle OTEC power system was shown in Fig. 1-2. A flow diagram of a typical closed cycle is presented in Fig. 1-6 (George and Richards, 1980). Water at 27.9°C (82.2°F), drawn from just below the ocean surface, warms a boiler (evaporator) that contains liquid ammonia. The liquid ammonia boils, producing vapor at a pressure of 940 kPa (9.3 atm). The vapor expands as it passes through a turbine, producing power to drive an electric generator. The vapor then

FIG. 1-4. Mini-OTEC deployed near Kailua-Kona, Hawaii (Courtesy Lockheed Missile and Space Co.).

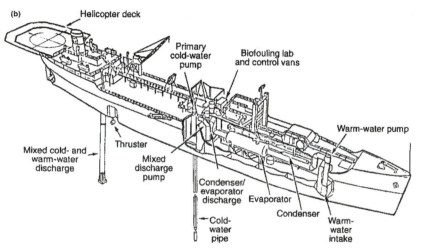

FIG. 1-5. (a) OTEC-1 heat exchanger test vessel deployed near Kalua-Kona, Hawaii (Courtesy C. Castellano). (b) Layout diagram for OTEC-1 subsystems (Castellano, 1981).

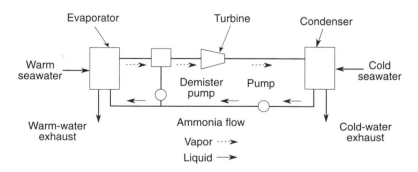

FIG. 1-6. Flow diagram of typical OTEC heat engine power cycle.

enters a condenser cooled by water at 4.0°C (39.2°F), which is drawn from a depth of approximately 1000 m (3300 ft).

The vapor pressure of ammonia at the temperature in the condenser, 10.3°C (50.6°F), is 620 kPa (6.1 atm). Thus, a significant pressure difference is available to drive the turbine and generate electric power.

As the ammonia condenses, the liquid is pumped back into the evaporator to complete the cycle, producing continuous power generation as long as the warm water and cold water continue to flow.

The closed OTEC cycle is basically the same as the conventional Rankine cycle employed in steam engines, in which the steam is condensed and returned to the boiler after driving a piston or steam turbine. OTEC differs by using a different working fluid and lower pressures and temperatures.

The closed OTEC cycle is also essentially the same as that used in commercial refrigerators, except that the OTEC cycle is run in the opposite direction so that a hot source and a cold heat sink are used to produce electric power, rather than the reverse. In refrigeration systems the temperatures and pressures of the working fluid are almost the same as those of the OTEC cycle, and the same working fluid may be used. Any of the common refrigerant fluids (freons) would be suitable for OTEC operation, but ammonia is the preeminent choice. Thus, much of the large body of technical information, engineering data, subsystems, and components developed over the years for the manufacture of refrigeration and cryogenic systems is directly applicable to OTEC and provides a good basis for the prediction of OTEC performance and costs.

1.2.1.2 OTEC Efficiency and Cost Implications

There is a theoretical limit, $\eta_{(max)}$, to the maximum efficiency of an OTEC system in converting heat stored in the warm surface water of the tropical oceans into mechanical work. As first defined by the French engineer, Sadi Carnot (1824),

$$\eta_{(max)} = \frac{T_w - T_c}{T_w},$$
(1.2.1)

where $\eta_{(max)}$ = Carnot efficiency; T_w = absolute temperature of the warm water; T_c = absolute temperature of the cold water.

For the ocean regions most suitable for OTEC operation, the annual average surface temperature is 26.7 to 29.4°C (80 to 85°F). Cold water at 4.4°C (40°F) or below is available at depths of about 900 m (3000 ft). Thus, the maximum OTEC thermal efficiency will be 7.5 to 8% (22/295 to 25/298); however, this ideal efficiency, even without unavoidable reductions caused by friction and heat losses, could be attained only at infinitesimal rates of power production.

Maximum power production per unit water flow requires both a high rate of heat transfer in the heat exchangers and a large pressure difference across the turbine. Without discussing details at this point, we may state that maximum power output will occur when the total temperature difference (ΔT) between the warm and cold water is divided roughly equally between (1) the ΔT required to promote a large difference in vapor pressure in the ammonia between the evaporator and condenser, and (2) the ΔTs required to induce a high heat transfer rate from warm water to ammonia in the evaporator and from ammonia to cold water in the condenser. An analysis is presented in Wu (1987).

Only the ΔT associated with the pressure difference across the turbine is associated with the production of mechanical work; therefore, in practical operation of an OTEC power system, the gross power efficiency is only about half the Carnot limit. This requirement reduces the maximum practical efficiency of OTEC gross power production to 3.5 to 4.0%.

Finally, analysis shows that 25 to 30% of the gross electric power generated by a floating OTEC plant operating with a ΔT of 22 degrees Celsius (40 degrees Fahrenheit) will be required to operate the water and ammonia pumps and to supply power to meet auxiliary needs for plant operation and ship station keeping. Thus, the net efficiency of conversion of the thermal energy stored in the ocean surface waters to net electric energy available at the on-board bus bar will be between 2.5 and 3%.

The small value of the OTEC net efficiency compared with efficiencies typical of heat engines that operate at high temperatures and pressures has led to assumptions by some energy planners that OTEC power would be too costly to compete with other methods of power production; however, design studies supported by research and development testing show that commercial production of OTEC will be cost effective, for the following reasons:

1. OTEC involves no costs for fuel or for its preparation and storage.
2. The low pressures and temperatures of the OTEC cycle permit major reductions in component costs compared with conventional power systems, which are dependent on high temperature, high pressure, and special materials for efficient operation.
3. OTEC reliability of operation and freedom from maintenance will be comparable to commercial refrigeration systems, which typically operate continuously for many years without shutdown. Operating factors of 85 to 95% are projected, compared with 50 to 70% for coal and nuclear plants.
4. OTEC operation will be safe and environmentally benign. Ocean basing frees OTEC from the taxes and delays imposed on conventional power plants by local governments opposed to power generation in their neighborhoods.
5. OTEC construction will employ facilities, procedures, and components that are standard in marine construction. Construction time will be 2.5 to 3 years, rather than the 6 to 10 years typical of coal and nuclear plants.

The preceding reasons combine to make the predicted costs of OTEC power attractive compared with other practical energy supply alternatives of comparable safety and environmental acceptability.

Because the thermal efficiency is low, OTEC operation requires the flow of large amounts of ocean water through the power system per kilowatt of power generated. This requirement in turn implies large heat exchangers, which must have a low unit cost if the total system is to meet competitive cost standards. The technical and design requirements to satisfy this criterion are discussed in the next section.

1.2.1.3 Basic Heat Exchanger Design Factors

The rate of heat transfer from one fluid through a dividing wall to another fluid is governed by the following general expression:

$$\dot{Q} = UA(T_1 - T_2), \qquad (1.2.2)$$

where $\dot{Q}$ = heat transfer rate; U = specific heat transfer coefficient (kW/m² °C); A = surface area (m²); T_1 = temperature of first fluid (°C); T_2 = temperature of second fluid (°C).

For the heat transfer rate from warm water to ammonia in the evaporator,

$$\dot{Q}_e = U_e A_e (T_{ww} - T_{ae}), \qquad (1.2.3)$$

where $\dot{Q}_e$ = evaporator heat transfer rate; U_e = heat transfer coefficient for the evaporator; A_e = heat transfer area of the evaporator; T_{ww} = warm-water temperature; T_{ae} = ammonia temperature in the evaporator.

For the heat flow from ammonia to cold water in the condenser,

$$\dot{Q}_c = U_c A_c (T_{ac} - T_{cw}), \qquad (1.2.4)$$

where $\dot{Q}_c$ = heat transfer rate for the condenser; U_c = heat transfer coefficient for the condenser; A_c = condenser area; T_{ac} = ammonia temperature in the condenser; T_{cw} = cold-water temperature.

Since the heat exchangers (HX) are of finite size and heat is transferred as water flows over the HX surfaces, the temperature of the water leaving the evaporator is lower than the temperature at the inlet. The ammonia boiling and condensation temperatures are determined by the pressures in the evaporator and condenser and are approximately constant.

The rate of heat transfer from water to ammonia varies with position in the HX; therefore, appropriate average values must be used for these quantities. Further elucidation is provided in Chapter 4.

The value of the coefficient U is dependent on the thermodynamic properties of the fluids, the flow velocities, the fraction of ammonia vaporized at a particular point in the heat exchanger, the pressures, the composition and construction of the separator material that forms the bounding surface between the fluids, the character of the surfaces on both sides of the separator (roughness, presence or absence of scale, biofouling, etc.), and the detailed design and dimensions of the heat exchanger. An analysis of the heat transfer process is given in Section 4.1.

As may be seen from Fig. 1-7, if a quantity of heat, Q, flows from water to ammonia, it must pass first from the water interior to the separating surface.[1] This requires a temperature difference, $T_w - T_{sb}$, the magnitude of which depends on the

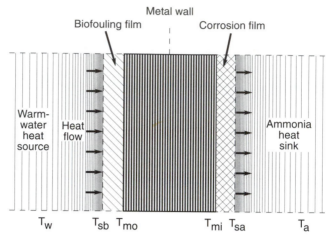

Fɪɢ. 1-7.   Heat transfer process in OTEC heat exchanger.

character of the flow at that point. This, in turn, depends on the local geometry of the flow passage. The heat then flows through any buildup of scale plus biofouling on the water side of the surface, where a further temperature drop, T_{sb}–T_{mo}, occurs. Similarly, the heat passes through the other resistances until it is finally absorbed by the bulk ammonia flow. We then have:

$$\dot{Q} = h_w(T_w - T_{sb}) = h_{sb}(T_{sb} - T_{mo}) = (k/x)(T_{mo} - T_{mi})$$
$$= h_{sa}(T_{mi} - T_{sa}) = h_a(T_{sa} - T_a) = U(T_w - T_a),$$

$$(1.2.5)$$

where T_w = bulk temperature of water; T_{sb} = surface temperature of biofouling-scale layer; T_{mo} = temperature of outside HX metal surface; T_{mi} = temperature of inside HX metal surface; T_{sa} = surface temperature of scale on ammonia side; T_a = bulk temperature of ammonia; h_w, h_{sb}, h_{sa}, and h_a = corresponding heat transfer coefficients; k = thermal conductivity of HX wall separating water and ammonia; x = thickness of HX wall.

Also, since

$$T_w - T_a = (T_w - T_{sb}) + (T_{sb} - T_{mo})$$
$$+ (T_{mo} - T_{mi}) + (T_{mi} - T_{ac}) + (T_{ac} - T_a),$$

$$(1.2.6)$$

it follows that

$$1/U = 1/h_w + 1/h_{sb} + x/k + 1/h_{sa} + 1/h_a.$$

$$(1.2.7)$$

Because the physical processes are so complex, it is generally not possible to predict accurate values of U from basic principles; however, many empirical formulas have been derived from experimental data, which may be used to relate experimental results and to provide a basis for design comparisons and performance predictions. These are discussed in detail in Chapter 4. It is of interest here to use

Eq. (1.2.7) to assess the relative importance of the terms on the right-hand side of the equation. From this assessment, one can estimate the effect of changes in the coefficients that could result from design improvements, changes in operating conditions, or degraded heat transfer due to biofouling or scaling.

As a baseline, values of the heat transfer coefficients in SI units typical of the current state of the art are listed here (English units are in parentheses):

h $\quad$ $= $ W/m^2 °C $\quad = $ (Btu/ft^2 h °F)

x $\quad$ $= $ m $\qquad\quad = $ (ft)

k $\quad$ $= $ W/m °C $\quad = $ (Btu/ft h °F)

$1/h_w$ $\quad = 0.00018$ (0.001)

$1/h_{sb}$ $\quad = 0$ to 0.000088 (0 to 0.0005)

x_{Al} $\quad = 0.0015$ (0.005); $\; x_{Ti} = 0.00076$ (0.0025)

$k_{Al\,5052}$ $\quad = 138$ (80); $\; k_{Al\,3003} = 156$ (90); $k_{Ti} = 17(10)$

$1/h_{sa}$ $\quad = 0.0000$ (0.0000)

$1/h_a$ $\quad = 0.000088$ (0.0005)

Ti $\quad\;\;\; = $ titanium

Al $\quad\;\;\; = $ aluminum (numerical subscripts are alloy designations).

With these values, we have for an aluminum HX with no biofouling,

$1/U$ $\quad = 0.00018 + 0.00000 + 0.00001 + 0.00000 + 0.000088$

$\quad\quad\;\; = 0.000278$

U $\quad\;\; = 3600$ W/m^2 °C (630 Btu/ft^2 h °F).

Substitution of more optimistic or more pessimistic values in the formula leads to the changes in performance shown in Table 1-1. The table shows that major improvements in power output will be possible with foreseeable improvements in heat transfer rates, and that biofouling effects will be minor if the fouling coefficient is maintained below an average value of 0.0001, as expected from experimental results.

1.2.1.4 Heat Exchanger Dimensions

The low heat transfer rates typical of OTEC systems impose a requirement for large surface areas in the heat exchangers. For typical values of the heat transfer

Table 1-1. Sensitivity of OTEC power system performance
to changes in baseline heat transfer coefficients

Change in overall heat transfer rate relative to baseline	% gain
Double water-side heat transfer coefficient	47
Double ammonia-side heat transfer coefficient	19
Double both coefficients	92
Double both coefficients (fouling factor 0.0001)	71
Double both coefficients (fouling factor 0.0005)	19
Replace aluminum with titanium	−12
Allow fouling factor to reach 0.0001 (time average)	− 3
Allow fouling factor to reach 0.0005 (time average)	−16

coefficients and heat exchanger ΔTs of approximately 5.6 degrees Celsius (10 degrees Fahrenheit), the net power delivered is about 0.16 kWe per square meter (0.015 kWe per square foot) of total heat transfer surface area. Thus, a 10-MWe state-of-the-art OTEC heat exchanger must have a total heat transfer surface area of approximately 60,000 m² (650,000 ft²) for both evaporator and condenser.

The principal determinant of OTEC heat exchanger size is the cross-sectional area and length required to accommodate water and ammonia flow through the passages that divide water flow from the working medium. For a reasonable compromise between pressure loss and heat transfer rate, the water flow velocity in the heat exchanger will be in the range of 1 to 2 m/s (3 to 6 ft/s). Thus, to accommodate the 23 m³/s flow (800 ft³/s) required for a 10-MWe HX module (see Section 1.2.2), the inlet cross-sectional area of the evaporator water passages will be 12.5 to 25 m² (135 to 270 ft²), and a similar area will be needed for the condenser passages.

The flow area required for ammonia is more difficult to characterize because ammonia enters the heat exchanger as a liquid and is progressively converted to a vapor, with a large volume increase as it passes through the HX. Details of the calculation, which is complex and specific to each particular HX concept, are presented in the references in Chapter 4. For this overview, we may assume that the HX will be designed to make the ammonia-flow passages parallel to and approximately equal in cross section to those of the water passages. The cross-sectional area of the evaporator and condenser passages to accommodate the combined flow will then be 25 to 50 m² (270 to 540 ft²) for a 10-MWe (gross) module. Additional area will be required for passage walls and structural features.

The depth, z (or dimension parallel to the flow), of an HX is determined by the time required for heat being transferred from the water to the working fluid (or vice versa) to change the bulk temperature of the water by the design amount, for example, 3.5 degrees Celsius (6.3 degrees Fahrenheit), in its passage through the HX.

Let us consider a narrow water channel with parallel walls, as shown in Fig. 1-8, with channel thickness, y, channel width one unit, and depth, z, in which heat flows through walls on both sides to the ammonia working medium. The average rate at which heat is transferred from the water to ammonia, in its passage through the HX, is

$$\dot{Q} = 2U(T_w - T_a)z = 2U(T_w - T_a)vt , \qquad (1.2.8)$$

where v is the flow velocity and t is the time. The quantity of heat transferred during flow through the heat exchanger is then

$$\dot{Q}t = \rho C_p zy(T_{w\,in} - T_{w\,out}) = 2U(T_w - T_a)zt , \qquad (1.2.9)$$

where ρ is the liquid ammonia density and C_p is the specific heat. Since $t = z/v$,

$$z = \frac{\rho C_p yv(T_{w\,in} - T_{w\,out})}{2U(T_w - T_a)} . \qquad (1.2.10)$$

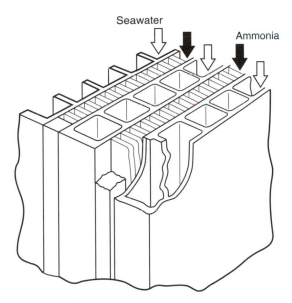

FIG. 1-8. Fluid flow paths in plate heat exchanger.

For example, if

$(T_{w\,in} - T_{w\,out})$	$= 3.50°C \ (6.3°F)$
$(T_w - T_a)$	$= 5.56°C \ (10.0°F)$
ρ	$= 0.824 \ kg/m^3 \ (64 \ lb/ft^3)$
C_p	$= 0.96 \ kcal/kg \ (0.96 \ Btu/lb)$
y	$= 0.0254 \ m \ (1/12 \ ft)$
v	$= 0.91 \ m/s \ (3 \ ft/s)$
U	$= 3620 \ W/m^2 \ °C \ (638 \ Btu/ft^2 \ h \ °F),$

then

$$z \qquad\qquad = 8.23 \ m \ (27 \ ft).$$

Furthermore, since the volume rate of water flowing is yvx and x is the total length of flow passage normal to the flow, $x = 990$ m (3250 ft).

Finally, if the inlet areas of water and ammonia passages are equal and the wall thicknesses are $(1/20) y$, the total cross-sectional area of each heat exchanger for this nominal 10-MWe (gross) module will be approximately 106 m² (1140 ft²). The platform area required just for the heat exchangers in a 54-MWe (40-MWe [net]) OTEC vessel will then be approximately 572 m² (6200 ft²) for the values of the parameters given.

1.2.2 OTEC Power Optimization

The principal results of the studies discussed in later chapters, which deal with optimization of closed-cycle OTEC power system design parameters, are summarized in the following. A 40-MWe (net) floating plant is used as the baseline.

1.2.2.1 Power Output

The optimum power output will occur when the total ΔT between the warm and cold water entering a closed-cycle OTEC plant is divided so that roughly half of the ΔT is used to promote heat transfer in the evaporator and condenser and the other half is used to achieve a difference in vapor pressure between the evaporator and condenser, that is, a pressure drop across the turbine. This requirement leads to a maximum heat engine efficiency of approximately 4% for a 22.2°C (72°F) warm-water temperature and a 4.4°C (40°F) cold-water temperature. Wu (1987) shows that the theoretical maximum power efficiency is given by the expression

$$P_{max}(\text{theor}) = 1 - \sqrt{\frac{T_{cw}}{T_{ww}}}.$$

For the temperatures cited, $\eta_{P(max)} = 3.77\%$.

Dependence of Power on Water Flow.   If the total ΔT is 22.2 degrees Celsius (40.0 degrees Fahrenheit), the optimum power output will occur when the temperature change in the water flowing between the inlet and exit of the evaporator is in the range of 2.2 to 5 degrees Celsius (4 to 9 degrees Fahrenheit). The optimum temperature change in the cold water flowing through the condenser will be in the same range. Since the cost of the cold-water ducts is significantly higher than the ducts for warm water, it is advantageous to design the power system to have lower cold-water flow in the condenser than in the evaporator. This implies a larger ΔT in the cold water than in the warm water.

For an average warm-water temperature change of 3.33 degrees Celsius (6 degrees Fahrenheit), 13.9 kJ (thermal) will be transferred to the ammonia in the evaporator per kilogram per second of water flow (6 Btu/s for a water flow rate of 1 lb/s).

With the assumption that the ΔT across the turbine is 11.1 degrees Celsius (20 degrees Fahrenheit) and the inlet water temperature is 26.7°C (80°F), the theoretical power generated by the turbine will be 0.49 kWe per kilogram per second (0.222 kWe per pound per second) of warm-water flow. For current state of the art, we may assume a combined turbine-generator efficiency of 82%. Thus, the electric output from the generator will be 0.40 KWe per kilogram per second (0.18 kWe per pound per second) of water flow.

Approximately 2350 kg/s (5170 lb/s) of warm-water flow will be required per megawatt of OTEC power generated. Approximately the same flow will be needed in the condenser. (If other variables were held constant, the heat flow from the condenser to the cold water would be 96.3% of the heat transferred to the evaporator, i.e., 1.00–0.037.) The estimated total seawater flow volume is then about 4.7 m³/s (165 ft³/s) per megawatt of gross electric power generated.

Water Pumping Power.   For a hydraulic head of 3.05 m (10 ft) and water pump efficiency of 75% (including electric motor drive), the estimated pumping power, P_{we}, for the evaporator is 9600 kg m/s (69,000 ft lb/s) = 94 kWe, per megawatt of gross electric power.

The condenser requirement is approximately 20% higher; therefore, the total water pumping power requirement is approximately 21% of the value of the gross power delivered.

Ammonia Pumping Power.   The total heat transferred to the ammonia per megawatt of electric power developed is 32.7 MW (thermal) (31,000 Btu/s).

The heat transferred to ammonia in the evaporator first raises the temperature until vaporization begins and then is absorbed as heat of vaporization at constant temperature; therefore, per megawatt of gross power,

$$P_a = \rho v A[(T_e - T_c)C_p + H_v],\tag{1.2.11}$$

where P_a = ammonia pumping power; ρ = liquid ammonia density = 625 kg/m^3 (39.0 lb/ft^3); v = ammonia inlet velocity; A = inlet area; T_e = evaporator temperature; T_c = condenser temperature; C_p = specific heat = 4.72 kJ/kg °C (1.13 Btu/lb °F); H_v = latent heat of ammonia vaporization at 21.1°C (70°F) = 1460 kJ/kg (509 Btu/lb); rvA = 27 kg/s (59 lb/s); and vA = 0.043 m^3/s (1.50 ft^3/s).

For the example shown in Fig. 1-6, the pressure difference across the ammonia pump is 397 kPa (57.6 psi). The calculated ammonia pumping power, P_a, if the pump efficiency is 72%, is then 23 kWe.

An accurate estimate of the power required to pump ammonia from the condenser to the evaporator exit requires detailed design data on ammonia pump efficiency and information about the requirements for recirculating the liquid ammonia that is trapped in the demister, which is discussed later. Anticipating the results of that discussion, we may assume that the ammonia pumping requirement will be about 3% of the gross power output of the OTEC electric generator.

Other Parasitic Power Requirements.   Power must also be provided for platform station keeping, for on-board personnel accommodations (hotel load), and for miscellaneous operational needs. The total is estimated at about 20 kW per megawatt or 2% of gross power.

OTEC Net Power Output.   The summary presented shows that approximately 26% of the gross power developed by an OTEC plant operating with a total ΔT of 22.2 degrees Celsius (40 degrees Fahrenheit) will be required for plant operating needs. Thus, the operating efficiency, defined as net power/gross power, will be 74%. The estimated net efficiency for this sample is then $0.74 \times 3.7 = 2.7\%$

Dependence of Net Power on ΔT.   Because OTEC power is linearly dependent on the ΔT between the water and the heat exchanger working fluid (ammonia), as well as on the ΔT between the inlet and exit of the ammonia turbine, the gross power varies as the square of the total ΔT; however, the parasitic power will be nearly independent of ΔT, since the pumping power and other power needs are not affected by a change in surface water temperature. For the typical case discussed here we have then, per megawatt (net),

$$
\begin{aligned}
P_{net} &= a\,(\Delta T)^2 - 0.26\\
&= a\,(22.2)^2 - 0.26 = 1.00 \text{ MWe}\\
a &= (0.26 + 1.00)/493.8 = 0.02252.
\end{aligned}\tag{1.2.12}
$$

Thus, for the range of surface temperatures between 23.9°C (75°F) and 30.0°C (86°F), corresponding to ΔTs of 19.4 to 25.6 degrees Celsius (35 to 46 degrees Fahrenheit), the output varies from

$$P_{net} = 0.000788 \times 352 - 0.26 = 0.70 \text{ MWe at } 23.9°\text{C } (75°\text{F})$$

to

$$P_{net} = 0.000788 \times 462 - 0.26 = 1.40 \text{ MWe at } 30.0°\text{C } (86°\text{F}).$$

Therefore, within the range of surface temperatures of interest for OTEC operation, the net power output of a typical plant will vary approximately as

$$P_{net(1)} / P_{net(2)} = (\Delta T_1)/(\Delta T_2)^n, \quad n = 2.54. \tag{1.2.13}$$

This steep dependence of net power on water surface temperature shows that OTEC cost per megawatt of net power output will have a pronounced minimum at ocean sites of high average surface temperature. This provides a strong impetus for focusing economic analyses of OTEC relative to other energy alternatives on the construction costs and costs of delivered power of OTEC plants designed for operation at sites of maximum ocean water temperature.

1.2.3 Water Ducting Subsystems

This section is concerned with the equipment and structures needed for drawing warm and cold water from the ocean, directing it to the heat exchangers, and returning it to the ocean. Water ducting is a major part of the OTEC system because of the large volume of water that must be handled and the size and length of the required cold-water pipe (CWP).

1.2.3.1 OTEC Cold-Water Pipe

General Requirements. Inherent in the OTEC concept is the use of cold water drawn from depths to 1000 m (3300 ft) to cool and liquefy the vapor emerging from the power plant turbine. The pipe and its attachment to the platform must be designed to withstand the static and dynamic loads imposed by the pipe weight, the relative motions of the pipe and platform when subjected to wave and current loads of up to 100-year-storm severity, and the collapsing load induced by the water pump suction. The CWP must be large enough to handle the required water flow with low drag loss; it also must be of a material that will be durable in seawater and not capable of forming corrosion or erosion products that would induce HX corrosion or be environmentally unacceptable.

A brief discussion is presented of the factors that must be considered in arriving at an optimum design for the CWP. The exposition is based on the analyses made in connection with the preliminary design of a 40-MW floating OTEC plant (George and Richards, 1980). The CWP requirements for a 50-MW plant resting on the sea bottom near the shore (shelf mounted) are also discussed briefly (Ocean

Thermal Corporation, 1984). A detailed discussion of CWP requirements and potential design solutions is presented in Chapter 4.

Systems Engineering Considerations. For the baseline 40-MWe floating OTEC plant, the general requirements for the CWP are the following:

length (L)	= 914 m (3000 ft)
water flow rate (w)	= 133 m³/s (4700 ft³/s)
	= 136,000 kg/s (300,000 lb/s)
diameter (D)	= 9.14 m (30 ft)
maximum angle between	
platform and CWP at joint	= 20°

The CWP length is defined by the need to draw water from a depth where the temperature is approximately 4.4°C (40°F). The water flow rate is determined by the desired power output and OTEC power plant efficiency in converting thermal energy to mechanical and then to electrical energy. The diameter of the CWP selected for a given power plant output results from a trade-off between CWP cost (roughly proportional to diameter, wall thickness, and length) and pumping power to overcome flow drag, which increases with the cube of the water velocity in the CWP. The allowable angle between the CWP and platform is determined by the ship and platform motions projected to occur under 100-year-storm conditions at the selected operating site.

The water flow requirements derived from the desired power output and plant efficiency were explained in Section 1.2.2. Short discussions of the remaining factors involved in CWP design will be presented here.

Cold-Water Pipe Diameter Determination. The power required for pumping cold water to the OTEC condenser system is given by the product of the water mass flow and the total hydraulic head. The hydraulic head is the sum of several factors:

1. The power needed to compensate for the depression below sea level of the height of the water in the CWP. Since the water in the CWP is denser than the average of the ambient seawater, the equilibrium position of the upper surface of the water in the cold-water column is approximately 0.6 m (2 ft) below sea level.
2. The hydraulic head of up to 3.0 m (10 ft) needed to overcome the pressure drop in the condenser.
3. The power (P_d) that must be expended to overcome the pressure loss caused by friction of the water flowing through the CWP.
4. The minor losses caused by screens and by loss of dynamic pressure at points in the flow by sudden area changes, that is, inlets, exits, or turns.

Items 1, 2, and 4 are not dependent on the diameter of the CWP. The friction loss, item 3, increases as the cube of the flow velocity and ultimately becomes the determining factor in the selection of pipe diameter.

We note that

$$\text{flow rate} = w = \rho A v = \rho D^2 (\pi / 4) v \tag{1.2.14}$$

flow drag force per unit flow = hydraulic head

$$= h_d = f(v^2 / 2g)(L/D), \tag{1.2.15}$$

where f is the friction coefficient.

The pumping power to overcome flow drag $= P_d$

$$= \text{flow rate} \times \text{hydraulic head}$$
$$= (fv^3 / 8g) L \rho D \pi. \tag{1.2.16}$$

Since

$$h_d = P_d / w$$
$$v = 4w /(D^2 \pi \rho)$$
$$h_d = f[4w /(D^2 \pi \rho)]^2 (1/2g)(L/D),$$
$$h_d = \frac{8 fw^2 L}{D^5 \pi^2 \rho^2 g}. \tag{1.2.17}$$

The quantitative implications of these relationships may be illustrated by an example using data from the baseline preliminary design of a floating 40-MWe (net) plant (George and Richards, 1980). For this example,

flow rate (w)	= 136,100 kg/s (300,800 lb/s)
seawater density (ρ)	= 1023 kg/m³ (64 lb/ft³)
CWP cross-sectional area (A)	$= D^2 \rho/4$
CWP diameter (D)	= 9.14 m (30 ft)
flow velocity (v)	= 2.0 m/s (6.7 ft/s)
drag coefficient (f)	= 0.0094 (smooth concrete pipe)
CWP length (L)	= 915 m (3000 ft).

Substitution of these values in Eqs. (1.2.17) and (1.2.16) gives:

hydraulic head to overcome pipe drag (h_d)	= 0.197 m (0.65 ft)
power to overcome CWP drag (P_d)	= 271 kW
baseline design total hydraulic head	= 3.2 m (10.5 ft)
baseline cold-water pumping power	= 4.27 MWe.

Since the hydraulic head to overcome pipe drag is a small fraction of the total head for this baseline design, it is apparent that the CWP diameter is larger than necessary.[2]

Selection of the optimal CWP size depends on the relative importance of a lower CWP cost as diameter is reduced versus increased plant cost per net kilowatt due to loss of net power if pumping power is increased. This requires the detailed analysis presented in Chapter 4; however, a rough estimate, made by using Eq. (1.2.17), is instructive. The pertinent data are presented in Table 1-2.

For the 40-MWe (net) baseline design, the cold-water flow is 136,100 kg/s (300,000 lb/s). The total hydraulic head is 3.2 m (10.5 ft), and the cold-water pumping power is 4.28 MWe. The hydraulic head associated with pipe drag as a function of pipe diameter is shown in column two of Table 1-2. The power to overcome flow drag in the pipe is shown in column three. The total pumping power is shown in column four. Column five presents the net power of the plant.

As discussed later (Section 1.4), the estimated cost for CWP construction and deployment is $18.4 million, and the estimated total deployed plantship cost is $144.4 million for the baseline plantship (1980 dollars).

If the CWP cost is assumed to vary directly with pipe diameter, the power costs per kilowatt vary as shown in column six of Table 1-2. If the cost varies as pipe diameter squared, the power costs per kilowatt are as shown in column seven.

Table 1-2 shows that the optimal CWP diameter is between 6.4 and 7.3 m (21 and 24 ft) for this 40-MWe (net) design. A more important result is the indication that the cost per kilowatt of net power from OTEC barge-mounted plants will be insensitive to moderate variations in the CWP diameter and cost. This mitigates a major perceived source of risk in the plantship investment. The conclusion must be modified for shore- or shelf-mounted plants. For these installations, the length needed for the CWP to reach a depth where the water temperature is 4.4°C (40°F) is approximately 3730 m (12,000 ft) at proposed sites. Also, pipe deployment costs are estimated to be much larger. Discussion is deferred to Chapter 4.

Cold-Water Pipe Structural Requirements.   The diameter and length of the CWP and the need for it to be able to withstand static and dynamic loads associated with ship motions, currents, and internal water flow make its design and fabrication a major technical challenge. Fortunately, the necessary engineering design principles have been developed in the offshore oil and shipbuilding industries, and

Table 1-2. Effect of CWP diameter on net OTEC power cost
[40-MWe (net) demonstration plant]

CWP diameter (m)	h_d (m)	P_d (MWe)	P_c (MWe)	Plant power (MWe net)	$/kW^a	$/kW^b
9.1	0.20	0.26	4.28	40.0	3610	3610
8.2	0.34	0.45	4.47	39.8	3582	3540
7.3	0.60	0.80	4.82	39.5	3563	3488
6.4	1.17	1.57	5.59	38.7	3589	3489
5.5	2.50	3.33	7.41	36.9	3714	3594

[a]CWP cost proportional to CWP diameter.
[b]CWP cost proportional to CWP diameter squared.

construction projects of similar magnitude have been successfully completed. Thus, experienced engineers concur that OTEC CWP construction and deployment are feasible and can be carried out at a reasonable cost (Cichanski and O'Connor, 1983; Det Norske Veritas, 1981).

A short summary of the requirements and potential approaches for CWP engineering is presented here. Full details are given in Section 4.2.

Cold-Water Pipe Dynamics. A CWP may be visualized as a long, thin, suspended tube of mass M, length-to-diameter ratio L/D, and wall thickness t. In the floating plant option the CWP will be supported at the top by a flexible joint fixed to a platform. (A schematic of the platform with a CWP is shown in Fig. 1-9.) The joint will experience angular deflections, which in 100-year-storm conditions are calculated to reach 18° in equatorial sites and 20° in the Caribbean, for the 40-MW demonstration barge. Angles will be smaller in commercial-size vessels. Heave and slewing excursions of the platform must be accommodated as well. Thus, the pipe will exhibit the lateral vibrational modes characteristic of a stretched string, coupled with forced vibrations and displacements induced by platform motions. Damping of the vibrations will be caused by friction of the water surrounding the pipe and flowing through the pipe. Local deflections will be caused by currents at various depths, and local, periodic stresses may be induced by eddy shedding.

Analysis of the motions of an actual CWP is complex, and discussion is postponed to Section 4.2; however, the analytical results have been compared with measured data from at-sea experiments with good agreement between theory and experiment. Thus, a good basis exists for calculating the limiting stresses and lateral excursions for CWPs made with practical construction materials and designs. The results of the test programs provide a firm basis for estimating the dynamic behavior expected in full-scale CWPs and give confidence that structural requirements may be estimated with sufficient accuracy to ensure satisfactory performance during deployment and operation of full-scale CWP designs.

The analysis and tests, presented in detail in Section 4.2, indicate that it will be feasible to build and practical to deploy suspended CWPs with existing technology. Conceptual design studies have been made of two specific designs:

1. An articulated CWP formed from pipe sections of light-weight concrete. The pipe sections are transported to the platform where they are joined together one after the other and progressively lowered from the OTEC platform until the full 910-m (3000-ft) length is deployed.
2. A compliant pipe made of reinforced plastic material, which would be constructed onshore as a single 910-m (3000-ft) tube and deployed by towing the tube to the OTEC platform, then sinking one end and inserting the other from below into the mounting joint on the ship.

Engineering data on two 9.1-m- (30-ft-) diameter CWP designs of the types discussed here are given in Section 4.2.

The first design employs 15-m- (50-ft-) long cylindrical sections (modules) made of post-tensioned lightweight concrete. The modules are connected through flexible joints having elastomeric pads of the type used between and under highway bridge sections.

The density of concrete can be lowered by replacing the sand with lighter materials such as fly-ash, without significant degradation in strength or suitability for use in marine applications. This feature allows the concrete specific gravity to be selected at an optimum value, 13% above that of seawater, which makes the deployed weight of the pipe only that required to hold the pipe angle at small values in the presence of ocean currents.

The second design uses a 910-m- (3000-ft-) long pipe made of concentric fiber-reinforced plastic tubes with an internal low-density plastic (syntactic foam) spacer. In this case, a weight is attached to the bottom of the slightly buoyant pipe to maintain the pipe at small angles from the vertical in the presence of lateral currents or platform motions.

The dynamic behavior of the CWP is determined by the longitudinal tension in the pipe wall, the bending strength (stiffness) of the wall, the longitudinal mass distribution, and the varying stresses imposed by the platform motions and ocean currents. In the first approximation, the CWP may be represented as a stretched string of uniform mass, which is supported at the top and subjected to a periodic force produced by wave action. The tension in the pipe decreases linearly with depth. If such a system is excited by a periodic force that is slowly increased from zero frequency, at a particular frequency the pipe will begin to vibrate, and, with no damping, the amplitude of the motion will increase at every cycle. This frequency is called the *resonant frequency* for the first vibrational mode. If the frequency of the periodic forcing function is now increased, the pipe motion will diminish to nearly zero; however, with further increase in the forcing frequency to a value about twice the first frequency, another resonance condition will occur, called the *first overtone* or *second normal mode*. Further overtones will be displayed as the forcing frequency is progressively increased.

The vibrational amplitude depends on the frequency of the forcing function and, with no damping, will assume extreme values for every instance in which the forcing frequency equals the frequency of one of the normal modes. With damping, energy is absorbed in proportion to the lateral vibrational velocity of the string, so that a limiting amplitude of vibration is reached at resonance conditions.

For a CWP suspended from a platform that is subject to wave motion, the forcing frequency will not be a single value but will be distributed over a spectrum of frequencies defined by the sea state, which will also be time dependent; however, for a given sea state, a narrow frequency range will dominate, and the CWP will vibrate in an overtone mode that most closely matches the exciting frequency. For the 100-year-storm condition in the equatorial Atlantic Ocean, the dominant frequency will be 1/10 to 1/13 Hz, that is, a wave period of 10 to 13 s. In Puerto Rico and Hawaii, the wave period during the 100-year-storm will be about 13 s.

1.2.3.2   Warm-Water Ducting

Warm water is drawn from the mixed layer near the surface at a depth selected to minimize effects of waves. The ducting arrangement depends on the platform configuration. Details are presented in Sections 4.2, 6.3, and 6.4.

1.2.3.3   Discharge Ducts

Discharge pipes that would conduct the HX flow to a depth below the mixed layer were considered necessary for moored OTEC power plants in early plans; however, experiments show that the discharge plume descends rapidly, indicating that external discharge pipes will not be needed in the absence of special environmental restrictions. The discharge pipes used in the OTEC-1 experiment failed due to inadequate attention to the engineering of their attachment to the ship; thus further development would be needed if these pipes were considered to be desirable.

1.2.4   *OTEC Energy Storage and Energy Transfer Subsystems*

The electrical energy generated on board an OTEC vessel may be stored in the form of chemical energy for periodic transfer to on-land users, or at favorable sites may be transmitted ashore via underwater power cables. Both options may be commercially attractive, depending on the cost of energy or fuel from alternative sources at the site of intended use.

The process of electrolysis provides a mechanism for efficient conversion of electrical energy to energy stored as internal energy of gaseous hydrogen, which may be recovered as heat of combustion or, with fuel cells, as electricity. Unlike heat engines, the energy conversion efficiency of an electrolytic process can approach 100% at ambient temperatures, and overall efficiency of a process involving electrolysis, transfer of the product to shore, and reconversion to electricity via fuel cells can be 60% or higher. Because grazing OTEC plantships can operate in ocean regions of maximum ΔT, the energy conversion losses can be compensated by higher net power output compared with shore-based OTEC plants. Furthermore, since the energy carrier is a liquid fuel, this mode of energy transfer has direct applicability to the most pressing U.S. energy need: motor vehicle fuel.

Electrochemical cells for water electrolysis have been manufactured commercially for many years to provide a convenient source of hydrogen for applications requiring a reducing atmosphere or to furnish pure oxygen for chemical or biological applications. Until recently, there has been no significant requirement for large units. The production of electrolytic hydrogen as an alternative fuel to supply the potentially large demands of vehicle transportation or power production has not been commercially attractive; however, research and development to improve efficiency, increase size, and reduce the cost of water electrolysis systems was conducted from 1975 to 1983, inspired by the predicted rise in hydrocarbon fuel prices. This work resulted in significant improvements in cell design and efficiency (Nuttal, 1981; Kincaide, 1981; Taylor et al., 1986).

Engineering design information for construction of efficient multi-megawatt electrolysis systems is now available. Plans have been announced for a 10-MW installation in Quebec that will use low-cost hydroelectric power to produce hydrogen for ammonia synthesis, and possibly for long-range distribution of gaseous hydrogen to consumers via pipeline (Taylor et al., 1986).

The Quebec plant will use unipolar cells, but bipolar cells with either acidic or alkaline electrolytes would be attractive alternatives. All types would be adaptable to shipboard installation and would be similar in cost. Preliminary design studies indicated that the General Electric solid polymer electrolyte (SPE) design would be of minimum weight and volume (Nuttal, 1981). The requirements are discussed in Section 4.3.

1.2.4.1 Hydrogen

Hydrogen is an excellent fuel that burns with high efficiency and produces only water as a combustion product. These characteristics have caused it to be advocated as the most desirable fuel to replace hydrocarbons in the future; however, hydrogen has two severe drawbacks: a very low density, both as a gas and as a liquid, and an extremely low boiling point ($-265°C$, $-445°F$). Cryogenic storage is required if hydrogen is to be used in liquid form. The estimated costs of liquefaction, storage, and transport of hydrogen from an OTEC plantship to users on land are high enough to make it unattractive in the near future to produce OTEC hydrogen as a fuel for motor vehicles or for power plants; however, preliminary design estimates indicate that OTEC liquid hydrogen could compete with other production options for applications that require liquid hydrogen per se, such as the space shuttle or hypersonic aircraft.

1.2.4.2   Ammonia

An attractive method of storing and transporting OTEC hydrogen is to combine it with nitrogen on the plantship to form ammonia, NH_3, which can be easily liquefied, transported, and stored by standard methods and equipment. Liquid ammonia is nine times as dense as liquid hydrogen and can be stored at ambient temperatures. Combustion of a liter of liquid ammonia produces 30% more heat than combustion of the same volume of liquid hydrogen. When ammonia burns, only water and nitrogen, the principal constituents of air, are formed as products; therefore, ammonia, like hydrogen, can be a fuel for the future because no pollutants are produced when it burns. Engine tests have demonstrated that ammonia is an excellent fuel for internal combustion engines, burning with high efficiency with an octane number of 111 (Starkman et al., 1966).

Ammonia is now produced commercially in multimillion-ton quantities by processes that use natural gas as a source of hydrogen. Ammonia is used worldwide as a fertilizer and is a major feed stock for the production of other nitrogen-rich fertilizers or plastics such as polyurethane. Thus, facilities and distribution systems are available for ammonia that would be suitable for direct introduction of OTEC ammonia into world markets if natural gas prices were to rise to values that would make OTEC ammonia competitive (Avery, 1988).

In the preliminary design of a baseline 40-MWe OTEC plantship, the electric power generated is used for on-board ammonia synthesis. The demonstration plant produces 103 metric tons (114 tons) per day of ammonia, which is stored on board for shipment to shore at approximately monthly intervals. A conceptual design was also prepared for a 325-MWe commercial plant that would produce 1100 metric tons (1200 tons) per day of ammonia. It was estimated that ammonia from such plants would be cost competitive with ammonia from natural gas if the natural gas cost rose to about $6.00 per GJ ($5.70 per million Btu). Details are presented in Chapter 4.

1.2.4.3 Methanol

The energy transfer method that offers the most promise for OTEC to become an early contributor to world energy supplies is the production of methanol (CH_3OH) on OTEC plantships (Avery et al., 1985). OTEC methanol could be a competitively priced substitute for petroleum-based fuels, even at 1989 gasoline prices. Methanol is a liquid at ambient temperatures. It burns cleanly and has an octane number of 110. Extensive tests with standard automobiles have shown that it can replace gasoline as a fuel with only minor modifications in operational factors and give a lower cost per kilometer than premium gasoline fuel (Fisher, 1984).

Methanol is made by combining two volumes of hydrogen with one volume of carbon monoxide (CO) in the presence of a suitable catalyst. CO is formed by the reaction of carbon with oxygen. The methanol synthesis reactions use the oxygen as well as the hydrogen of OTEC water electrolysis, with a corresponding reduction in the product cost. The low cost of coal shipment makes it commercially attractive to transport coal to the OTEC plantship to supply carbon for CO production. A design of an OTEC plantship with nominal 160-MWe output power that produces 1750 tonnes (1925 tons) per day of methanol is described in Chapter 4 (Ebasco and Rockwell, 1984).

1.2.4.4 Energy Transfer from Grazing OTEC Plants

OTEC plantships designed to produce ammonia, methanol, or other energy products will cruise slowly (graze) in regions of the tropical oceans where ΔT exceeds 22 degrees Celsius (40 degrees Fahrenheit). Energy transfer will involve shipboard storage of the energy product for periods of about 1 month, followed by transfer of the product to a tanker for transport to world ports where the product will be stored for transshipment to the ultimate consumer. The basic equipment and procedures required for this method of energy transfer are in general commercial use. Some development will be necessary for shipboard operation. The requirements are discussed in Chapter 4. In brief,

1. Electrolyzers now in commercial use must be scaled up to a suitable size and efficiency for economical production of hydrogen on OTEC plantships.
2. Commercial air-separation plant subsystems must be arranged for shipboard operation as integral units in OTEC ammonia or methanol plants. For methanol production, scale-up of the Rockwell molten carbonate gasifier is necessary (see Section 6.4).

3. Modifications in commercial ammonia and methanol plant controls and operating procedures will be needed to optimize the processes for OTEC operation.
4. Storage tanks and filling and transfer equipment must be adapted to the OTEC requirements.
5. Specific designs and procedures for product transfer at sea must be demonstrated.

Ammonia and methanol are now produced commercially in plants readily adaptable to OTEC shipboard operation. The on-board storage of OTEC products will employ tanks of conventional design for liquid ammonia or methanol, configured for efficient use of available space in the hull. Appropriate venting and other safety measures will be provided. Ammonia transfer will adapt procedures developed and employed for commercial at-sea ammonia transfer to liquid ammonia tankers. Methanol transfer will use procedures employed in refueling of Navy ships. Details of the transfer process proposed in the preliminary design of the 40-MWe demonstration ammonia plantship and the proposed 160-MWe methanol plantship are given in Chapter 6.

1.2.4.5 Volume Requirements for OTEC Product Production and Transfer

Energy transfer as discussed here involves water electrolysis followed by conversion of the gaseous products into a form suitable for storage and transfer. For the 40-MWe baseline design (George and Richards, 1980), the weight and volume requirements for electrolysis were defined by the size and weight of a system using electrolyzers of the GE SPE type. The electrolyzers would be housed in vessels of $1 m^2$ (10 ft^2) cross-sectional area containing 290 individual electrolysis cells, with a total voltage requirement of 524 V DC and a current of 10,000 A.

The components of the pilot ammonia synthesis plant, consisting of compressors, catalysis unit, water distillation and purification, and ammonia storage are accommodated on other decks at the same ship location. The complete ammonia plant is housed in a space 32.9 m (108 ft) wide, 21.3 m (70 ft) fore and aft between the 11.3-m (37-ft) and 31.4-m (103-ft) levels. Ammonia storage tanks occupy space above the keel 11.0 m (36 ft) high, 32.9 m (108 ft) wide, and 11.3 m (37 ft) fore and aft.

A 1750-tonne/day (540,000-gal/year) methanol plant housed on a 160-MWe (nom) commercial OTEC methanol plantship will occupy a space 43 m (140 ft) wide and 30 m (100 ft) long. The process and space requirements for the energy storage methods described here are discussed in Chapter 4.

1.2.4.6 Power Transmission from Moored OTEC Plants

Electric power transmission from OTEC plants moored offshore can be commercially attractive at sites that satisfy two requirements: availability of a suitable water temperature difference within a distance from shore small enough to make power transmission practical, and an electric power demand on shore large enough to support an OTEC plant of cost-effective size. The islands of Oahu in Hawaii and Puerto Rico have attractive sites for moored OTEC plants that could contribute to U.S. needs. Similar opportunities exist on other islands in the Caribbean and throughout the Pacific. If underwater power transmission for distances of about 160 km (100 miles) can be proved feasible, OTEC plants could also be placed near the edge of the continental shelf in the Gulf of Mexico and at other sites bordering the Atlantic, Pacific, and Indian oceans (Morello, 1978; Garrity and Morello, 1978; Garrity et al., 1980).

Successful deployment and operation of a moored OTEC plant requires development of a mooring or station-keeping system that will hold the plant on station at offshore sites where the water depth exceeds 1100 m (3600 ft), which is needed to leave ample clearance for the CWP intake. The mooring must also be compatible with the power-cable system designed to bring electric power from the OTEC plant to the substation on land. Several methods have been devised that seem suitable for mooring or position control of an OTEC platform sited where the water depth is 1100 m (3600 ft) or more. The mooring studies have included a number of platform types and sizes and have addressed wave and wind conditions predicted for the worst storm expected to occur at the site during a 100-year period. Methods of rotating the platform to the optimal heading during the storm were included. Differential traction of the mooring cables and use of thrusters mounted at the corners of the platform were investigated (Trimble and Robidart, 1978). A "watch circle"[3] no greater than 5% of the mooring depth was imposed as a design requirement to ensure compatibility with the power-cable requirements.

Two potential designs of mooring lines for a multiline mooring system that combines features of earlier studies are shown in Fig. 1-9 (Ross and Wood, 1981). Three potential power-cable support methods are shown in Fig. 1-10 (Schultz et al., 1981). Several concepts have been proposed that address the combined mooring and power transmission requirements, but an optimal system has not been defined. The perceived technical difficulties of moored floating plants led to a diversion of Department of Energy (DOE) program emphasis after 1981 to shelf- or shore-mounted concepts for this OTEC application.

The equipment for the transfer of OTEC power to shore includes the power-cable mounting, power-conditioning equipment, switching gear, operator's control station, and safety equipment. In the 40-MWe demonstration vessel design, the total electric power transfer system occupies a space 33 m (108 ft) wide and 21 m (70 ft) long.

1.2.5 *Propulsion Requirements for OTEC Grazing Operation*

Energy transfer via an OTEC-produced fuel involves the use of plantships that maneuver slowly in tropical ocean areas of maximum ΔT. Analysis of the propulsion needs of the baseline 40-MWe ammonia plantship shows that the value of the maximum thrust will be determined by the requirements for ship survival during the most severe 100-year storm (George and Richards, 1980). For this storm, extreme waves of significant wave height of 8.8 m (29 ft) are predicted to occur for periods

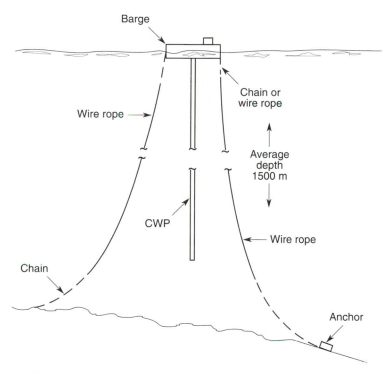

FIG. 1-9. Diagram of mooring options for an OTEC barge (Ross and Wood, 1981).

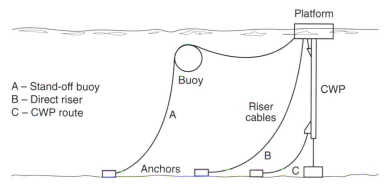

FIG. 1-10. Diagram of power cable deployment options for moored OTEC power plant (Schultz et al., 1981).

totaling several hours per year at sites within 10° of the equator in the Atlantic Ocean. For a severe storm, when the wave height reaches 3.5 m (12 ft), the ship power plant is shut down and the propulsion power is supplied by standby diesel-electric generators. Diesel-electric-driven thrusters mounted at the four corners of the OTEC platform, rated at a maximum power of 2 MWe each (2500 hp), that is, 8 MWe total, are used to maintain heading and ship stability under these extreme conditions. Under the normal grazing operation, at an average 0.4-knot speed, a total thruster power of 1.4 MWe supplied by the OTEC power plant suffices. Thrusters with the desired power and ability to direct the thrust in any desired horizontal direction (azimuthing capability) are commercially available. A discussion of the requirements is given in Chapter 4.

1.2.6 *OTEC Platform*

The size, configuration, and method of construction of an OTEC platform (or building) will depend on many factors: power output, site, type of installation (i.e., shore-based, bottom-mounted, floating but moored, or cruising [grazing]), method adopted for minimizing interaction between platform motions associated with sea conditions and CWP motions, method of energy delivery, and perceived optimal system characteristics that will maximize efficiency and minimize cost and risk of construction, deployment, and long-term operation.

The discussion in the following parts of this section presents an overview of basic engineering considerations that furnish quantitative guidelines for determining platform type, dimensions, and method of construction. This provides a context for a summary of the conclusions drawn from analyses and ocean tests regarding platform size and configuration.

Design data for floating plants of 40 and 160 MWe at the bus bar are selected for this overview. Discussions of platforms for other system options, including smaller, shore-based systems are presented in Chapters 4 and 5.

1.2.6.1 General Requirements

The OTEC platform is the structure that supports and houses all of the subsystems and components of an operational OTEC system, whether floating, shelf mounted, or shore mounted. Many alternative designs have been proposed and subjected to varying degrees of study that have reflected changing views of the program objectives. The platform designs may be grouped into three categories.

1. Platforms for OTEC systems that would be produced in large enough numbers to make significant impacts on U.S. or world energy needs. Of necessity, these systems are floating platforms, or "plantships," maneuvering slowly on the tropical oceans and designed to produce fuels and other energy-intensive products for shipment to world ports.
2. Platforms for OTEC systems designed to supply electric power to public utilities having a large enough power demand to warrant construction of 100- to 400-MWe installations. Of necessity, these are large, offshore, moored installations or shelf-mounted or shore-based installations, where warm and cold water are available within a reasonable distance from shore.
3. Small OTEC systems (1 to 40 MWe) designed as demonstration units or to meet a special tropical island demand. Of necessity, the potential contribution

of these systems to the major energy needs of the United States or other developed countries is small, except for their inputs to the technical and commercial development of OTEC as a major energy source; however, such small systems can be cost effective at many island sites and could induce a total demand large enough to generate commercial interest. Indeed, such interest appears to be attracting Japanese involvement.

This overview and the remaining chapters focus on defining the technology that is needed for OTEC to meet the requirements of Category 1; however, funding support in the United States, after initial emphasis on Category 2, turned to Category 3 and has been directed to the investigation of specific component or subsystem designs. Thus, conceptual designs of complete OTEC systems are available for only eight U.S. systems. Of these, three are for moored installations, three are for shelf-mounted systems, and two are for plantships. Preliminary engineering designs were completed for only one 40-MWe plantship, one moored 40-MWe power plant, and one 50-MWe shelf-mounted system.

Land-based (shelf-mounted or onshore) OTEC installations, which are site specific, are discussed in Chapters 4 and 5.

1.2.6.2 Floating OTEC Platform Types

Parametric studies of six platform concepts for floating OTEC plants ranging in power from 50 to 500 MWe (gross) were made under DOE direction during 1977–1978 (Waid, 1978; Scott and Rogalski, Jr., 1978).[4] The types considered were:

1. Ship (or barge)
2. Cylindrical platform
3. Semi-submersible
4. Spar
5. Submersible
6. Sphere

Sketches of the designs are shown in Fig. 1-11.

Each type was evaluated at each power level with regard to technical feasibility, component accessibility and maintenance, cost, time to operational use, and risk. Types 1, 3, and 4 were rated as attractive candidates, but the marine engineering background and facilities were more extensive for Type 1. Subsequent work on floating platforms concentrated on barge configurations.

1.2.6.3 OTEC Barge Design

Marine engineering principles and technology developed for conventional barges are applicable to the design of platforms for moored OTEC plants and to cruising plantships as well, since these will maneuver at an average speed of only 0.2 m/s (0.4 knot).

By DOE direction, the evaluations by Waid and by Scott and Rogalski were based on OTEC power plants designed for offshore generation of power for transmission to onshore utilities.

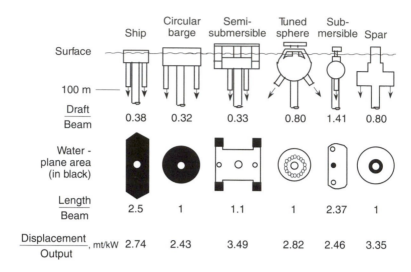

FIG. 1-11. Alternative platform configurations for moored OTEC power plants (Waid, 1978).

Construction and deployment of floating OTEC platforms under 200 MWe in size would use existing technology and construction facilities; however, departures from conventional barge practice would be necessary to meet two special requirements of OTEC:

1. A large volume along the longitudinal centerline of the platform must be provided for the heat exchanger bays, CWP, and cold-water storage bay (moon pool). Fore-and-aft bending and twisting loads must therefore be taken by structures at the sides and ends of the platform.
2. Efficient integration of all of the plant subsystems is required, particularly the CWP, power plant, and energy delivery subsystems, to achieve an optimal structural design of minimal cost. Incorporation of the large heat exchangers and their associated water ducting in an efficient total design requires particular consideration.

The total plant or plantship must conform to the standard criteria for seaworthiness, habitability, safety, and durability expected in any vessel designed for long-time operation at sea.

The platform design proceeds in a series of steps:

1. A preliminary estimate of gross volume and weight is made on the basis of the dimensions and weights of the power system components, CWP and associated plenum and water ducting, and energy transfer system. A rough assessment of the space requirements for operational and crew facilities is also made.
2. A preliminary conceptual arrangement of the subsystems is made and iterated to define interface and space requirements for an efficient integration of the subsystems into a total design.

3. A hull configuration is chosen that will provide space and support for the arrangement of subsystems defined in Step 2 and that satisfies naval architecture and marine engineering requirements such as the length-to-beam ratio, buoyancy, pitch and roll angles, angular excursions of the joint between the CWP and platform, and survival strength. Materials, facilities, and construction procedures are considered from the standpoint of minimizing cost and schedule, and a provisional design concept is selected.
4. Weights and volumes of the light ship and fully loaded platform are assessed, reflecting the inputs from Step 3. Ballasting requirements are calculated that will adjust the displacement and metacentric height to appropriate values in operational and survival conditions. If incompatibilities are found, the design process is recycled until a satisfactory design is achieved.

The platform–CWP combination must be designed so that angular motions at the joint between the CWP and platform will be small in normal operation and will not exceed safe limits, even with waves of the magnitude predicted for a 100-year storm.

The volume of water flowing to the heat exchangers implies a design in which the water pumps are mounted so that the water exhausts at an elevation above sea level that balances the head required to overcome pressure drops in the heat exchanger cycle. To minimize pumping power costs, water flow velocity in the ducts should be moderate. The ducts and heat exchanger bays for a power module are designed to form a compact unit associated with the corresponding pump outflow, so that they become an integral part of the structural configuration of the ship.

As in other aspects of OTEC design, thorough systems engineering is required to achieve a platform design that will be optimal for performance, durability, and maintainability at minimal cost. The major aspects will be summarized here. The design details are discussed in Chapter 4.

Platform Dimensions. The platform dimensions are defined by the volume requirements of the major subsystems and the ways in which they may be combined to form a practical total design. The dimensions and volumes of these subsystems may be estimated by considering the size requirements defined for the baseline 40-MWe plantship. These resulted from a systems integration and optimization program (George and Richards, 1980).

The platform is a box structure 135 m (443.5 ft) long, 43 m (140 ft) wide, and 27 m (89 ft) deep. A sketch of the baseline barge and the general layout of the baseline plant are shown in Figs. 6-56 and 6-57 in Chapter 6. Details of the design are presented in Section 6.5.1.

1. The baseline power plant consists of four modular units of nominal 10-MWe net output. Each module includes heat exchangers and associated structures, ammonia pump, demister and turbine, and electric generator.
2. The water ducting subsystem includes the CWP assembly with a pipe-to-platform joint and supporting structure, warm-water intakes and pumps, and

ducts to carry water to the heat exchangers and afterward discharge it to the ocean in a manner that will be environmentally satisfactory. The cold-water pumps direct water into a moon pool at the top of the CWP.

3. Transfer of OTEC energy from the bus bar on the plantship to shore is by conversion of electrical energy to stored chemical energy in the form of ammonia. The baseline design assigns an area of 700 m² (7500 ft²) for this subsystem. The electrolysis system, ammonia plant, and ammonia storage and transfer gear are included in this volume. The area and volume requirements of the power subsystem for direct transfer of electric power to shore in the moored plant configuration, including mooring and power conditioning space, are estimated to be the same for the 40-MWe moored baseline vessel as for the energy conversion plant.

4. Crew accommodations are provided for 40 people, including space for visitors and supporting facilities. The personnel are housed in a five-story hotel. Area for shops and additional facilities is provided on lower decks.

5. Wing walls extending the length of the platform are required to provide strength to resist bending and twisting loads. For the baseline design these are box beam structures.

The total platform area of the baseline barge is 5770 m² (62,100 ft²) not including the sponsons. The main deck is 23.5 m (77 ft) above the keel. Hull volume not occupied by the power plant equipment, water ducting, and energy transfer equipment is allocated to machinery, fuel storage, ballast, electrical gear, positioning control equipment, safety equipment, and miscellaneous items. The positions of the various subsystems are shown in Fig. 6-56.

The 40-MWe ammonia plantship design has been used as a baseline for preliminary design studies of a 182-MWe (160-MWe [nom]) plantship (Ebasco and Rockwell, 1984). A layout drawing of the methanol plantship is shown in Fig. 6-39. The vessel is 275 m long, 100 m wide, and has a draft of 20 m (Richards et al., 1984; Avery, 1984). Details are presented in Section 6.4.7.

Comparison of the dimensions of the 160-MWe (nom) plantship with the 40-MWe (nom) baseline shows that more efficient use of the platform area assigned to the cold-water ducting and wing wall structures is achieved in the larger plantship; however, no specific effort was made to minimize packaging for the hotel and miscellaneous structures, in scaling to 160 MWe. The power plant area was based on the use of the 10-MWe modules of the baseline design and remained constant. Thus, the area assigned is much larger than necessary.

Construction Materials. The technology and materials for construction of OTEC floating platforms of either steel or concrete are available at many shipyards in the United States and other countries. Although steel is the predominant choice for ship construction, concrete with reinforcing steel and/or post-tensioning is more suitable for OTEC barges because of several factors:

1. The hull cost per unit of power output is significantly less for concrete construction.

2. The greater weight and volume of concrete, which increase ship drag, are not a significant drawback for moored or slowly moving OTEC barges.
3. Since considerable ballast is needed to compensate for the buoyancy of submerged OTEC heat exchangers, the greater weight of the concrete hull simply reduces the amount of ballast needed.
4. Concrete is compatible with aluminum, but steel requires special precautions to avoid corrosion induced by electrolytic action or by deposition of corrosion-inducing particles of iron or rust on aluminum heat exchanger surfaces.
5. Concrete is durable in seawater, whereas steel ships must be returned to drydock at intervals of approximately 5 years for rust removal and reapplication of rust-prevention coatings.

These factors have led to the selection of concrete hull designs for almost all of the proposed OTEC installations.

Shipyards suitable for the construction of floating OTEC plants are available on the east and west coasts of the United States, along the Gulf of Mexico, and in other maritime countries. New graving docks would usually be required but would be a small addition to the total investment. A listing of suitable facilities is presented in Chapter 4.

1.3 OTEC RESEARCH AND DEVELOPMENT STATUS

After the initial evaluations of OTEC in 1974 showed its potential to become a significant energy resource, attention was directed in the U.S. program to the technical areas that would need expansion before a firm judgment could be made about OTEC practical feasibility. Critics questioned whether a Rankine cycle could be operated efficiently with the small ΔT available for OTEC, whether any net power could be produced, whether low-cost heat exchangers could be built that would be durable in seawater, whether fouling of the heat exchanger surfaces would rapidly degrade performance to unacceptable levels, whether a CWP could be built and deployed, whether interactions between ship motions and the CWP would be too severe for its survival in storms, whether OTEC operation would be unacceptable environmentally, whether energy could be transferred from an OTEC plant to shore, whether suitable sites for OTEC existed, whether one conceptual design of OTEC should be favored over others, and, finally, whether OTEC plants could be built at a low enough cost to make OTEC energy cost competitive with other energy alternatives. All of these questions were addressed and answered with positive results in the research and development programs conducted before funding was terminated.

The initial OTEC research and development program was organized to develop a technology base specifically related to OTEC from which answers to the critical questions would be expected to emerge. Research programs were established with the general goals and program direction listed in Table 1-3.

Table 1-3. OTEC research and development programs

Topic	General objective	Program direction[a]
Heat engine performance	Analyze OTEC power cycles	ANL—closed cycles
	predict net output	SERI—open cycles
Heat exchanger design	Relate heat transfer rates to	
	geometry and flow variables	ANL
Heat exchanger materials	Determine durability of commercial	
	HX alloys in OTEC environment;	
	estimate life-cycle costs	ANL
Biofouling	Determine rate under OTEC conditions	ANL
	Investigate biology;	
	investigate control methods	LBL/UC
CWP design	Determine CWP-hull dynamics	NOAA, JHU/APL
	Evaluate construction methods	
	and materials	NOAA, JHU/APL
Energy transfer	Define compatible mooring and	
	power cable for floating plant	NOAA
Energy product	Design complete plantship systems	
	for ammonia, methanol, and hydrogen	JHU/APL
Environmental interactions	Oceanographic data	NOAA, LBL/UC
	Biofouling of structures	LBL/UC
Environment on OTEC	100-year storm	NOAA, JHU/APL
OTEC on environment	Biological effects;	
	· temperatures and flow patterns;	
	miscellaneous impacts	LBL/UC

[a] ANL, Argonne National Laboratory, Chicago, Illinois; SERI, Solar Energy Research Institute, Golden, Colorado; LBL/UC, Lawrence Berkeley Laboratory, University of California, Berkeley, California; NOAA, National Oceanic and Atmospheric Administration; U.S. Dept. of Commerce, Washington, D.C.; JHU/APL, The Johns Hopkins University Applied Physics Laboratory, Laurel, Maryland.

A brief summary of the principal results is given in the following paragraphs. Detailed information is provided in the references. The results are also discussed in Chapters 4 and 5.

1.3.1 *Thermodynamic Performance of Heat Engines for OTEC*

The goal of this program was to identify heat engine cycles applicable to OTEC and to select those of most promise for further study and development. Factors considered included theoretical efficiency, working fluids, background technology, availability of materials and components, technical unknowns, projected development time, and cost.

The initial review of potential closed-cycle heat engine types for OTEC application showed that Rankine closed cycles employing ammonia, Freon R-22, and propane were attractive candidates (Olsen et al., 1973). The Claude open cycle (Claude, 1930) and two new open-cycle concepts, one employing foam (Zener and Fetkovitch, 1975; Zener and Kay, 1980) and the other mist (Beck, 1975), to lift water to operate a hydraulic turbine merited investigation. Subsequent studies led to the concentration of effort on the ammonia closed cycle and the Claude open cycle.

The initial analyses showed that closed-cycle efficiency is almost independent of the choice of working fluid if optimum conditions are chosen for each cycle. (The theoretical maximum efficiency is independent of the heat transfer medium.) Therefore, the choice must be based on other factors: working pressure, compatibility of the working fluid with heat exchanger and turbine materials, heat exchanger size, turbine and ducting size, safety, and estimated risks and costs. Evaluation of these factors has led to the choice of ammonia for the working fluid in most of the conceptual designs of OTEC systems. Freon R-22 has been favored in a few studies.

1.3.2 Heat Engine Design and Performance

Although designs and engineering data were available in 1974 for the analysis of various heat exchanger types, the applicability to OTEC requirements was not known. A research and development program to investigate this subject was established at the Argonne National Laboratory of DOE. The program included construction of a test facility specifically designed to supply warm- and cold-water flows at temperatures and velocities typical of OTEC (Thomas et al., 1979). Tests were conducted of seven designs of evaporators and condensers that analysis indicated would be suitable for OTEC, including one type in which the water flow was external to the tubes containing the ammonia working fluid. In no case was the performance degraded from predicted levels because of difficulty in establishing nucleate boiling with the low ΔT, which had been raised as a potential problem by power plant engineers familiar with high-temperature systems. Overall heat transfer rates ranged from values typical of conventional shell-and-tube designs (~2270 W/m^2 °C, 400 Btu/ ft^2 h °F) to three times those levels with advanced concepts (~6800 W/m^2 °C, 1200 Btu/ft^2 h °F). Quantitative data directly applicable to large-scale OTEC designs were obtained. Fresh water was employed, which eliminated the possibility that biofouling would complicate interpretation of the results. The work also included the development of computer codes for analyzing the data and predicting performance in large systems where different operational conditions might apply (Thomas et al., 1979). The program proved that OTEC heat exchanger performance would be satisfactory for practical designs.

1.3.3 Heat Exchanger Materials

Heat exchangers for commercial use have been constructed from alloys of aluminum, copper, stainless steel, and titanium. For particular applications, the choice of material has depended on tradeoffs among suitability for fabrication and packaging, durability, thermal conductivity, and cost. It was recognized that OTEC would require unusually careful selection of heat exchanger materials and fabrication techniques because of the large area per net kilowatt of power output and the special demands imposed by the seawater heat source. The volume of the OTEC heat exchangers and associated water ducting imposes a need for minimal cost materials and construction, as well as careful integration of the power system into the total platform. If this were not done, the heat exchanger cost could become the major factor in total OTEC system cost. The selected heat exchanger materials must not only withstand the corrosive action of seawater but also must not contaminate the outflow with dissolved chemicals hazardous to marine life.

The initial review of this program element led to the establishment of a research and development program under Argonne National Laboratory direction to furnish test data on the durability of candidate aluminum alloys and to provide further data on copper and stainless steel alloys, which might offer intermediate cost advantages (Kinelski, 1979; LeQue, 1979). Test programs were set up, principally at the Naval Coastal Systems Center, Panama City, Florida; at the LeQue Center for Corrosion Research, Wrightsville Beach, North Carolina; at the Center for Energy and Environmental Research, San Juan, Puerto Rico; and at the Sea Coast Test Facility of the University of Hawaii. The last two had access to warm ocean water typical of OTEC operating conditions. Five years of testing in Hawaii proved that several aluminum alloys showed corrosion rates low enough to qualify them for 30-year life in OTEC heat exchangers. Since the life-cycle cost of aluminum heat exchangers will be much lower than the costs of other alternatives, the OTEC system cost estimates discussed later are based on the use of aluminum (Panchal et al., 1985). The demonstration that high-cost titanium heat exchangers would not be required was a major result of this program. The data are presented in Chapter 4.

1.3.4 Biofouling Control

The possibility that biofouling of the heat exchangers would quickly degrade OTEC performance was raised as a critical issue at the beginning of the OTEC program. Investigation has shown that the seriousness of the problem was overestimated. Fouling rates are much lower in tropical open-ocean waters suitable for OTEC operation than in coastal waters, which are rich in marine life; therefore, biocontrol methods are much more effective for OTEC than for typical marine heat exchangers. Chemical agents can be used in concentrations that are environmentally safe, and physical methods can be effective at lower intensities or longer time intervals.

One reason for apprehension about the effects of biofouling was a misconception by many evaluators that the low thermal efficiency of OTEC would make its performance particularly sensitive to a reduction of heat transfer by biofouling. This is not so. As shown earlier (Section 1.2.2), the percentage reduction in the overall heat transfer coefficient due to fouling is the same whether ΔT is 20 or 1000°C. It is correct to state that as heat transfer coefficients are improved by better design, the sensitivity of the performance to biofouling will increase.

Beginning in 1978, experiments were conducted in the Department of Energy program to determine rates of biofouling under conditions typical of OTEC operation in Hawaii (Pandolfini et al., 1980; Liebert et al., 1981), in the Gulf of Mexico (Little, 1978), and in Puerto Rico (Sasscer et al., 1980). It was found that fouling would reach unacceptable levels in about 6 weeks without fouling controls; that is, the fouling coefficient, R_f, would exceed a value of 0.000088 m^2 °C/W (0.0005 ft^2 h °F/Btu). [The fouling coefficient is the reciprocal of the heat transfer coefficient, h_{sb}, defined in Eq. (1.2.7).] The research also included tests of a wide variety of control methods, including both physical and chemical means. As a result of this work, a practical method of OTEC biofouling control is now available. It has been shown that injection of 70 parts per billion of chlorine for 1 h/d into the warm-water flow to the evaporator effectively prevents film formation. As of July 1986, heat exchanger test samples had been exposed to seawater with this chlorine concentration for more than 1000 d, with no significant reduction in the heat transfer coefficient due to biofouling. The chlorine added is one-twentieth of the amount considered environmentally acceptable by the U.S. Environmental Protection Agency. There was no biofouling in the condenser test samples exposed to cold-water flow; therefore, there was no need for chlorine addition (Berger and Berger, 1986).

Figure 1-12 displays biofouling test results for a typical experiment using warm seawater with chlorine injection, and for a control run in which fouling was allowed to build without chlorine injection until R_f reached the value 5×10^{-4} m^2 h °F/Btu, after which the heat exchanger surface was cleaned and the process repeated.

The research and development program also demonstrated that biofouling can be controlled by physical means, including brushing and ultraviolet and ultrasonic radiation; however, these methods are less attractive to use and are not likely to be adopted unless chlorine or other biocides are arbitrarily, or as the result of new evidence, ruled to be unacceptable.

The biofouling program also included studies of the biology of slime formation on surfaces exposed to seawater. The process involves an initial phase in which a bacterial film is deposited on the heat exchanger surface. This film provides sustenance for marine organisms, which gradually build up a layer that covers the surface. When this stage is reached, growth proceeds at a relatively fast, approximately constant, rate. Experiments show that the value of R_f is directly proportional to the film thickness (Liebert et al., 1981).

Another important result of the biological investigations is that no significant differences are found in the organisms that cause fouling at sites in Hawaii, Puerto Rico, and the Gulf of Mexico. This result gives confidence that biocontrol measures that are effective in Hawaii will be applicable throughout the tropical ocean area (Sasscer, 1981; Liebert et al., 1981).

1.3.5 *Cold-Water Pipe Research and Development*

The projected size and length of suspended OTEC CWPs, as well as the apparent problems of designing suitable structures that can be constructed, deployed, operated successfully, and able to survive tropical storms, have led to this aspect of OTEC development being described as a "challenge" by advocates and an "impossibility" by critics. Fortunately, the advocates' views have been substantiated in ocean tests at reasonable scale.

Early review showed that concrete pipes of sizes suitable for OTEC CWPs had been constructed and emplaced in a number of offshore oil installations, in diameters of 9.1 to 15.2 m (30 to 50 ft) (Haynes and Rail, 1977). A concrete pipe 21 ft in diameter and 8 miles long carries water from Lake Havasu to the Colorado River in Arizona. Fiberglass-reinforced plastic storage vessels up to 24.4 m (80 ft)

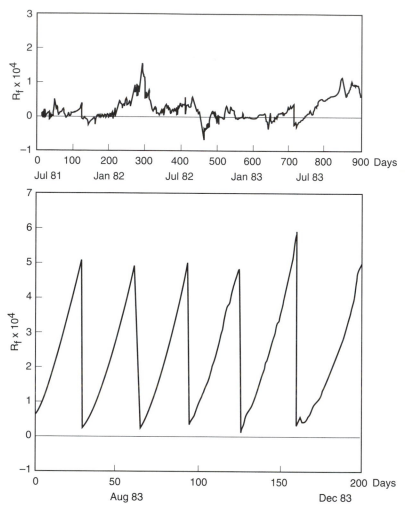

Fig. 1-12.   Change of fouling coefficient with time in HX tubes exposed to warm seawater. (Top) With chlorine addition of 70 ppb/h per day. (Bottom) Without chlorine but cleaned when R_f reached 5×10^{-4} ft^2 h °F/Btu (Berger amd Berger, 1986).

in diameter were in use on land. Thus, facilities were available or could readily be assembled that could be used to build pipe segments of these materials that could then be joined to form pipes of the required length. The major questions that arose were the following:

1. How do we determine the dynamics of the CWP- platform system as a function of the design variables?
2. Would the CWP-platform loads be a determining factor in the total system design? For example, would a submerged or spar design for the platform be required for a feasible design of the CWP to be achieved?

3. Would the cost or risk of construction and deployment of the CWP be intolerably high?
4. What CWP structural design and what materials of construction would be suitable in terms of systems compatibility, cost, and endurance?

Conceptual design answers have emerged for all the items listed.

By drawing on the extensive background established for offshore oil drilling, computer codes for analysis of CWP dynamics were prepared. The calculated motions and stresses have been compared with observations made at sea in scale tests of flexible and jointed pipes of diameter 1.5 to 2.4 m (5 to 8 ft) and 150 m (500 ft) long (Donnelly et al., 1979; Hawaiian Dredging and Construction Company, 1982; Paulling, 1959). The agreement between theory and experiment provides a good basis for predicting both the strength requirements and the angular motions of pilot and commercial-scale CWPs, under operational and 100-year-storm conditions. Design analyses and tests of a 3.05-m- (10.0-ft-) diameter jointed section indicate that CWPs made of rigid sections of lightweight concrete connected by flexible joints will satisfy the design requirements and will be feasible to build and deploy at a reasonable cost (Cichanski and O'Connor, 1983; O'Connor, 1981). Flexible CWPs employing concentric tubes of fiberglass-reinforced plastic separated by a low-density plastic (syntactic foam) spacer are also predicted to meet the design requirements and to be of reasonable cost to construct and deploy (Hove and Grote, 1980). Other options also appear attractive.

The design estimates indicate that the costs of construction and deployment of suspended concrete or fiberglass-reinforced-plastic CWPs for plantships ranging in size from 40 to 160 MWe will be in the range of 10 to 15% of the total ship cost. Thus, a large underestimate of the cost of the CWP system would have a small effect on the unit power cost. This conclusion does not apply to shore- or shelf-mounted OTEC plants, for which the long lengths required and the difficult deployment and emplacement of supporting structures cause the CWP cost to be a large part of the total cost.

Detailed engineering design information on both suspended and bottom-mounted CWPs is presented in Sections 4.2 and 6.5.

An acceptable solution to the CWP dynamics and structural problems permits selection of a barge design for the platform. As discussed in Section 1.2.4, conceptual and preliminary design studies indicate that the barge platform will be optimal from the standpoints of accessibility, component packaging, constructability, maintenance, and cost.

1.4 ENERGY TRANSFER

1.4.1 *Energy Transfer Via OTEC On-Board Fuel Production*

The problems of mooring and direct power transmission from an OTEC plant may be avoided by using the power generated on board to manufacture a fuel, which is shipped to customers on land, where it may be used for vehicle transportation or to generate electric power. This option permits the OTEC plant to operate at design capacity 24 h/d, a major advantage compared with other solar systems.

Investigation of this topic proceeded through the following phases:

1. A survey was made to identify potential energy products, to assess the technical feasibility of producing them by OTEC, and to furnish information on costs and economic factors (Konopka et al., 1977). Preliminary data were developed for hydrogen, ammonia, aluminum, methanol, gasoline, methane, lithium, and chemicals derived from saline solutions. These studies indicated that the commercial processes used for the production of the products listed could be adapted to OTEC plantships; however, only aluminum, ammonia, and perhaps sea chemicals appeared to offer low enough costs to be worthy of further investigation.

2. The conclusions of the studies in Phase 1 were strongly influenced by the assumption that feedstocks and energy products would be shipped to and from the OTEC plantship by barges, and that the plantships would be sited in the Gulf of Mexico. Accordingly, shipping costs were high, and plant efficiencies were 30 to 40% lower than those expected for the maximum surface water temperatures. Subsequent studies showed that for commercial-sized plantships designed for operation in tropical Atlantic Ocean sites and serviced by commercial ammonia tankers, OTEC ammonia production could be profitable (Avery et al., 1976).

3. The favorable results of Phases 1 and 2 led to the preliminary design of a 40-MWe demonstration OTEC ammonia plantship and a conceptual design of a 325-MWe commercial plantship. The 40-MWe design provided a baseline for estimates of equipment requirements and costs for the production of OTEC hydrogen, aluminum, and methanol. Because of platform size requirements and transportation costs, hydrogen and aluminum were not deemed favorable for early development. Methanol was found to warrant additional study (BARDI, 1982).

4. Further investigation of OTEC production of methanol led to the design of a 160-MWe (nom) methanol plantship that would produce 1750 tonnes (1925 tons) of methanol per day. The cost of OTEC methanol delivered to a U.S. port from this plant was estimated to be 20 to 30% below the projected cost after 1990 of methanol produced from natural gas in new plants in the United States (Avery et al., 1985). A layout diagram for this OTEC plantship is shown in Fig. 1-13 (Ebasco and Rockwell, 1984).

1.4.2 Energy Transfer Via Underwater Cable

Power cables for underwater transmission of electric power over long distances are in use at many sites throughout the world, and facilities for their construction and deployment are available commercially (Garrity and Morello, 1979; Pieroni et al., 1979); however, OTEC power transmission poses two requirements that go beyond present technology: operation at depths over 900 m (3000 ft) and ability to withstand continual flexing associated with attachment of the cable to a moored but bobbing platform.

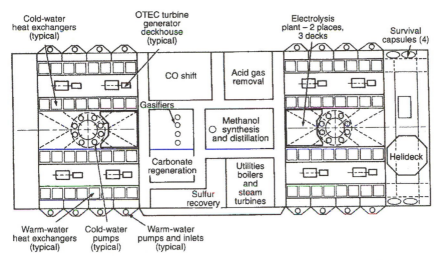

FIG. 1-13. Design of platform of 160-MWe (nom) OTEC methanol plantship (EBASCO and Rockwell, 1984).

An extensive research and development program, established to address these issues, involved the following phases:

1. A survey was made of cable types and characteristics to define limiting capabilities of existing systems with regard to size, operable depth, flexure capabilities in angle and number of cycles, and cost. Self-contained oil-filled (SCOF) and cross-linked polyethylene (XLPE) cables were selected as promising candidates (Garrity et al., 1980).

2. An examination was made of design options for the platform, mooring, and power-cable system. The study showed that spar or tension-leg moored systems would reduce cable design problems; however, the benefits were not judged to be sufficient in floating plant designs to overweigh the total system advantages of the moored barge configuration.

3. Studies were made of methods for supporting the flexible riser section of the cable and attaching it to the platform and the bottom-mounted section of the cable. The studies showed that basically different designs would be required for:

   a. Moored plants that could be sited within a few miles of shore, as in Hawaii, Puerto Rico, and other islands.

   b. Moored plants sited at the edge of the continental shelf in the Gulf of Mexico, 120 to 240 km (75 to 150 miles) from the coast.

For category (a), 400-kV AC cables would be feasible. The cables would be deployed in a trench on the sea bottom from a facility onshore to an anchoring point within the watch circle of the moored plant and then would rise to a flexible attachment point on the vessel. If cables of this type and length needed to be replaced during the life of the plant, the cost would be acceptable.

An extensive test program was completed in which sections of cables of the SCOF and XLPE types were mounted in a fixture and subjected to continuous flexing simulating 115 years of mechanical and electrical stresses under seawater pressure at a depth of 1830 m (6000 ft) and power levels above 100 MW (9.5×10^4 Btu/s). The tests indicated that riser cables of the XPLE design would meet the design requirement for 30-year life (Soden et al., 1982).

For category (b), power losses would become excessive if AC transmission were used over distances of more than about 16 km (10 miles); therefore, a DC cable would be necessary. The cost of laying the bottom-mounted section of the DC cable for the long distance and the low probability of failure of the entrenched cable make it undesirable to specify a design for this part of the cable that would incorporate features to allow cable replacement. This suggests the use of a "permanent" bottom-mounted cable joined to a replaceable, flexible riser cable; however, the use of a separate section for the riser introduces some problems that have not been resolved. It would be difficult to design, test, and ensure adequate performance of a junction box between the two sections of cable at a depth of 1200 m (4000 ft) or more. If the junction box were mounted on a buoy held at a depth of 30 to 60 m (100 to 200 ft), where it would be accessible to divers but would be relatively unaffected by surface waves, it would introduce new design factors for both sections of the cable as well as for the mooring system. Use of a spliced cable would mitigate some problems but would bring difficulties in ensuring satisfactory performance of the spliced section.

A decision to focus attention on shelf-mounted installations terminated further investigation of long-power-cable design issues in the OTEC program supported by the U.S. Department of Energy; however, effort directed to power transmission between islands in Hawaii may provide the desired design information, if interest develops in OTEC electric power transmission to the Gulf states or similar situations in other parts of the world (Traut et al., 1982).

1.5   INVESTIGATIONS OF ENVIRONMENTAL INTERACTIONS OF OTEC

Early consideration of the environmental effects of OTEC operation identified two major topics for further investigation (Lewis and Duncan, 1978; Lewis, 1979):

1. The effects of the biological and physical marine environment on the continuing operation of an OTEC system. This included the selection of appropriate materials and structures, development of environmentally acceptable methods of biofouling control, and assessment of the design requirements for the system to withstand the worst storms predicted at the operating site (Bretschneider, 1977, 1978; Bretschneider and Tamaye, 1977; Kinelski, 1982; Panchal et al., 1985).

2. Effects of OTEC operation on the marine environment. This involves documentation of the marine environment at potential OTEC sites and assessment of impacts of the large warm- and cold-water flows of OTEC on local ocean structure, flow patterns, currents, temperatures, and marine biology (Ditmars and Paddock, 1979; Hartwig, 1981).

1.5.1 *Impacts of the Environment on OTEC*

1.5.1.1 Survey of Oceanographic Data at Sites of OTEC Interest

Many teams of oceanographers and biologists were recruited early in the OTEC program to compile information that could serve as a data base for assessing OTEC performance and environmental interactions. Regions investigated in some detail included the Gulf of Mexico, Hawaii, Puerto Rico, and the tropical Atlantic Ocean. Available data were assembled on temperature, salinity, winds, currents, and marine life, from the surface to a depth of 1000 m (3280 ft). To the extent that funds and personnel were available, additional measurements were made at specific points of interest. This information has been invaluable in judging the potential interactions of OTEC with the environment.

1.5.1.2 Impacts of the Marine Biological Environment on OTEC Operation

The initial program in this field was designed to provide the basic marine biological information needed for OTEC design and performance evaluation. Topics assigned for investigation included the identification of marine organisms involved in biofouling, description of the fouling process, effect of operational site on fouling, measurement of biofouling rates, measurement of the reduction of heat transfer as a function of biofouling buildup, effectiveness of various chemical and physical methods of biocontrol, and development of facilities, instruments, and evaluation procedures required for the different investigations.

The program has reaffirmed the observation that the quantitative description of the marine world and its interaction with ocean-based structures, processes, and materials is an infinite task. Nevertheless, the needed basis for the design of OTEC power systems that will operate with high performance in the tropical ocean regions suitable for OTEC has been established.

The results of the studies of heat exchanger biofouling were discussed in Section 1.3.

In summary,

1. Fouling of heat exchangers occurs slowly under OTEC evaporator operating conditions and does not occur in the cold environment of condenser operating conditions. Biofouling can be prevented by injecting chlorine 1 h/d into the inlet water at a concentration of 70 parts per billion. Physical methods of biocontrol are also available but are less cost effective.
2. Tests in Puerto Rico, the Gulf of Mexico, and Hawaii, which show similar results, provide evidence that biofouling is not site specific at deep-water tropical sites suitable for OTEC operation.
3. Concerns that biofouling could seriously degrade OTEC performance and operating cost have been allayed.
4. Materials have been identified and tested that would satisfy at reasonable cost the OTEC requirements for the 30-year life of the power system components, platforms, and CWPs.

1.5.1.3   OTEC Design to Withstand Storm Conditions

OTEC systems must have long operating life (~30 years), including the ability to withstand the winds, waves, and currents of the most severe storm predicted to occur over a period of 100 years at the site of plant operation (the 100-year storm). Formulas developed by Bretschneider (1979) show that the 100-year storm would produce waves of 8.8 m (29 ft) significant wave heights[5] in the Atlantic Ocean near the equator, 11.3 m (37 ft) in the Caribbean near Puerto Rico, and 9.1 m (30 ft) at Hawaiian sites. The baseline 40-MWe barge and CWP system is designed to withstand these conditions. A maximum angle of 20° between the platform and CWP at the junction is allowed in the design. Water tunnel tests of a 1/30-scale model of the baseline configuration confirmed the suitability of the grazing plantship design for the Atlantic site. The tests also showed that minor fairing and bulkheads would need to be added to resist hurricane waves at the Puerto Rico site. The evaluation confirmed the soundness and practicality of the concrete barge design (George and Richards, 1980, 1981).

1.5.2   *Effects of OTEC Operation on the Marine Environment*

1.5.2.1 Biological Effects

Although it is recognized that the cold water flowing through an OTEC plant is rich in nutrients and could well add to the economic value of OTEC, it is desirable for the first OTEC plants to operate in a way that will cause minimal environmental impact, either beneficial or adverse. OTEC operation could affect the biological environment by adding stimulants to marine growth (for example, nitrogen) or deterrents such as traces of chlorine or other contaminants. Temperature changes and currents induced by water leaving the heat exchangers could also have positive or negative effects on the ambient ecology.

A substantial program was established to explore the many questions raised and to provide a basis for an environmental impact assessment. The general conclusion is that the operation of a demonstration OTEC plant of 40-MWe output in accord with present design plans will produce no significant ecological changes. Plants of 100 to 500 MWe also appear innocuous if suitably separated from each other in coastal sites (Meyers et al., 1981; Jirka et al., 1981). It is recognized that subtle effects will not become evident until demonstration plants are built and are in operation (Quinby-Hunt, 1981).

The initial designs of moored and shelf-mounted plants call for the outflow to be discharged in a way that will minimize its mixing with the water near the surface and will facilitate its sinking to a depth that will prevent its subsequent rise to the surface. This is more important for moored plants, which could affect the ecology in local fishing or recreational areas, than for grazing plants that will operate over a wide range of high seas.

A moored OTEC plant is designed to draw surface water from a depth of 9.1 to 19.8 m (30 to 65 ft). The water is cooled 2.8 to 3.9 degrees Celsius (5 to 7 degrees Fahrenheit) in passage through the evaporators. Cold water drawn from the ocean depths is warmed an equal amount in passage through the condensers. The two flows are mixed and are then discharged at an appropriate depth. For the baseline 40-MWe moored plant design, the mixed flow emerges downward at a depth of about 76 m (250 ft) through eight pipes 4.6 m (15 ft) in diameter. The temperature of the discharged water is $15.5 \pm 1.7°C$ ($60 \pm 3°F$), and the exhaust velocity is 2.1 m/s (6.8 ft/s). The momentum of the flow and its lower density than the surrounding water cause the water column to descend to a depth of about 200 m (650 ft) before it equilibrates with the surrounding water (see Chapters 4 and 5). No measurable effect on the water temperature at the surface is expected from operating a single OTEC plant under actual conditions.

If moored plant operation were established and continued in a completely still environment, the water emerging downward would gradually induce a large-scale circulation pattern that would ultimately get to the surface; however, since currents of fluctuating intensity and direction are always present in the ocean, it appears that steady circulation patterns would not have a chance to become established with plants of commercial size. Surveys of the flow from actual plants will be needed to provide a detailed picture of the impact of moored plant operation on the hydrodynamic and biological environments of the plants.

Grazing OTEC plantships will be sited in regions of the tropical oceans where an adequate ΔT is available. The plants will cruise within an ocean region of several hundred square kilometers per plant at speeds of about 1.26 m/s (0.5 knot). If constructed in accord with the present designs, these plants will discharge the outflow at a depth of 20 m (65 ft), and the outflow will slowly descend to a depth of about 200 m (600 ft) before equilibrating. A small part of the outlet flow will remain in the mixed layer near the surface and could stimulate the production of marine life, as is characteristic of tropical regions near Peru where upwelling brings nutrient-rich deep water to the surface. The lack of oceanographic analysis of the flow patterns and the absence of experimental data prevent quantitative prediction of the effects.

The potential increase in marine life associated with OTEC plantship operation is expected to be beneficial, although it could lead to the need for somewhat more vigorous methods to control biofouling of the OTEC heat exchangers. The needed data can be acquired when the first commercial OTEC plantships begin operation (Laurence and Roels, 1977).

1.5.2.2 Temperature and Flow Effects of OTEC Operation

Early surveys of oceanographic data provided information on the temperature difference between the surface and a depth of 1000 m (3280 ft) in the world's oceans (Miller, 1978; Wolff, 1978). Wolff's data are presented in Fig. 1-1. Data on current profiles in regions of interest to OTEC were provided by Molinari (1979). A comprehensive examination of modeling techniques was given by Allender et al. (1978).

There is detailed discussion of possible environmental impacts of OTEC operation in Section 9.2. The available geophysical data and modeling information can be used for estimates of the long-term effects of OTEC plant operation on the distribution of the plant outflow and on ocean surface temperatures. Using a one-dimensional model, Martin and Roberts (1977) derived the temperature and flow impacts of operation of 100 and 1000 moored 200-MWe plants, uniformly distributed at spacings of 107 and 34 km in the 1 million square kilometer area of the Gulf of Mexico. Operation of 1000 200-MWe plants in the Gulf of Mexico would cause changes in the mixed layer that would stabilize after about 2 years with a decrease of the annual average surface temperature of about 1.3°C. With 100 plants, the decrease in surface temperature would be only 0.05°C. (The electric power production of 1000 200-MWe plants would be equal to 25% of the total U.S. electric power demand in 1990.) In this one-dimensional model, the principal effect of the OTEC operation is to produce a flat oval of mixed effluent water centered at about 850-m depth, which steadily increases in temperature. The predicted gain in temperature in this zone is about 0.3°C/year for the 1000-plant case and 0.03°C/year for the 100-plant siting. In actual operation, cold water flowing into the Gulf of Mexico would displace the warmer water at the 1000-m level, so that net plant output would be determined by the rate of flow of cold water into the Gulf. Similar conclusions appear reasonable for dense siting of OTEC plants in other tropical ocean areas; however, definitive studies have not been done. A study by Brin of the effect of continuing OTEC operation on the cold-water reservoir below 1000 m in depth showed that the impact would be insignificant if OTEC plants were sited to produce no more than 0.2 MWe/km^2 of ocean surface area (Brin, 1981).

Refinement of the models to include three-dimensional analyses and experimental data from at-sea plant operations are required for precise estimates (Allender et al., 1978). It is clear from the present studies that the thermal and hydrodynamic impacts will be small for operation of individual plants and that large numbers of OTEC plants can be operated without causing environmental concerns. Decisions on optimal plant spacing for maximum long-term power production can await data from multiplant operational experience.

A more extensive discussion of environmental impacts of OTEC operation is given in Chapter 9.

1.5.2.3 Miscellaneous Impacts

The on-board power output from ocean-sited plantships will be converted to a fuel or energy-intensive product for transport to on-land users. The conversion processes and product transport procedures will give in-depth consideration to possible environmental impacts and will be designed to be in full compliance with national and international environmental guidelines. Proposed plantships will be designed to meet those guidelines. Since construction of OTEC plants and plantships will be gradual on a global scale, there will be time to assess the actual environmental impact of OTEC operation and to take appropriate countermeasures, if necessary, before significant adverse consequences could occur. Further details are presented in Chapter 9.

1.6 OTEC COSTS AND COMMERCIALIZATION

1.6.1 *Costs of Electric Power and OTEC Products*

The investment requirements and operating costs for OTEC systems of demonstration size were determined in the OTEC preliminary design and conceptual design programs funded by the DOE. This information and data from French, Japanese, and Taiwanese programs provide a base for estimating investment and operating costs for commercial plants and plantships. The most complete cost assessments were made in the preliminary design of a land-based OTEC power plant by the Ocean Thermal Corporation, and in the programs resulting in baseline designs of barge-mounted OTEC plants that were done by the Applied Physics Laboratory of The Johns Hopkins University. The land-based plant would be sited in Hawaii on the island of Oahu. The floating plants include a moored plant to be sited off Punta Tuna, Puerto Rico, and grazing OTEC plantships sited near the equator in the Atlantic Ocean. The cost data for OTEC systems are presented in Chapter 7.

1.6.2 *OTEC Energy Prices Relative to Prices of Alternative Systems*

From the plant investment and operating cost data for an energy production system, an investor can estimate the sales price of power or products that will return an attractive profit to a private investor or allowable return to a publicly owned enterprise. This sales price represents the result of an in-house assessment of factors that include: perceived risk, projections of future markets, government subsidies (for example, government-supported research and development, government-supplied facilities, depletion allowances), tax incentives, financing options, and other factors. Without knowledge of these proprietary factors, unbiased judgments cannot be made in comparing the quoted prices of power or fuels from alternative energy systems. A financial analysis procedure devised by Mossman (1988), based primarily on plant investment, that uses the same guideline to estimate the sales price that would return an acceptable profit for each energy production system allows a valid comparison to be made of the relative commercial attractiveness. This procedure eliminates distortions and misunderstandings caused by failure to include the effects of government subsidies that differ widely among the various options.

The financial analysis procedure and investment data to rate systems proposed to meet future requirements for fuels and power are presented and compared in Chapters 7, 8, and 9. The studies show that OTEC systems will be commercially attractive when the product costs are compared with the costs of supplying motor vehicle fuels from conventional systems. An investment in OTEC methanol plantships with total capacity sufficient to replace all U.S. gasoline derived from imported petroleum is estimated to provide investors an annual return after taxes of 38% on their equity. The assessment is based on a comparison of OTEC investment needs with the investment requirements, without subsidies, of existing commercial and proposed nonsolar options that must satisfy long-term requirements for environmental acceptability and safety.

NOTES

1. Heat is not a substance. Therefore, it is not correct to speak of heat flowing from one position to another; however, it is pedagogically useful to describe the process in this manner. For an interesting discussion of this point, see Fenn (1982).
2. In the floating 40-MWe baseline design, the CWP diameter was not chosen to minimize system costs but to demonstrate a CWP in the pilot plant test that would be the maximum size considered feasible to construct with facilities available in 1979.
3. The tension in the mooring lines must be adjusted so that the ship cannot move from the design surface position a distance more than 5% of the mooring depth.
4. Parallel studies were made by Basar et al. (1978); however, they assumed a fixed connection between the CWP and the platform, which made their conclusions invalid for platforms with a flexible CWP joint.
5. Significant wave height is the average height of the highest 10% of the waves occurring during a storm of specified duration.

REFERENCES

Allender, J. H., J. D. Ditmars, R. A. Paddock, and K. D. Saunders, 1978. "OTEC physical and climatic environmental impacts: An overview of modeling efforts and needs." *Proc. 5th Ocean Thermal Energy Conversion Conf.,* Miami Beach, Fla., Feb., III-165.

Avery, W. H., 1984. "Methanol from ocean thermal energy." *Johns Hopkins APL Tech. Dig.,* **5**, 159.

——, 1988. "A role for ammonia in the hydrogen economy." *Int. J. Hydrogen Energy,* **13**, 761.

——, R. W. Blevins, G. L. Dugger, and E. J. Francis, 1976. *Maritime and construction aspects of ocean thermal energy conversion (OTEC) plant ships.* Johns Hopkins Univ. Applied Physics Lab., SR 76-1B.

——, J. D. Richards, and G. L. Dugger, 1985. "Hydrogen generation by OTEC electrolysis and economical energy transfer to world markets via ammonia and methanol." *Int. J. Hydrogen Energy,* **10**, 727.

BARDI, 1982. *Coal to methanol study using OTEC technology.* Brown and Root Development, Inc. (BARDI).

Basar, N. S., J. C. Daidola, and N. M. Maniar, 1978. "OTEC ocean systems evaluation." *Proc. 5th Ocean Thermal Energy Conversion Conf.,* Miami Beach, Fla., Feb., IV-15.

Beck, E. J., 1975. "Ocean thermal gradient power plant." *Science,* **189**, 293.

Berger, L. R., and J. A. Berger, 1986. "Countermeasures to microbiofouling in simulated ocean thermal energy conversion heat exchangers with surface and deep water in Hawaii." *Solar Energy Update: Final Issue,* **SFU-86-12**, 31.

Bretschneider, C. L., 1977. "Operational sea state and design wave criteria: state of the art of available data for U.S.A. Coasts and equatorial latitudes." *Proc. 4th Annual Conf. on Ocean Thermal Energy Conversion,* New Orleans, March, IV-61.

——, and E. E. Tamaye, 1977. "Hurricane wind and wave forecasting techniques." Univ. of Hawaii Look Lab., **6**, No. 1.

——, 1978. "Operational sea state and design wave criteria for potential OTEC sites." *Proc. 5th Ocean Thermal Energy Conversion Conf.,* Miami Beach, Feb., IV-17 & IV-236.

——, 1979. "Hurricane design winds and waves and current criteria." Univ. of Hawaii Look Lab. Report No. 79-45.

Brin, C., 1981. *Energy and the oceans*. Ann Arbor Publishers, Inc., Butterworth Group, Chapter 8.

Carnot, S., 1824. "Reflexions sur la puissance motrice du feu et sur les machines propres a developper cette puissance." Paris, *Bachelier, Annales Scientifiques de l'Ecole Normal Superieure*, **2**, No. 1, 383 (1872).

Castellano, C. C., 1981. "Overall OTEC-1 status and accomplishments." *Proc. 8th Ocean Energy Conf.*, Washington, D.C., **2**, 971.

Cichanski, W., and R. F. Mast, 1979. "Design of a concrete cold water pipe for ocean thermal energy conversion (OTEC) systems." *Proc. Offshore Technology Conf.*, Houston, Tex., **11**, 1653.

Cichanski, W., and J. S. O'Connor, 1983. "Development of a concrete cold water pipe for ocean thermal energy (OTEC) systems." in *Alternate energy sources III*, Washington, D.C.: Hemisphere Publishing Corp., p. 321.

Claude, G., 1930. "Cold water pipe for ocean thermal energy conversion (OTEC) systems." *Mechanical Engineering* **52**, 1039.

Det Norske Veritas, 1981. "Rules for the design, construction and inspection of fixed offshore structures." Oslo.

Ditmars, J. D., and R. A. Paddock, 1979. "OTEC physical and climatic impacts." *Proc. 6th OTEC Conf.,* Washington, D.C., June, 13.11.

Donnelly, H. L., J. T. Stadter, and R. O. Weiss, 1979. "Cold water pipe verification test." *Proc. 6th OTEC Conf.,* Washington, D.C., **1**, 6.2-1.

Ebasco Services, Inc., and Rockwell International Corp., 1984. "Coal to methanol OTEC plantship study using the Rockwell molten carbonate gasification system." Johns Hopkins Univ. Applied Physics Lab., APL Contract No. 601863-L, March.

Fenn, J. B., 1982. *Engines, energy and entropy*. New York: W. H. Freedman and Co.

Fisher, M., 1984. *Methanol fuel program*. Bank of America, San Francisco, Cal. (internal publication).

Garrity, T. F., and A. Morello, 1979. "A theoretical study of technical and economic feasibility of submarine cables for OTEC plants." *Proc. 6th OTEC Conf.*, Washington, D.C., **1**, 7.1.1.

Garrity, T. F., R. Eaton III, T. Dalton, C. Pieroni, and J. P. Walsh, 1980. "Design, test and commercialization considerations of OTEC pilot plant riser cables," *Proc. 7th Ocean Energy Conf.*, Washington, D.C., **3**, 3.5-1.

George, J. F., D. R. Richards, and L. L. Peroni, 1978. *A baseline design of an OTEC plantship. Vol. A: Detailed report*. Johns Hopkins Univ. Applied Physics Lab., SR-78-3A.

George, J. F., and D. Richards, 1980. *Baseline designs of moored and grazing 40-MWe OTEC pilot plants*. Johns Hopkins Univ. Applied Physics Lab., JHU/APL SR-80-1A.

——, 1981. "Model basin tests of a baseline 40 MWe OTEC pilot plant." *Proc. 8th Ocean Energy Conf.*, Washington, D.C., June, 77.

Gershunov, E. M., Y. H. Ozudogru, and J. M. Betts, 1981. "Analysis of the up-ending problem of OTEC-1 cold water pipe." *Proc. 8th Ocean Energy Conf.*, Washington, D.C., June, **2**, 991.

Hartwig, E. O., 1981. "OTEC environmental biological oceanographic program." *Proc. 8th Ocean Energy Conf.*, Washington, D.C., June, 507.

Hawaiian Dredging and Construction Co., 1982. *OTEC cold water pipe at-sea test program*. Hawaiian Dredging and Construction Co. Design Report No. 1.

Haynes, H. H., and R. D. Rail, 1977. "Concrete structures—fabrication methods." *Proc. 4th Annual Conf. on Ocean Thermal Energy Conversion*, New Orleans, La., V-70.

Hove, D. T., and P. B. Grote, 1980. "OTEC CWP baseline designs." *Conf. Proc. 7th Ocean Energy Conf.*, Washington, D.C., June, **1 & 2**, 3.3.1.

Jirka, G. H., J. M. Jones, and F. E. Sargent, 1981. "Intermediate field plumes from OTEC plants: Predictions for typical site conditions." *Proc. 8th Ocean Energy Conf.*, Washington, D.C., June, 475.

Kincaide, W. J., 1981. "Advanced alkaline electrolysis systems for OTEC." *Proc. 8th Ocean Energy Conf.*, Washington, D.C., 685.

Kinelski, E. H., 1979. "Biofouling, corrosion and materials overview." *Proc. 6th OTEC Conf.*, Washington, D.C., June, **2**, 12.1.

———, 1982. "OTEC biofouling, corrosion and materials program." *Conf. Proc. 7th Ocean Energy Conf.*, Washington, D.C., June, **1**, 5.1.

Konopka, A., N. Biederman, B. Talib, and B. Yudow, 1977. "Alternative forms of energy transmission from OTEC plants." *Proc. 4th Annual Conf. on Ocean Thermal Energy Conversion*, July, **3**, 47.

Laurence, S., and O. A. Roels, 1977. "Potential mariculture lead of sea thermal power plants," *Proc. 4th Annual Conf. on Ocean Thermal Energy Conversion*, New Orleans, La., March, III-21.

LeRoy, R. L., 1983. "Industrial water electrolysis: past and future." *Int. J. Hydrogen Energy* **6**, 401.

LeQue, F. L., 1979. "Qualifying aluminum and stainless alloys for OTEC heat exchangers." *Proc. 6th OTEC Conf.*, Washington, D.C., June, **2**, 1.

Lewis, L. F., 1979. "OTEC environmental and resource assessment program," *Proc. 6th OTEC Conf.*, Washington, D.C., **1**, 13.1.

———, and C. P. Duncan, 1978. "OTEC environmental and resource assessment program," *Proc. 5th Ocean Thermal Energy Conversion Conf.*, **III**, 1.

Liebert, B. E., J. Larsen-Basse, J. A. Berger, and L. R. Berger, 1981. "Biofouling and corrosion studies at Keahole Point, Hawaii." *Proc. 8th Ocean Energy Conf.*, Washington, D.C., June, **1**, 421.

Little, T. E., 1978. "Selection of seawater pumping systems for OTEC power plants." *Proc. 5th Ocean Thermal Energy Conversion Conf.*, Miami Beach, Fla., Feb., **1–5**, 199.

Martin, P. J., and G. O. Roberts, 1977. "An estimate of the impact of OTEC. operation on the vertical distribution of heat in the Gulf of Mexico," *Proc. 4th Annual Conf. on Ocean Thermal Energy Conversion*, New Orleans, La., March, IV-26.

Meyers, E. P., S. M. Sullivan, and J. R. Donat, 1981. "Programmatic environmental impact statement on commercial OTEC licencing." *Proc. 8th Ocean Energy Conf.*, Washington, D.C., **1**, 453.

Miller, A. R., 1978. "A preliminary comparative study of historical sea surface temperatures at potential OTEC sites." *Proc. 5th Ocean Thermal Energy Conversion Conf.*, Miami Beach, Fla., **III**, 214.

Molinari, R. L., 1979. "Thermal and current data for the Gulf of Mexico and South Atlantic relative to placement of OTEC plants." *Proc. 6th OTEC Conf.*, Washington, D.C., June, **2**, 13.3-1.

Morello, A., 1978. "Bottom power cables connecting floating plants to shore." *Proc. 5th Ocean Thermal Energy Conversion Conf.*, Miami Beach, Fla., Feb., II-5.

Mossman, B. J., 1989. "OTEC methanol plantship proposal: A financial evaluation." Project in fulfillment of the requirements for the M.B.A. degree from the Wharton School of the University of Pennsylvania. (Document available from the author.)

Nuttal, L. J., 1981. "Utilization of ocean energy." *Proc. 8th Ocean Energy Conf.,* Washington, D.C., **1**, 679.

Ocean Thermal Corp., 1984. *40-MWe OTEC power plant.* Ocean Thermal Corporation, Preliminary design engineering report, **1-4**.

O'Connor, J. S., 1981. *Lightweight concrete OTEC cold water pipe tests, phase II.* Johns Hopkins Univ. Applied Physics Lab., SR-80-5A.

Olsen, H. L., G. L. Dugger, W. B. Shippen, and W. H. Avery, 1973. "Solar sea power plant conference and workshop." A. Lavi, ed., Carnegie Mellon Univ., Pittsburgh, June, p. 185.

Panchal, C. B., J. Larsen-Basse, L. R. Berger, J. A. Berger, B. J. Little, H. J. Stevens, J. B. Darby, L. E. Genens, and D. L. Hillis, 1985. *OTEC biofouling control and corrosion protection study at the sea coast test facility 1981-83.* Argonne National Lab., Report No. OTEC-TM5.

Pandolfini, P. P., J. L. Keirsey, G. L. Dugger, and W. H. Avery, 1980. "Alclad aluminum, folded-tube heat exchangers for OTEC." *Proc. 3rd Miami International Conf. on Alternative Energy Sources,* University of Miami, Dec.

——, E. J. Francis, and L. L. Perini, 1986. *OTEC system.* Johns Hopkins Univ. Applied Physics Lab. Internal report.

Paulling, J. R., 1979. "Frequency-domain analysis of OTEC CWP and platform dynamics." *Proc. 6th OTEC Conf.,* Houston, April, 1641.

——, 1959. "Frequency domain analysis of OTEC CWP and platform dynamics." *Offshore Technology Conf.,* Houston, OTC paper 3543.

Pieroni, C. A., R. T. Traut, D. O. Libby, and T. P. Garrity, 1979. "The development of riser cable systems for OTEC power plants." *Proc. 6th OTEC Conf.,* Washington, D.C., June, **1**, 7.2-1.

Quinby-Hunt, M. S., 1981. "Nutrient and dissolved oxygen studies at OTEC sites." *Proc. 8th Ocean Energy Conf.,* Washington, D.C., **1**, 537.

Resales, L. A., C. Dvorak, M. M. Kwan, and M. P. Bianchi, 1978. "Materials selection for ocean thermal energy heat exchangers." *Proc. 5th Ocean Thermal Energy Conversion Conf.,* Miami Beach, Fla., Feb., **4**, VIII-231.

Richards, D., W. H. Avery, E. J. Francis, M. Jones, B. Heyer, A. L. Kohl, and J. K. Rosemary, 1984. "OTEC methanol-from-coal process plantship studies." *Amer. Soc. Mech. Eng.,* 84-WA/Sol-32, 2.

Ross, J. M., and W. A. Wood, 1981. "OTEC mooring system development: Recent accomplishments." *Proc. 8th Ocean Energy Conf.,* Washington, D.C., June, 85.

Sasscer, D. S., 1981. "Microbiology of aluminum and titanium ocean thermal energy conversion (OTEC) evaporator tubes at a potential OTEC site." *Proc 8th Ocean Energy Conf.,* Washington, D.C., June, **1**, 403.

——, T. Morgan, and T. R. Tosteson, 1980. "Biofouling measurements from a moored floating platform off Punta Tuna, Puerto Rico." *Conf. Proc. 7th Ocean Energy Conf.,* Washington, D.C., June, **1**, 10.3-1.

Schultz, J. A., T. C. Dalton, and R. Eaton, 1981. "The development design status of riser cable systems for OTEC pilot plants." *Proc. 8th Ocean Energy Conf.,* Washington, D.C., June, 607.

Scott, R. J., and W. W. Rogalski, Jr., 1978. "Considerations in selection of OTEC platform size and configuration." *Proc. 5th Ocean Thermal Energy Conversion Conf.,* Miami Beach, Fla., Feb., 77.

Soden, J. E., R. Eaton, and J. P. Walsh, 1982. "Progress in the development and testing of OTEC riser cables." *Proc. Oceans '82 Conf.,* Washington, D.C., Sept., **1**, 587.

Starkman, E. S., H. K. Newhall, D. Sutton, T. Maguire, and L. Farber, 1966. "Ammonia as a spark ignition fuel: theory and application." *Soc. Automotive Engineers*, SAE paper #660155, 765.

Sverdrup, H. U., M. W. Johnson, and R. H. Fleming, 1942. *The oceans: Their physics, chemistry, and general biology*. Englewood Cliffs, N.J.: Prentice-Hall.

——, 1940, "On the annual and diurnal variation of the evaporation from the oceans." *J. Marine Research,* **3**, 93.

Taylor, B. J., and B. Shih, 1979. *An OTEC cold water pipe hydroelastic experiment*. National Oceanic and Atmospheric Admin., DOE/NOAA/OTEC-19, 1.

——, and R. McHale, 1984. "An OTEC slope mounted cold water pipe experiment." *Oceans 84*, **C6**, 345.

——, J. E. A. Anderson, K. M. Kalyanam, A. B. Lyle, L. A. Phillips, and J. E. Nitsh, 1986. "Technical and economic assessment of methods for the storage of large quantities of hydrogen." *Int. J. Hydrogen Energy*, **11**, 5.

Thomas, A., J. J. Lorenz, D. L. Hillis, D. T. Yung, and N. F. Sather, 1979. "Performance tests of the 1-MWt shell and tube heat exchangers for OTEC." *Proc. 6th OTEC Conf.*, Washington, D.C., **2**, 11.1.

Traut, R. T., J. E. Soden, J. P. Kurt, C. A. Chapman, and G. N. Okura, 1982. "Design of Hawaii deep water 250 kV DC power cable." *Proc. Oceans '82 Conf. Records*, Marine Technology Society, Washington, D.C., Sept., 1265.

Trimble, L. C., and C. M. Robidart, 1978. "A power system module configuration using aluminum heat exchangers." *Proc. 5th Ocean Thermal Energy Conversion Conf.*, Miami Beach, Fla., Sept., V-19.

——, and W. L. Owens, 1980. "Review of mini-OTEC performance." *15th Intersociety Energy Conversion Conf.*, Seattle, **1**, 1331.

U.S. Department of Energy, Assistant Secretary Energy Technology, Division of Solar Technology.

Waid, R. L., 1978. "OTEC platform design optimization." *Proc. 5th Ocean Thermal Energy Conversion Conf.*, Miami Beach, Fla., Feb., IV I-14.

Wolff, P. M., 1978. "Temperature difference resource." *Proc. 5th Ocean Thermal Energy Conversion Conf.*, Miami Beach, Fla., Feb., III, 11.

Wu, C., 1987. "A performance bound for real OTEC heat engines." *Ocean Engineering* **14**, 349.

Zener, C., and J. Fetkovitch, 1975. "Foam solar sea power plant." *Science,* **189**, 294.

——, and M. I. Kay, 1980. "Results of foam OTEC studies at Carnegie-Mellon University and University of Puerto Rico." *Proc. 3rd Miami International Conf. on Alternative Energy Sources*, 462.

2

OTEC HISTORICAL BACKGROUND

2.1 THE DEVELOPMENT OF HEAT ENGINES

As in other branches of technology, the understanding of the physical and chemical principles underlying the operation of heat engines followed long after such systems were in commercial use. Apparently both the ancient Egyptians and Chinese were able to use steam or combustion gases to do work in special applications; however, the first practical use of a heat engine was the steam-driven piston engine for pumping water from mines, invented in 1698 by the Englishman Thomas Savery. This was followed by a better device invented in 1712 by Newcomen and further developed by Smeaton, which was widely adopted for mining operations in the tin mines of Cornwall and the British coal mines. In 1763, James Watt invented his greatly improved steam engine, which laid the foundation for the industrial revolution based on steam power. Interesting accounts of these developments are presented in Fenn (1982) and Callendar and Andrews (1958).

By 1800, there were nearly 500 engines of Watt's design emplaced throughout England for pumping water, working metal, or other uses. Steam use in ships was successfully demonstrated by Fulton on the Hudson River in New York in 1807. Railroad transportation based on steam-driven locomotives was introduced by Stephenson in 1812 following small beginnings in 1801 by Trevithick.

As the steam engines of Newcomen were manufactured and installed, their performance was measured by the amount of water that could be pumped to a given height per bushel of coal burned. The heating value of the coal being used was approximately 1 million Btu per bushel. The data of Table 2-1 show how the thermal efficiency of steam engines improved with time. It is interesting to note that the industrial revolution began with engines of less than 1% efficiency and blossomed with the development of Watt's engine of 2.7% efficiency.

2.2 CARNOT, D'ARSONVAL, AND CLAUDE— EARLY FRENCH EXPERIMENTS

Watt and his predecessors related the performance of their engines in pumping water to what could be accomplished by horses engaged in the same task. An average value of the power capability of a horse was estimated by Watt, who

Table 2-1. Steam engine efficiencies (Forbes, 1958)

Date	Builder	Thermal efficiency (%)
1718	Newcomen	0.5
1767	Smeaton	0.8
1774	Smeaton	1.4
1775	Watt	2.7
1792	Watt	4.5
1816	Woolf compound engine	7.5
1828	Cornish engine	12.0
1834	Improved Cornish engine	17.0
1878	Corliss compound engine	17.2
1904	Triple-expansion engine	23.0

established the unit of one horsepower as the power needed to raise 33,000 pounds 1 foot in 1 minute. Thus, a determination of the efficiency of steam engines in service could be defined in terms of bushels of coal burned per horsepower delivered; however, no basis existed for estimating what performance might be attained by further engine improvements. The answer to this question was first stated by the French engineer, Sadi Carnot, in a paper written in 1824. Carnot recognized that the production of work by the thermal expansion of a fluid (or any substance) required the transfer of heat from a source at one temperature (T_1) to a heat sink at a lower temperature (T_2). He showed that the efficiency of such a process had a theoretical limit defined by the relationship:

$$\frac{\text{Work performed}}{\text{Heat consumed}} = \frac{T_1 - T_2}{T_1}. \tag{2.2.1}$$

Implicit in Eq. (2.2.1) is the recognition that heat and work are both forms of energy and should be expressible in the same units; however, the relationship was not established until 1843, when careful experiments by the English scientist James Prescott Joule gave a value of 770 foot-pounds of work as being equivalent in energy to 1 Btu of heat. More accurate measurements have established the mechanical equivalent of heat at its present value of 778.5 ft lb/Btu. Carnot's paper apparently went unnoticed by engineers until 1849, when William Thompson (later Lord Kelvin) presented to the Royal Society of England an analysis of the implications of Carnot's paper with respect to the theoretical performance of heat engines. Two years later Thompson and Rudolph Clausius, in Germany, independently enunciated the law of conservation of energy, which was established as the First Law of Thermodynamics. Carnot's equation became recognized as the Second Law of Thermodynamics, although it had been formulated 25 years earlier than the First Law.

Steam engine technology developed rapidly after 1850, aided by an understanding of the benefits of high temperature and pressure in achieving improved efficiency, which resulted from thermodynamic analyses by Lord Kelvin, Rankine, and others. A necessary supplement to the theoretical work was the concomitant development of precision manufacturing techniques, improved mechanical design, instruments for measurement and control of temperature and pressure, and stronger materials, all of which were applied to the design and construction of steam engines to meet the more stringent operating conditions of the more efficient engines.

The successful development of steam engines led to the recognition that the heat engine cycle could be run backwards to produce cooling if a gas such as ammonia or sulfur dioxide, which could be liquefied under pressure at room temperature, was substituted for water in the boiler of a heat engine. The process would involve bringing the material to be cooled into thermal contact with the boiler and then allowing the liquid to vaporize and expand at room temperature, doing mechanical work in driving a piston against atmospheric pressure. Jacob Perkins patented the first successful closed cycle using such a process in 1834. In 1856 Alexander Twining demonstrated the first commercial production of ice using liquefied gas expansion. Use of refrigeration equipment for food processing was begun by James Harrison in 1857, which provided a major impetus for further development.

As refrigeration applications evolved, various working fluids were tried and their thermodynamic properties measured and documented. Thus, by 1880, a good technological base existed for the design of engines, based on the Rankine cycle, that could use liquids other than steam as working fluids. Some attempts had been made to use compressed gases, such as air and carbon dioxide, for tramway propulsion. These were unsuccessful because of the cooling that accompanied the work produced by the expanding gas. Nevertheless, the potential simplicity of the process led a French scientist, Le Bon, to publish a paper in 1881 in which he proposed the use of these compressed gases as the motive power of the future because they offered a convenient method of storing energy and transporting it to a distance. Le Bon also stated that this source of energy, obtainable from natural sources, could provide power for trains and ships when supplies of oil became exhausted, possibly before the end of the century!

Le Bon's paper prompted a compatriot, Arsene D'Arsonval, to publish an article later in the same year that proposed heat engines with liquefied gases as the working medium to derive power from low-temperature sources of heat available in nature. D'Arsonval explained that such engines could produce appreciable power with a temperature difference between boiler and condenser as low as 15°C, and that such temperature differences were widely available in nature. In particular he noted that a suitable temperature difference existed in the oceans at the equator, where a temperature of 4°C could be found at a depth of 1000 m. Thus, engines could be designed to use the warm water at the surface to boil a fluid and the cold water at depth to condense it and provide the pressure difference for engine operation.

D'Arsonval's recognition of the tropical oceans as a potential source of world energy, through use of the temperature difference between surface water and deep water, entitles him to recognition as the father of the technology now known as ocean thermal energy conversion, or OTEC.[1]

D'Arsonval listed the data shown in Table 2-2 to illustrate the pressure difference that would be available for some potential liquefied gases that could be

Table 2-2. Pressure difference for potential working fluids (D'Arsonval, 1881)

Working fluid	Pressure difference (mm Hg)
Sulfur dioxide	137
Dimethyl ether	170
Methyl chloride	181
Ammonia	328
Hydrogen sulfide	550
Nitrous oxide	1390
Carbon dioxide	1646

used as working fluids. The table assumes a boiler temperature of 30°C and a condenser temperature of 15°C.

Apparently no attempt was made to implement D'Arsonval's proposal for power generation based on ocean temperature difference until 1926, when Georges Claude, a French engineer and former student of D'Arsonval, embarked on a program to demonstrate that ocean thermal power generation could be practical. Claude believed that the cost of constructing heat exchangers of the large area required for closed-cycle OTEC operation as envisioned by D'Arsonval would make the process uneconomical, and that corrosion and biofouling of the heat exchangers would pose further difficulties. He proposed to avoid those problems by using the warm seawater itself as the working fluid in an open cycle, now known as the Claude cycle. The basic process has been described in Section 1.2.

Claude demonstrated the feasibility of this concept in 1928 by an experiment at Ougree-Marhaye in Belgium in which the cooling water at 30°C from a steel plant was used as a source of warm water for the boiler (evaporator) and the water of the Meuse River at 10°C as the condensing fluid. The setup incorporated a turbine of 1 m diameter mounted in a pipe between an evaporator chamber evacuated to a pressure of 0.04 atm (the vapor pressure of water at 30°C), which induced boiling of warm water and production of steam, and a condenser vessel in which cold water with vapor pressure of 0.02 atm was sprayed to condense the vapor. The pressure difference between the two vessels induced rapid steam production and a high mass flow rate that could generate significant power in the turbine despite the very low pressure and temperature of the steam. In the test, the turbine attained a speed of 5000 rpm and generated a power output of 50 kW.

The success of this test enabled Claude to get financial backing for the next step: demonstration of the use of cold deep water from the ocean in combination with warm surface water to generate ocean thermal power.

For this purpose, Claude chose a site at Matanzas Bay in Cuba where the water temperature at the surface was 25 to 28°C and cold water was available at a depth of 700 m at a distance of 1600 m from shore. The experimental plant used at Ougree was moved to the new site. The major task was now the construction and deployment of the long pipe that would carry the cold water from depth to the plant on shore. The water flow needs of the 50-kW plant would have been met by use of a 0.6-m diameter pipe and cold-water flow velocity of 2 m/s; however, Claude estimated that heat flow through the pipe walls during the long transit through the 1600-m pipe would warm the water almost to the surface temperature before it arrived at the plant. Therefore, he chose to use a 2-m pipe, which would supply ten times the required flow of cold water and increase the pumping power tenfold but would reduce the warming to two or three degrees. Deployment of the large pipe was beset with severe problems, partly due to the site characteristics, partly to stormy weather, and partly to inadequate supporting facilities. Two pipes were lost in deployment, but Claude continued and on the third try was successful in installing a 2-km-long pipe, and operating the plant successfully. The results were in accord with his earlier engineering estimates and demonstrated the technical feasibility of the open-cycle concept. The cold-water pipe for this test failed in a storm after 11 days of plant operation and caused the project to be terminated; however, the successful operation proved for the first time the technical feasibility of conversion of ocean thermal energy to electric power.

For one of the tests in which 22 kW of power was produced, 200 liters/s of cold water entered the condenser at 13°C and left at 15°C. Warm water entered at the same rate at 27°C and left at 25°C. Thus the theoretical Carnot efficiency was 4.0%. The thermal efficiency was 1.3% (ratio of work output to heat taken from the water). Because of the power needed to pump the water flowing through the oversized cold-water pipe, the plant produced no net power (Claude, 1930).

In 1933, Claude installed a complete open-cycle plant, designed for ice production off the coast of Brazil, on the 10,000-ton barge *Tunisie* (Claude, 1933). The open-cycle power system was mounted in an enclosure 8 m in diameter and 25 m long, and included a turbine designed to produce 2000 kW of shaft power, of which 1200 kW would be used for the ice machine. The ship was to be connected by a flexible tube to a 2.5-m-diameter cold-water pipe suspended from a semi-submersible float. Unfortunately, during deployment, heaving of the float prevented successful attachment of the pipe, and in the process the pipe was lost, forcing abandonment of the project from depletion of funds. A schematic diagram of the Brazil plant from Marchand (1979) is shown in Fig. 2-1.

Despite his severe financial losses, Claude maintained his interest in OTEC. In 1940, he proposed a plan to the French government for a 40-MWe plant intended for installation at Abidjan, Ivory Coast, where a favorable site with a shallow lagoon of solar-heated water could be used as the warm-water source. A deep canyon filled with cold water approached to within 5 km of the coast. Claude proposed to tap the cold water through a tunnel from shore. The project encountered problems and proceeded slowly, but in 1948, a French government company, "Energie de Mer," was formed with specific objectives for development of the concept. The program led to several significant accomplishments (Marchand, 1979):

1. An optimum evaporator design was selected after many tests.
2. A de-aerator of French commercial design was shown to be effective and of low power consumption.
3. A low-pressure turbine of 8 m diameter was successfully constructed.
4. A complete, compact plant was designed with projected cold-water consumption of 1 m³/s per MW of gross power. A 2-m-diameter cold-water pipe was

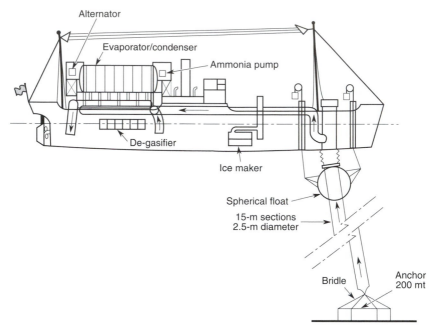

FIG. 2-1.   Design of Claude OTEC plant deployed in Brazil (Marchand, 1979).

designed and tested. This was made of rigid sections of steel pipe 6.25 m long, joined by 2.5-m-long flexible sections of rubber reinforced with steel springs.

5. A pipe 150 m long of this construction was successfully deployed at a depth of 300 m.
6. A photograph of a section of the cold-water pipe of the 3.5-MWe OTEC plant designed for the Abidjan installation is shown in Fig. 2-2.

After the development program was completed, plans were made in 1956 for construction of a 5-MWe plant. Cost estimates showed that the plant would be economical if a 30-year life could be attained and construction costs and financing assumptions were valid; however, the project was abandoned when it was announced that a large dam and hydroelectric plant were to be built at Abidjan.

Studies were made in 1958 to determine whether the plant would be economically attractive for installation in Guadaloupe, but the site was found to be unsuitable. This ended active French interest in OTEC until the 1970s.

2.3   AMERICAN INVESTIGATIONS PRIOR TO 1973—
THE ANDERSONS

There is no record of any interest in ocean thermal energy development after the end of the French work until 1963, when James H. Anderson, Jr., presented his thesis entitled "A Proposal for a New Application of Thermal Energy from the Sea" to the Massachusetts Institute of Technology for his B.S. degree. In 1964, Anderson

Fig. 2-2. CWP ready for deployment at Abidjan, Morocco (courtesy, P. Marchand).

and his father described the concept at a meeting of the American Society of Mechanical Engineers. A study of Claude's work convinced them that the open-cycle concept was impractical because the very low pressure of the steam in the Claude cycle required turbines too large to be feasible, and expensive and complex equipment would be needed for removing the gases dissolved in the seawater, which would otherwise accumulate and block the operation. Instead, the Andersons proposed to develop the original Rankine closed-cycle concept of D'Arsonval with improvements available in the background technology of the refrigeration and cryogenic industries.

Features of the Andersons' design, further refined in 1966, included the use of a floating plant with a power system employing propane as the working fluid, heat exchangers mounted at a depth at which the water pressure would approximately balance the pressure of the propane, and selection of construction materials and techniques that would minimize cost. A schematic diagram of the plant is shown in Fig. 2-3. A cost estimate indicated that their proposed system would be able to deliver power from sites in the Caribbean at lower rates than oil, coal, or nuclear alternatives. Although the Andersons' proposal was published in the journal of the American Society of Mechanical Engineers, it appears to have created no visible interest during the 1960s among energy planners, who assumed that nuclear power and fossil fuels would meet future energy demands at minimum cost.

2.4 SURVEY OF U.S. DEVELOPMENTS 1970–1985

The complacent view that U.S. energy demands would be met by the orderly development of nuclear power to provide electricity, and continued U.S. and world

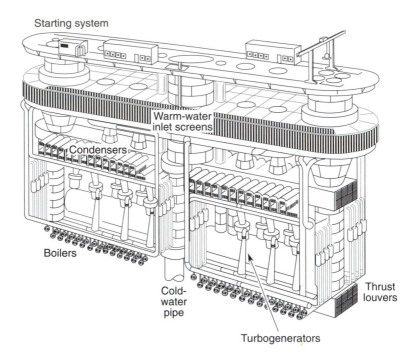

Starting system

Warm-water inlet screens

Condensers

Boilers

Cold-water pipe

Turbogenerators

Thrust louvers

FIG. 2-3. Design of the Andersons' 100-MWe OTEC power plant (Anderson, 1985).

production of low-cost oil and coal for other energy needs, was replaced by apprehension and drastic revision of energy forecasts in 1973. This resulted from two events totally unanticipated by the energy planning community:

1. A groundswell of concern developed about the effects on human health and quality of life of pollution of air, water, and land by automobiles, power plants, and industry. This apprehension, augmented by public concern about the hazards of nuclear developments, began in the late 1960s and grew to political power by 1973.
2. Arab potentates gained control of Mideast oil production, which led to formation of OPEC and worldwide escalation of petroleum prices, with severe impacts on U.S. and world economies.

The first event led to vigorous public opposition to the construction of nuclear and coal power plants; the second to rapid increases in the price of fuel for automobiles, power plants, and industry. In this situation, strong interest arose among scientists and engineers not affiliated with the energy establishment in the development of pollution-free, safe, and inexhaustible sources of energy that could replace the conventional sources now publicly unacceptable. Attention was focused on the direct use of solar energy to produce electric power and heat, and on energy sources indirectly dependent on solar energy such as winds, waves, tides, heat stored in the surface waters of the tropical oceans, and wood, kelp, and other forms of so-called "biomass" that could be processed to produce fuels.

The time was appropriate for a resurgence of interest in ocean thermal energy. A major contribution to this interest was made in an article in *Physics Today* written by the distinguished physicist, Clarence Zener (Zener, 1973). He called attention to the Andersons' design of an ocean thermal plant and its estimated costs, and stated that the rising costs of nuclear plants would make them more expensive than ocean thermal plants. After a careful review of positive and negative factors for ocean thermal plants versus nuclear and coal plants, Zener concluded that ocean thermal plants would be superior from both cost and environmental standpoints.

Zener's paper inspired interest by many research organizations in the United States and led the U.S. National Science Foundation (NSF) to expand its support of study programs on ocean thermal energy. (A small contract on the subject had been awarded to the University of Massachusetts in 1972.) Under NSF sponsorship, the first OTEC workshop was held at Carnegie–Mellon University (CMU) in June 1973 and attracted approximately forty attendants, representing NSF, CMU, the University of Massachusetts, The Johns Hopkins University Applied Physics Laboratory (JHU/APL), and representatives of government and industry.

Interest in OTEC, along with other solar-based energy sources, grew rapidly with strong support from the U.S. Congress. A sense of urgency was provided by the oil embargo of 1974. Funds for new energy alternatives were appropriated in response to public demands for the development of energy options that would be safe, environmentally acceptable, and nonpolluting. From this impetus, the Energy Research and Development Administration (ERDA) was established in January 1975 with the mission to evaluate all potential energy options for the U.S. future and to support development of the most promising forms. To provide foxes to guard the fledgling products being hatched, personnel for the new agency were drawn from the existing agencies concerned with energy research and development, in rough proportion to their budgets: namely, 90% from the Atomic Energy Commission (AEC), 10% from the Bureau of Mines, and 0.1% from the NSF, which acquired responsibility for all solar energy programs.

After its preliminary review, ERDA endorsed the development of renewable energy options but contended in public statements that OTEC, along with other solar energy options, could not be expected to contribute significantly to U.S. energy needs before the year 2000 and that emphasis in ERDA programs should be placed on development of nuclear and coal options. It was therefore stated that the funds supplied by Congress for OTEC should be used to support research programs on ocean energy subsystems and components that would provide a technical base for later OTEC development, if and when needed. This position accurately represented the opinions of the public utilities, the nuclear power industry, the oil companies, and the coal companies. It did not reflect the views of the scientists and engineers who had studied energy options in some depth, nor of the general public and their representatives in Congress. A situation developed in which the Congress every year appropriated more money for ocean energy development than ERDA [and its successor in 1977, the Department of Energy (DOE)] requested, and the funds were then distributed by DOE to support, for the most part, unfocused efforts on energy projects. In the OTEC program, the funds were used initially to establish a broad program to investigate the ocean sciences, engineering disciplines, and components basic to OTEC development, including two small contracts to consider potential OTEC system designs.

Reports on these projects were issued in 1975, with the conclusion that OTEC systems were feasible and projected costs were attractive. (The goals and accomplishments of the research and development programs are discussed in Section 1.2 of this volume.) As the work progressed, it became apparent that the technology was ready for the design and construction of pilot systems that would provide valid engineering data on which to base commercial development; however, reorientation of the work toward pilot plant construction was opposed by the DOE management. Instead, a major program was initiated to provide a large-scale test of one component of an OTEC system, the heat exchanger, and lesser efforts were directed to further studies of other components and subsystems.

The OTEC program of ERDA inherited a research orientation from the NSF and was devised with the long-range interests of the electric power industry in mind. An alternative base for OTEC support existed in the research and development program of the Maritime Administration (MARAD) of the U.S. Department of Commerce. MARAD's research and development goals were directed to enhancement of the commercial success of the U.S. Merchant Marine and were focused on efforts that would provide commercially attractive opportunities for new ship construction and products that would improve the U.S. balance of trade. The MARAD evaluation of OTEC, based on their strong background in marine engineering and the orientation of the division head, Marvin Pitkin, toward solving problems rather than studying problems, convinced them that OTEC should be oriented toward system development and gave MARAD a positive interest in supporting a research and development program that could lead to early commercialization of the concept.

With this objective, MARAD awarded a contract to JHU/APL in 1974 to investigate the "Maritime and Construction Aspects of Ocean Thermal Energy Conversion (OTEC) Plantships." The program considered site selection, plantship design, on-board production of ammonia and other OTEC products, maritime impacts, and legal and political implications. The final report on the program, issued at the conclusion of the contract (Avery et al., 1976), affirmed the feasibility of OTEC and stated that OTEC plantships should be capable of profitable production of ammonia and other products. With government support during development, it was projected that ammonia plantships could be in commercial operation within 10 years.

The positive conclusions of the report led MARAD to support a follow-on program to explore the investment requirements for commercial development of OTEC plants. A favorable report on this study was issued in 1977.

In April 1977, the results of these studies led JHU/APL to propose to MARAD the preliminary engineering design of a 20-MWe OTEC pilot plant that, after construction and testing at sea, could supply baseline engineering information on performance and costs needed by industry before commercial construction could be considered. After consultation between DOE and MARAD, the proposal was jointly funded by MARAD and DOE. Technical direction was assigned to the Ocean Energy Division of the Central Solar Technologies branch of DOE. The report on this project (George et al., 1979) led to an increase in the scope of the effort to include preparation of baseline preliminary design studies of a 40-MWe moored power plant for a site at Punta Tuna, Puerto Rico, and for a 40-MWe plantship for ammonia production, which would graze in an Atlantic Ocean region 500 to 1500 miles east of Brazil.

An important development during this period was the recognition by faculty members at the University of Hawaii and by State of Hawaii officials that unique facilities for OTEC research and development and other solar energy programs could be provided in Hawaii on the island of Hawaii (the "Big Island"). Under the leadership of John Craven, Dean of Marine Engineering at the University of Hawaii, and with the active support of Senator Matsanaga, the Hawaii legislature established the Natural Energy Laboratory of Hawaii (NELH), dedicated to the support of OTEC and other solar energy programs including mariculture, at Kea-hole Point near Kailua-Kona on the Big Island. Since the laboratory was located on a old lava flow that extended into the sea, the water directly offshore was typical of seawater in the open ocean and therefore nearly ideal for study of the effects of seawater biofouling on OTEC heat exchangers. This was considered to be a critical issue in judging OTEC practicality. The first studies of biofouling rates of OTEC heat exchangers were made here in 1975. The NELH has continued to play an active role in OTEC and is now the major contributor to OTEC development (see Section 2.7).

In 1977, frustrated by the Administration's reluctance to support development of a complete OTEC system, representatives of the Lockheed Corporation, the Dillingham Corporation, and the Hawaii State government decided to proceed without Federal support in an at-sea test of a complete OTEC system at a site near Kea-hole Point on the island of Hawaii. Loan of a U.S. Navy scow to serve as a platform was arranged. Funding for design and engineering of the power plant was provided by Lockheed. Operations and facilities support and environmental assessments were provided by the Hawaii State government. Funding for modification of the barge and for design, construction, and deployment of a polyethylene cold-water pipe was furnished by Hawaiian Dredging and Construction Co., a subsidiary of Dillingham Corporation. Alfa-Laval Corporation provided heat exchangers, instrumentation, and test support. Worthington Pump supplied seawater pumps, and Rotoflow Corporation provided a turbine generator. Arrangements for the cooperative program were completed in January 1977, and design and analysis began on February 15, 1977. Seventeen months later, installation and at-sea deployment of the complete system, christened "Mini-OTEC," were completed, including construction and deployment of a 0.7-m-diameter (28-in.), 670-m-long (2200-ft) cold-water pipe. After checkout and initial testing, a milestone in OTEC development was reached on August 28, 1978, with the first-time generation of net

OTEC power at sea. From program inception to demonstration, the generation of net OTEC power had required only 19 months!

Operation of Mini-OTEC continued with complete success until its scheduled shutdown on November 18, 1978. (A photograph of Mini-OTEC is shown in Fig. 1-4.) Over 600 h of operating time were logged. Attempts to gain DOE support for an extension of the test period to provide longer-term information on heat exchanger performance and operational reliability were unavailing; however, results of the test were in accord with the predictions and provided proof of the feasibility of OTEC deployment and operation at sea. A detailed description of the complete program is presented in Section 6.3.1.

Meanwhile, in the DOE program major effort was devoted to the design of a floating "heat exchanger test facility," later known as OTEC-1. As the program developed from 1975 to 1977, a plan emerged for modifying a "mothballed" T-2 tanker, *Chepachet*, to serve as a floating platform for "the sole purpose of testing scaled prototype heat exchangers and other equipment under a variety of operating and environmental conditions . . . over a 3½ year interval." A deliberate decision was made not to include a turbo-generator. Vigorous opposition to the plan was expressed by non-DOE members of the OTEC community on the grounds that the program would require funding comparable to a small pilot plant but would not produce data directly applicable to OTEC plant design. Specifically, it would not use any subsystem of a design proposed for OTEC large-scale plants, so it would not eliminate the need to build a pilot plant later. This objection applied to the platform, to the cold-water pipe (CWP), to the platform–CWP joint, to the mooring system, to energy transfer, to the power system, and even to the low-performance shell-and-tube heat exchanger planned for the first test series. Without a turbine, OTEC-1 could not generate any electricity and, therefore, could not produce information about total power plant performance. It also could not provide electric power to operate the water pumps, which had to be driven by diesel engines at a high fuel cost. The criticisms were ignored by DOE in the implementation of the program, which proceeded according to the announced plan, to deployment of OTEC-1 in November 1980, after some delays and cost overruns. A photograph is shown in Fig. 1-5.

The OTEC-1 test program was conducted until May 1981 when testing was terminated because of the high cost of providing diesel fuel to operate the water pumps, associated expenses for operating personnel, and funding restrictions. The test successfully accomplished its limited objectives (Castellano, 1981):

1. Deployment according to plan of the 2200-ft-long CWP, which was made of three 1.2-m-diameter (4-ft) polyethylene tubes 670 m long (2200 ft) bound together axially to form a single CWP (the pipe was assembled on shore, then towed afloat to the operational site 6 miles from Kea-hole Point, and finally upended and inserted into the gimbal support on the ship, where it operated satisfactorily until testing was discontinued);
2. Mooring of the vessel in water of 4500 ft depth, release from the mooring, operation in a grazing mode for several days, and reattachment to the mooring;

3. Successful operation of the CWP–platform connection during wind, wave, and current changes that resulted in CWP–platform angles up to 18°;
4. Operation of a shell-and-tube evaporator and condenser in a closed ammonia cycle at a 38-megawatt thermal (38-MWt) heat duty;
5. Demonstration of biofouling control with low-level chlorine injection. Tests were also made of the operation and effectiveness of the proprietary AMERTAP system, which circulates elastomeric spheres that scrub biofouling from the heat exchanger tubes as the balls pass through the shell- and tube-evaporator.

With the continuing rise in oil prices throughout the late 1970s, public support for solar energy development escalated. Many presentations on the status and promises of renewable energy development were made to members of Congress and their staff members to acquaint them with new energy opportunities offered by solar-based sources. This led to passage in 1980 of Public Law 96-310 that established a national goal to develop solar energy technologies that would be able to supply 1% of the nation's energy needs in the year 1990 and 20% in the year 2000. Goals for OTEC established in Public Law 96-310 of 1980 included:

1. Demonstration by 1986 of at least 100 MWe of electric capacity or energy product equivalent from OTEC systems and demonstrations of at least 500 MWe by 1989;
2. In the mid 1990s, achievement of an average cost of electricity or energy product produced by installed OTEC systems that is competitive with conventional energy sources for the Gulf Coast region of the United States and for islands in the United States, its territories, and its possessions; and
3. A national goal of 10,000 MW of OTEC electrical energy capacity or energy product equivalent by the year 1999.

A parallel bill, Public Law PL-320 of 1980, established guidelines for OTEC financial and regulatory assistance by Federal agencies.

With this support, a formal structure was in place for a U.S. program that could have led to demonstration of commercial-sized OTEC systems by 1990. The U.S. Congress appropriated $6.3 million of fiscal year 1980 funds to initiate development programs consonant with the goals of the legislation. In September 1980, in response to the Congressional directive, DOE issued a Program Opportunity Notice (PON) inviting industry to participate in a phased cost-sharing development program to produce a 40-MWe OTEC pilot plant that would demonstrate the cost and performance features necessary for reasonably sized, full-scale commercialization of the technology. The PON stated that DOE would fund five to eight contracts at $900,000 each for the first phase of a six-step program beginning with *conceptual design* of potential OTEC systems, and proceeding through *preliminary design, engineering design, construction–deployment–acceptance testing, joint operational test and evaluation,* and, finally, *transfer of ownership to private industry* for continuing operation. Federal cost sharing of 60 to 85% was promised. The stated purpose of the PON was to elicit a broad range of responses that would give maximum scope for innovative concepts, applications, and siting options, as well as management and cost-sharing proposals.

The response to the PON was enthusiastic. Consortia were formed by industry that included major representation from the shipbuilding, power plant, and refrigeration industries, pump manufacturers and others, who contributed many man-months of effort in formulating responses. Eight proposals were received representing three positioning options (two land-based, five moored, and one grazing); eight plant concepts; proposed sites in Puerto Rico, Hawaii, the Gulf of Mexico near Tampa, the Marianas Islands, and the Pacific Ocean south of Hawaii for the grazing plant; and several product and management options. However, with the change of Administration in 1980, the ground rules were changed. Only two conceptual-design contracts were awarded, both for shelf-mounted designs to deliver electrical power to the island of Oahu, Hawaii. The remaining $4.5 million of the appropriation was deferred to be used for a single second-phase contract award. This action of DOE was regarded by the OTEC community as a gross breach of faith, and dealt a body blow to major industrial interest in OTEC commercialization.

Two conceptual design contracts were awarded, one to General Electric (Lessard, 1982) and one to Ocean Thermal Corporation (1982). On the conclusion of these contracts, one contract was awarded by DOE to Ocean Thermal Corporation for the preliminary design phase, after which DOE terminated support for OTEC system development and announced that further development would have to be done exclusively with private funding. The preliminary design contract was completed in November 1984.

With installation of the new Administration in 1980, perceptions of the need for the alternative energy sources diminished. Funding was progressively reduced and, in 1984, the DOE OTEC system development program was terminated after the preliminary design of a land-based 40-MWe OTEC plant for siting at Kahe Point on the Island of Oahu, Hawaii, was completed. Plans were announced by Ocean Thermal Corporation for beginning construction of the 40-MWe plant in late 1986 with private financing but were put on hold when oil prices dropped precipitously in 1985.

No funds were requested for OTEC by DOE in fiscal years 1982, 1983, and 1984, but Congress appropriated modest funds specifically identified to support the research programs directed by the Solar Energy Research Institute and Argonne National Laboratory. In 1985, DOE announced a reorientation of its thinking about OTEC and requested funding for support of limited research efforts on open-cycle OTEC. The DOE interest was stimulated by a Japanese offer to support work in Hawaii through the Pacific International Center for High Technology Research (PICTR). Minimal funding was requested throught fiscal year 1989 for study of CWPs for small, shore-based plants and for research on power plant subsystems, but no funding was requested for 1990 and 1991.

2.5   JAPANESE PROGRAMS 1970–1985

In 1970, Japanese consideration of the strong dependence of their economy on imported energy led them to form a Committee on Investigation of New Power Generation Methods, which identified OTEC as one of the technologies to study.

Thus, Japanese interest in OTEC preceded the revival of interest in France and the United States. Preliminary results of the investigation, including a system concept, were presented by Kamogawa (1972) to the fifth general meeting of the Pacific Basin Economic Council in a paper entitled "Equatorial Industrial Marine Complex." The concept involved OTEC power generation and use of the nutrient-rich cold water for mariculture.

Interest in OTEC continued to develop and led to the inclusion of OTEC in a development program on new energy technologies called the "Sunshine Project," which was established in 1974 (Kajikawa, 1982). Concurrently, experimental research was begun at the Electrotechnical Laboratory of AIST in Tokyo and at the Science and Engineering Faculty of Saga University in Saga. Design studies of land-based OTEC power plants were also started in this period by the Tokyo Electric Power Service Co. (TEPSO).

The objectives and program accomplishments during the period 1974–1979 have been summarized by Kamogawa (1980) as follows: "The objectives of the OTEC feasibility study . . . were the evolution of technical and economic factors of the concept as a whole. As a first step, design work was conducted in 1974 on a 1.5-MW land-based experimental OTEC power plant. In the second year, 1975, a conceptual design was prepared for an ocean-based demonstration OTEC power plant rated at 100 MW and sited at a benign tropical location. The design was based on the current state of technology with minor improvements, particularly in the heat exchangers. In the third year, 1976, an improved conceptual design of a 100-MW power plant, to be located in the sea near Japan, was made. This later design was based on advanced technology. The results of the study were reported. In the following two years (1977 and 1978), additional work on the previous year's designs was carried out. Emphasis was placed especially on the optimum design of 100-MW plants, the platform structure and station-keeping at the representative sites, structure model experiments in a wave flume, and improvements in the heat exchangers."

The studies of 100-MW OTEC plants included estimates of total system investments and operating costs for floating and semi-submerged concepts, from which delivered power costs could be estimated. The results indicated that construction costs would be in the range of $2000–3000/kW, and busbar power cost would be in the neighborhood of 60 mills/kW h (1977 dollars) (Homma and Kamogawa, 1979). This estimate led to expectations that OTEC power could become competitive with other sources if commercial operation was established. Accordingly, a three-phase development plan was outlined that would involve major subsystem development and component testing and evaluation from 1978 to 1983, development of a 10- to 25-MW at-sea engineering test facility along with additional major component development during the period 1983–1989, and beginning of commercial construction in 1990.

Specific goals established for the first-phase program included:

1. Development of economically viable heat exchanger designs by laboratory tests, small-scale at-sea tests, and system simulation tests;

2. Development of the technology needed for the design of cold-water pipes and mooring systems through analysis, simulation, and small-scale model tests in water tunnels and at sea;
3. Investigation of oceanographic conditions and the thermal resources for candidate sites at sea; and
4. Development of the design methods to determine the optimum OTEC system from an economic standpoint for the proposed sites, recognizing the temperature distributions and seasonal variations in the thermal resource.

The first-phase program led to a number of significant accomplishments:

1. An on-shore, 100-kW closed-cycle OTEC test facility was installed on the seashore of the island of Nauru, in the equatorial Pacific Ocean at latitude $0.5°$S and longitude $167°$E. The plant employed a polyethylene CWP of 70 cm inner diameter and 1250 m length that drew cold water from 700 m depth. This gave a ΔT of 24°C. Titanium shell-and-tube heat exchangers were installed. A thin copper coating on the water side of the tubes in the evaporator and fluting of the working-fluid side in the vertical condensers were used to enhance heat transfer. The plant operated successfully for a series of experiments conducted over a period of about 1 year that confirmed predicted performance. A maximum power output of 120 kW and net power of 31.5 kW were demonstrated. Testing was concluded in September 1982 (Ito and Seya, 1985).
2. A Japanese "Mini-OTEC" was deployed successfully in October 1979 at a site off the coast of Shimane in the Japan Sea, where the temperature was 23.1°C at the surface and 0.5°C at a depth of 230 m (Uehara et al., 1980). Details of the test are given in Section 6.3.4.
3. A research and development program based on a 50-kW experimental facility was established at Saga University to investigate the performance of plate-type heat exchangers and other power system components, and to provide information for system optimization of OTEC power plants. The data obtained were used for analysis of an optimum power cycle for a 100-MWe OTEC power plant (Uehara, 1982).
4. Construction of a 50-kW closed-cycle OTEC plant was started by the Kyushu Electric Power Co. in Tokunoshima, on the southwestern coast of the island of Kyushu. The design employs plate-type heat exchangers and ammonia working fluid.
5. Conceptual designs of 1-MWe barge- and spar-type OTEC pilot plants were prepared. An OTEC simulator was developed to measure the sensitivity of the system performance to variations in the input conditions, for example, effects of heave, pitch, and roll on pump operations. Follow-up led to the preliminary design of a 1-MWe barge-type test facility for a site off Kumejima, near Okinawa. The steel barge (100 m × 30 m × 17 m, draft 12 m) would be attached to a mooring buoy that would support a steel CWP 1.8 m inner diameter × 700 m long (Kajikawa et al., 1978).
6. Experimental programs were initiated to support the design work with data on advanced materials, design methods, and procedures.

2.6   FRENCH PROGRAMS 1970–1985

Interest in OTEC was renewed in France in 1973 but did not become active until 1978, when studies were begun to explore OTEC opportunities in the Pacific islands under French supervision. The program was outlined in three phases:

1. A 2-year study program to evaluate the technical and economic feasibility of OTEC plants of 1 to 18 MWe. Subjects considered included closed- and open-cycle power systems, shore-based and floating options, CWP design including tunneling through the break-water zone, site selection, and economic tradeoffs among electric power production, supply of fresh water, and mariculture. This work led to the conclusion that both closed and open OTEC cycles were feasible and economically promising. Construction would be possible with available equipment and facilities in the size range of 1 to 10 MWe. Moored, floating plants, and land-based plants were judged to be feasible, but the last option was preferred for the pilot plant because of accessibility, a presumption of lower risk, and the fact that the land installation would provide options for mariculture and uses of cold water. Fresh-water production would not be needed in Tahiti but could be studied for applications at other sites. This phase was completed in 1982.

2. The second phase of the program began in 1982 with projected completion of the preliminary design of a 5-MWe pilot plant for a land site in Tahiti by 1985. Work in this phase included:

   a. A detailed documentation of the near-shore environment over a 3-year period was prepared, including bathymetry and photography of the sea floor, water temperatures and compositions, waves and swells, macrofouling of potential materials of construction, and seismic measurements.

   b. A consortium, "ERGOCEAN GIE," formed from a number of major French companies, was assigned responsibility for investigating the power plant design options involving the heat exchanger cycles and the CWP. The open-cycle studies focused on the degassing problem and selection of evaporator and condenser designs. At Grenoble, a facility of 400 kW capacity was designed for this purpose. Work on the closed cycle concentrated on the heat transfer characteristics of shell-and-tube heat exchangers of titanium, aluminum, and inox (stainless steel) with smooth and enhanced surfaces. This work was supported by investigations of corrosion, biofouling, and cleaning at the IFREMER facility at Brest.

      The cold-water pipe investigations were directed to definition of opti-mum materials and methods of construction for the pipe, identification of the most attractive procedures for its deployment and installation, and detailed study of the configuration of the sea shelf to indicate suitable trajectories for pipe emplacement. Composite constructions of fiber glass and resin for the CWP, and buoyant versus bottom-mounted designs for its positioning were studied. Mathematical analyses and simulations were conducted to evaluate the dynamic behavior of the pipe during deployment and when subjected to current forces after installation. It was specified that the pipe would be 3 m in diameter and 3 km long.

   c. Financial analyses and economic projections were made to evaluate the potential role of OTEC as a future energy source. Considerations of future costs of petroleum and expected technical development of OTEC led to a conclusion that 50-MWe OTEC plants could compete in the midterm with diesel-powered generation of electric power. Benefits of fresh-water production were included.

The results of these studies led to a provisional decision to proceed with the design of a 5-MWe land-based OTEC pilot plant that could be in operation in 1989.

2.7 OTEC PROGRAMS 1986–1990

The sharp drop in the price of crude oil in January 1986 led to a worldwide reduction in support of alternative energy research and severe cutbacks in funding. As a result, plans for private or cost-shared funding of OTEC development were greatly reduced or put on hold in the United States, Japan, and France; however, some earlier programs have continued at a small scale. As of mid-1991, renewed interest is being exhibited.

Research efforts on open-cycle OTEC have continued at the Natural Energy Laboratory of Hawaii (NELH) at facilities provided by the government of the State of Hawaii. The work has benefited by commercial development at NELH of several privately funded mariculture ventures, initially designed to use the nutrient-rich cold water available from the open-cycle OTEC experiment. NELH has also provided facilities for development of a new design for the aluminum heat exchanger that will greatly reduce OTEC heat exchanger costs.

Engineering studies and experiments have also continued at Saga University in Japan (Uehara, 1990). Designs have been announced for a 5-MW on-shore plant and for a 25-MW floating plant for Philippine Island sites. The projected cost of power from the floating plant is competitive with coal power plants and with the 5-MW plant with diesel electric generation.

It is reported that plans are being made for construction of an OTEC demonstration plant in Taiwan, based on the design studies of Liao et al. (1986).

Further discussion of OTEC development programs is presented in the following chapters.

NOTES

[1]In the same article D'Arsonval notes that the temperature difference between hot springs and local river water, or between rivers and ice in glaciers, could be used, and he also suggests the possibility of storing winter snow for summer generation of power in Paris, in conjunction with the use of warm water from the Seine. He comments that he could suggest other possibilities but does not want to step on the terrain of Jules Verne.

REFERENCES

Anderson, J. H., 1985. "Ocean thermal power: the coming energy revolution." *Solar and Wind Technology* **2**, 25.

Avery, W. H., R. W. Blevins, G. L. Dugger, and E. J. Francis, 1976. *Maritime and construction aspects of OTEC plantships*. Johns Hopkins Univ. Applied Physics Lab., SR76-1B.

Callendar, H. L., and D. H. Andrews, 1958. "Heat." *Encyclopedia Britannica* **11**, 315.

Castellano, C. C., 1981. "Overall OTEC-1 status and accomplishments." *Proc. 8th Ocean Energy Conf.,* Washington, D.C., **2**, 971.

Claude, G., 1930. "Power from the tropical seas." *Mechanical Engineering,* **52**, 1039.

——, 1933. "Sur une usine flottante Claude-Boucherot." *Academie des Sciences (France),* May.

D'Arsonval, A., 1881. "Utilisation de forces naturelles." *Revue Scientifique,* **2**, 370.

Fenn, J. B., 1982. *Engines, energy and entropy*. New York: W. H. Freeman.

Forbes, R. J., 1958. in *A history of technology*. Vol. 4, p. 164. New York: Oxford University Press.

George, J. F. , D. Richards, and L. L. Perini, 1979. *A baseline design of an OTEC pilot plantship*. Johns Hopkins Univ. Applied Physics Lab., SR 78-3A.

Homma, T., and H. Kamogawa, 1979. "An overview of Japanese OTEC development." *Proc. 6th OTEC Conf.,* Washington, D.C., June **1**, 3.3.

Ito, F., and Y. Seya, 1985. "Operation experiences of 100-kW OTEC and its subsequent research." *Marine Technology Society, Ocean Energy Workshop*. Washington, D. C., 35.

Kajikawa, T., T. Agawa, H. Takayawa, K. Nishiyama, M. Amano, and T. Homma, 1978. "Study on OTEC power cycle characteristics with ETL-OTEC-II experimental facility." *Proc. 5th Ocean Thermal Energy Conversion Conf.* **2**, V-164.

——, 1982. "Status of R&D activities in Sunshine project." *Marine Technology Soc., Ocean Energy Workshop*. Washington, D. C., 35.

Kamogawa, H., 1972. "Equatorial marine–industrial complex. A report for the Natural Resources Development Committee." *Proc. 2nd International Ocean Development Conf.,* Tokyo, Oct. 1.

——, 1980. "OTEC research in Japan." in *Energy*. Pergamon Press Ltd.

Lessard, D., 1982. "Conceptual design of a tower-mounted, 40-MWe pilot plant for Oahu, Hawaii." *Proc. 5th Miami Conf. on Energy Sources,* Miami Beach, Dec.

Liao, T., J. G. Giannotti, P. R. Van Mater, Jr., and R. A. Lindman, 1986. "Feasibility and concept design studies for OTEC plants along the east coast of Taiwan, Republic of China." *Proc. International Offshore Mechanics and Arctic Engineering Symposium,* **2**, 618.

Marchand, P., 1979. "The French OTEC program." *Proc. 6th OTEC Conf.,* Washington, D. C., June, **1**, **790631**, 3.2-1.

——, 1986, "Ocean renewable energy resources: A chance for the future?" *Conference SUT "Exclusive Economic Zones,"* London, 1.

Ocean Thermal Corp., 1982. *Ocean thermal energy conversion (OTEC) pilot plant conceptual design study*. Ocean Thermal Corp., internal report.

Paoloni, E. C., 1982. "Conceptual design of a 40-MWe OTEC pilot plant on an artificial island off Oahu, Hawaii: OTEC power for Hawaii." *Proc. 5th Miami Conf. on Alt. Energy Sources,* Miami Beach, Dec., 3.

Uehara, H., 1982. "Research and development on ocean thermal energy conversion in Japan." *Proc. American Institute of Chemical Engineers 17th IECEC Conf.,* Washington, D. C., June, **4**, 1454.

——, 1990. "Ocean energy." *Japan Society for Mechanical Engineers* **14.4**, 1.

——, S. Nagasaki, and H. Yokohama, 1980. "Deployment of cold water pipe in the Japan Sea." *Proc. 7th Ocean Energy Conf.,* Washington, D. C., June, **II**, 14.4.

——, H. Kusuda, M. Monde, T. Nakaoka, T. Yamashita, and N. Maeda, 1981. "The test of OTEC plant in Imari Bay." *Proc. 8th Ocean Energy Conf.,* Washington, D. C., June, **2**, 803.

Zener, C., 1973. "Solar sea power." *Physics Today*, January, 48.

3

OTEC SYSTEM CONCEPTS

3.1 GENERAL COMMENTS ON SYSTEMS ENGINEERING

Systems engineering is a top-down approach to program management and systems procurement. It optimizes the development process by ensuring that the operational, technical, and cost goals (and limitations) of a total proposed system are understood before development begins. The requirements for the "forest" are determined before the features of the "trees" are specified. It makes a basic assumption that a team endeavor under single-system management will be established with authority to define development goals and assign subsystem programs and funding. It recognizes that each system requires a unique management structure that is based on the qualifications of the people and organizations available for the total endeavor.

Systems engineering begins with an authoritative request or requirement for a system that would provide new capabilities or would reduce existing problems in a significant technical activity. After personnel and level of effort for a preliminary assessment of the need are identified, the initial effort then involves these steps:

1. A precise definition is prepared of the specific operational need for which the proposed system must provide a solution. For example, this book addresses the present national need for a new energy system that can provide a practical, timely, cost-effective, and nonpolluting alternative to petroleum-based fuels for transportation. The need arises from three factors:
    a. The perception that an alternative to dependence on petroleum fuels for transportation must be developed to avoid severe disruption of world economies in the early years of the twenty-first century;
    b. Evidence that combustion of fossil fuels is causing a significant increase in the carbon dioxide content of the atmosphere (if not reduced, this could eventually produce a "greenhouse effect," leading to large-scale changes in climate and an increase in sea level, with severe economic consequences); and
    c. The belief that solar energy can be used via OTEC to supply nonpolluting fuel in sufficient quantity, at low enough cost, and in time to become a practical alternative to dwindling or unavailable petroleum supplies.

Failure to define the system need with sufficient clarity is a root cause of most system development difficulties. It must not be assumed that the government or other sponsor's statement of a requirement is adequate. The need must be stated in systems engineering terms, including the criteria for determining the suitability of the system to be developed and the date when the new system must be operational.

2. A listing of alternative system concepts for meeting the need is prepared. For alternative energy sources, this includes fuel synthesis from coal, nuclear power, hydropower, photovoltaic generation of electricity, solar thermal heat production, energy from biomass, OTEC, wind power, and other possibilities.

3. A first-order evaluation is made of the concepts identified in (2) for technical feasibility, potential contribution, environmental impacts, manufacturing capability, operational practicality, rough cost, and estimated development time. The purpose of this evaluation is to select three or four options of major promise that fulfill the operational need for in-depth consideration. A common difficulty in this task is a desire to avoid choosing any option until it can be proved that the "best" concept has been identified, even though several options would apparently meet the system requirements; however, selection of the favored concepts should not be based on a rating system that incorporates inadequate data. Engineering judgment must not be replaced by decisions based on a matrix of faulty numbers.

4. An engineering study (conceptual design) is made of each of the attractive concepts to document component performance and cost where available, to identify needs for additional information, and to suggest research and development programs that would provide information required to assure technical feasibility and minimum cost for the concepts.

5. Development programs are outlined for one or more of the favored concepts. In each program, the research and development needs are identified in separate subprograms, each of which is evaluated according to the rationale outlined. A multiphase development plan is then prepared that outlines a work schedule for each subprogram, showing estimated completion dates and identifying efforts and funding critical to completion of the program on the desired time scale.

6. If the required funding is available and time scales are suitable, development programs are established that proceed through the phases of preliminary design, engineering design, construction, deployment, and operational testing, with review at the end of each phase. The final phase is commercial construction and enduring operation of the system.

The requirement for a new system to be designed to achieve a minimum cost deserves special comment. Since World War II, research and development of major new systems have been funded by government requirements for advanced technology to meet perceived threats to national security or public welfare. The goal of such efforts has been to move the science and engineering underlying the new concept beyond existing frontiers. Advanced performance became the criterion for excellence. Cost entered the picture as the final step in decisions among candidate systems, which were then required to adhere to pre-established design specifications. Such programs were regarded as high risk, were exciting to work on, and attracted the most competent program participants. Conversely, programs with a goal to reduce cost while maintaining established performance came to be regarded as dull, suitable for second-level personnel, and unimportant.

This attitude must be changed in planning renewable energy developments, which must be of reasonable initial cost if they are to attract industrial support as opposed to governmental support and become commercially successful.

Costs of the first production version of systems requiring development of new technology are usually underestimated, often by large amounts. A common cause is optimism about the expected performance of undeveloped components or subsystems, and failure to consider off-design features. The largest errors result from several faulty assumptions or procedures in the design process:

1. Bottom-up rather than top-down systems design. This procedure, favored by government program managers, leads to development and optimization of system components by different contractors as isolated items. In final assembly, it is found that components or performance is incompatible and that system redesign is required. This leads to extended schedules and large cost overruns. It is analogous to awarding individual contracts to makers of a jigsaw puzzle, one contract for manufacture of the red pieces, another for the blue pieces, and so on, with a final contract to a systems contractor to assemble the puzzle using the delivered pieces.
2. Failure to involve industrial contractors who would be potential producers and installers of the system in the initial phases of the systems design. Many systems at the end of development are faced with major redesign because the components or systems are not compatible with manufacturing facilities or procedures for cost-effective production.
3. Failure to identify design for minimum cost as a primary development goal. The desire of engineers to design for performance that exceeds the state of the art, with cost as a secondary objective, can introduce unforeseen problems and delays, with serious cost overruns.
4. Lack of a well-defined development, test, and production plan with appropriate, scheduled funding subject to timely reviews of progress.

3.2   OVERALL REQUIREMENTS AND OPTIONS FOR ALTERNATE ENERGY SOURCES

In 1987, 37% of the world energy production was provided by petroleum fuels (Statistical Abstracts, 1990). It is recognized that the petroleum resources of the earth are finite and that world demands for liquid fuels are depleting the reserves at a rapid rate. This could eventually make petroleum fuels so costly that U.S. and world economies would be critically affected. Major national security issues are also involved because 80% of the world oil reserves are in the Middle East. It is imperative that cost-effective alternatives to petroleum be developed in time.

Much effort has been given to the search for alternatives to the use of petroleum fuels. Coal may be substituted for some fraction of the liquid or gaseous fuels used by industry to provide process heat, but coal is impractical as an energy source for transportation or general purposes. Thus, it may be estimated that approximately 40% of the world's energy needs must be supplied by liquid or gaseous fuels, if the present forms of transportation and nonindustrial energy uses are to be continued.

Several alternative solutions to the problem of petroleum fuel depletion have been proposed:

1. Improvement in the efficiency of liquid fuel use. This is the most attractive immediate approach because most of the equipment and devices that use liquid or gaseous fuel that were designed before 1980 sacrificed efficiency for lower cost or did not consider efficiency important in the design. Thus, large gains can be possible with small changes in design. Implementation of this approach in the United States since the mid-1970s has significantly reduced fuel use with minor impacts on life styles or equipment costs. Major gains in efficiency would be possible if practical methods were developed for replacing internal combustion engines with electric motors powered by fuel cells (Flavin, 1992); however, it is clear that with continued increases in world demands for energy, particularly in the developing countries, efficiency improvements will not be sufficient.

2. Production of synthetic conventional fuels from coal in land-based plants. This approach has been shown to be technologically feasible but was not commercially attractive with projected costs of crude oil at $28 per barrel and natural gas at $5 per million Btu. The proposed equipment and processes require the use of approximately two atoms of carbon for each atom incorporated in the liquid fuel. Thus, the greenhouse problem would be doubled by using coal with these processes to replace petroleum fuel. The environmental concerns would add to the cost.

3. Substitution of liquid-fuel energy by electrical energy produced in nuclear or coal-fired power plants. Electrical heat can replace oil or gas used for heating, but at high cost. Heat from electrical energy at $0.06/kW h is equivalent in cost to heat from oil at $74/bbl. Higher efficiency can compensate to some extent. Electric energy can be used to charge storage batteries, which can provide motive power for vehicles; however, great reductions must be made in battery weights and volumes before battery-powered vehicles could be practical for ordinary transportation purposes.

4. Production of fuels from biomass, such as wood, organic waste materials, kelp, or agricultural products. This approach is technically feasible and has been successfully used for production of ethyl alcohol for motor vehicle fuel, particularly in Brazil (Renner, 1989). Large government subsidies have encouraged development of an infant industry in the United States. With this option, an equilibrium energy cycle could possibly be established in which the CO_2 produced by combustion of biomass-produced fuels would be consumed

in the production of new biomass, with no net increase in the concentration of CO_2 in the atmosphere. It is an interesting possibility, but much further investigation will be needed to assess the chemical and biological equilibria. The costs and energy supply possibilities have not been estimated on a worldwide scale. A fundamental difficulty with this approach arises from the fact that the fuel is produced in a dilute solution and the energy required to concentrate and purify it can approach or exceed the heating value of the fuel.

5. Production of alternate fuels using electric energy to electrolyze water to form hydrogen and oxygen, followed by conversion of the products into fuels suitable for transportation or other energy needs. This approach can be attractive if electric power can be made available at low enough cost to allow production of the fuel product at acceptable prices. It is the basis for the strong interest being shown internationally in a future "hydrogen economy," based on solar or nuclear generation of electricity as a primary energy source (Marchetti, 1987; Veziroglu, 1987). Public perceptions of hazards of nuclear power from accidents and radioactive wastes may bar large-scale development of low-cost nuclear power. Direct use of solar energy for electricity production via photovoltaic cells has become progressively more efficient but requires collection and storage equipment that may make it uneconomical or environmentally unacceptable. Hydroelectric, wind, geothermal, and wave power offer limited opportunities. OTEC is unique as a safe, environmentally benign, vast, and inexhaustible energy source. It appears to be the most attractive option to meet world total demands for liquid fuels safely, economically, and without atmospheric pollution or adverse environmental impacts.

These considerations make OTEC a leading candidate for systems engineering evaluation as a practical way to produce a significant alternative to petroleum-based liquid fuels. To be acceptable, the OTEC system, when commercially available, must satisfy the following requirements:

1. It must offer a practical potential to furnish a significant fraction of the world liquid fuel demands–1 million barrels per day petroleum fuel equivalent by about the year 2000, possibly 15 million bbl per day by 2025.
2. It must have low enough cost to withstand competition from remaining sources of petroleum-based fuels and synthetic alternatives. Programs and developments to attain this cost objective should have highest priority. Ideally, the cost liabilities of dependence on petroleum fuels, including environmental factors, subsidies, and national security, would be weighed in the competition.
3. It must be environmentally acceptable.
4. It must be safe and be perceived to be safe by the public.

Commercial development of OTEC would satisfy all of these requirements.

A further, highly important but nontechnical OTEC requirement must be recognized: Commercialization will not take place unless industrial and political support can be gained for a development and commercialization plan that will achieve and maintain financial backing for OTEC development. Funding must be available through successive phases of engineering design, pilot demonstration, full-scale construction and checkout, and finally, construction and deployment of cost-effective plants in sufficient numbers to provide a significant alternative fuel capability. The lack of such a plan has curtailed progress toward commercialization of OTEC technology in the United States.

3.3   OTEC PRELIMINARY SYSTEMS ENGINEERING

The major factors that govern the design and selection of OTEC systems that could meet the requirements listed are examined in the following sections.

3.3.1 *OTEC Energy Resource Evaluation*

A brief discussion of the resources of warm and cold water in the tropical oceans that could be used for OTEC operations was presented in Section 1.1. In this section, information needed for selection of sites suitable for operation of OTEC plants and plantships, for evaluation of environmental interactions, and for prediction of performance as a function of design variables is presented. References to reports containing further information for detailed design requirements are also included. The information available includes not only temperatures but also data on waves, currents, winds, and marine biology pertinent to OTEC design.

3.3.1.1 Ocean Temperature Distribution (ΔT)
Oceanographic data from worldwide compilations of bathymetric surveys have been used by Wolff (1978) to prepare maps presenting the annual ΔTs in the ocean areas of interest for OTEC. The isotherms are presented in Fig. 1-1. Wolff et al. (1979) tabulated detailed data for each month of the year for ten potential OTEC sites selected by the U.S. Department of Energy (DOE). A map showing the location of the sites is reproduced in Fig. 3-1. The OTEC plantships would use

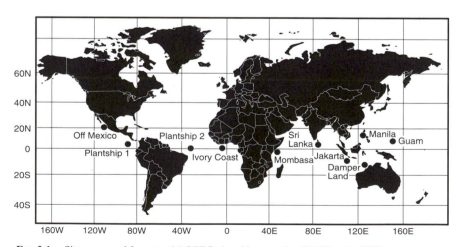

FIG. 3-1.   Sites surveyed for potential OTEC plantship operation (Wolff et al., 1979).

satellite observations of ocean surface temperatures as a basis for continually stationing the plantships throughout the year in the local regions of highest ocean surface temperature.

The available data show that the cruising procedure will improve the average annual ΔT available to OTEC by 1 to 2°C compared with fixed positioning. The average power output will be increased by 12.5 to 25%.

The ΔT data may be used for an estimate of the net OTEC power that will be produced at the ocean sites listed. As explained in Section 1.2, net OTEC power will vary as ΔT to about the 2.5 power.

The data show the advantages of operating OTEC at sites of high average ΔT; for example, the average net power of the grazing plant is 40% higher than the power of the land-based plant of the same design sited in Hawaii. The data also show the penalties that are associated with large seasonal variations in ΔT. For example, the net power at the Mobile, Alabama, site in the Gulf of Mexico would vary from 25% of the annual average value in February to 200% of the averaged value in August.

The reports of Wolff present temperature data at 100-m intervals from the surface to 1500-m depth for the sites listed. In addition, a temperature profile to 1500-m depth in Puerto Rico is presented by Goldman and Pesante (1979) and to the same depth near Kahe Point in Hawaii, by Wolff (1978). Data by Hill and Dugger (1979) are given for a 900-m depth. Limited data on temperature profiles have also been published by Uehara (1982) for the island of Nauru (0.5°S lat., 167° E long.), by Marchand (1980) for Tahiti, and by Munier et al. (1979) for the island of St. Croix (U.S. Virgin Islands). These data may be used to assess the tradeoffs at each site between improved plant output with a larger ΔT and larger system cost as the cold-water pipe is lengthened to draw in colder water available at greater depth.

The area in which the ΔT between the ocean surface and 1000 m depth exceeds 22°C, as shown in Fig. 1-1, is approximately 60 million square kilometers. An estimate of the bus-bar power that could be generated annually by deployment of OTEC plants in this region at a spacing that would be environmentally acceptable exceeds 10 million MWe (net) (see Chapter 9). This would provide on-board OTEC power for potential production of fuel energy of 200 quads/year of ammonia or 640 quads/year of methanol (Avery et al., 1985). These high-octane liquid fuels have been demonstrated to be good alternatives to gasoline. Gasoline consumption in the United States in 1987 was 14 quads of fuel energy.

3.3.1.2 Winds, Waves, and Currents versus Geography and Time

Structures intended for use in the ocean environment must be designed to withstand the most severe environmental conditions expected during the projected life of the installation. Allowance for fatigue of structures, induced by the fluctuating ocean environment in normal operation during the plant's life, must also be included in the design (Giannotti et al., 1981). The general requirements are common to all marine structures. Over the years, millions of items of data have been collected on the physical characteristics of waves, winds, and currents throughout the ocean areas and shorelines of the world (Churgin and Iredale, 1979). The data are most numerous for regions important to world commerce and may be sparse for many sites of potential OTEC interest; however, an adequate data base is available for

OTEC preliminary design purposes. These data have been used to estimate, by time of year and geographical location, the stresses that would be imposed on planned structures and to provide a basis for appropriate structural design. (The design requirements are discussed in Chapter 4.)

The most severe stresses are associated with the massive waves produced by hurricanes, which have reached record heights of nearly 33 m (100 ft) at latitudes above 30°. Hurricanes do not occur at latitudes in the ±10° zone centered on the equator; so the structural requirements are significantly less for ships and installations confined to that region.

Methods for predicting hurricane frequencies and intensities were developed and have been refined since 1959 by Bretschneider. His formulas are now accepted as the standard for design calculations (Bretschneider, 1975, 1977, 1978, 1979). For example, his formulas predict for the most severe storm in 100 years, maximum wave heights of 16 m for the tropical mid-Atlantic Ocean and 27 m for the southern coast of Puerto Rico (George et al., 1979). The damage potential to structures from waves of this magnitude with their associated currents has major significance for the design of floating OTEC platforms as well as shore-based installations.

Stresses imposed by winds on exposed structures must be considered in OTEC design but are small compared to the loads induced by wave action, and play a secondary role in OTEC systems engineering. Fatigue problems in normal operation are also expected to be of secondary importance for overall system design, and will be less severe in the more benign environment near the equator. There may be a need in some components to design around possible resonance effects associated with differences in average wave frequency at potential sites.

3.3.1.3 Conclusions Regarding the OTEC Resource

It is clear from the preceding discussion that the ocean energy resource that could be used by grazing OTEC plantships is ample to provide an alternative to U.S. dependence on petroleum fuels.

The major resource implications for OTEC systems engineering are that power output will be much higher for plants operating with the high ΔT available near the equator, and that structural requirements and costs will be appreciably less for installations sited in the less rigorous environment at low latitudes. The large waves and currents that develop in shallow water during storms place special structural demands on near-shore and shore installations. Thus, much lower cost per kilowatt of on-board power can be expected for plants designed for grazing operation at ocean sites near the equator. This conclusion is of fundamental importance in the selection of OTEC system alternatives.

3.3.2 *OTEC Power Systems Alternatives*

Two generic OTEC systems employing a gaseous working fluid to drive a prime mover were described in Chapters 1 and 2:

1. The closed-cycle (CC) process first described by D'Arsonval (1881) employing the vapor created by warming a liquefied gas;

2. The open-cycle (OC) process described by Claude and Boucherot (1928) and Claude (1930) in which low-pressure water vapor, generated by seawater boiling under reduced pressure, is the working fluid.

Other alternatives for OTEC include the use of water vapor produced by vigorous surface evaporation at low pressure to lift droplets of liquid spray or foam to an appreciable height where they may be collected. Power may then be produced by directing the water through a hydraulic turbine as it flows back to sea level. The mist-lift system of Ridgway (1977) and the foam-lift design of Zener and Fetkovitch (1975) employ this procedure. Power may also be produced through the use of hot and cold ocean water sources for thermoelectric generation (Kajikawa, 1982). The concepts are discussed in detail in Chapters 4 and 5.

Estimates of the performance, size, and cost requirements of OTEC OC power systems based on component testing at reduced scale are available. Less detailed estimates indicate that the other concepts mentioned previously would have similar net efficiencies and would have similar cold- and warm-water mass flows per kilowatt of net power delivered; however, there will be major differences in the systems engineering requirements for power system components and for the platform, both in size and configuration, which must be considered in evaluating the overall merits and costs for OTEC applications. Discussion of the systems engineering implications of these factors is restricted here to the first two concepts since complete systems designs have not been developed for the last three.

3.3.2.1 Differences between Closed-Cycle and Open-Cycle Systems

A fundamental, major difference between the CC and OC systems is in the sizes of the ducts and turbines, which are roughly proportional to the specific volume of the working fluid. In the closed cycle employing ammonia, the inlet pressure to the turbine is approximately 860 kPa (125 psi). In the Claude open cycle, the water vapor pressure is 3.1 kPa (0.45 psi at 25°C). Allowing for differences in thermodynamic properties, flow velocities, and design variables, the duct area must be over 100 times as large in the latter case. A duct and turbine about 15 m in diameter (50 ft) would be required for the OC system to accommodate the water vapor flow of a 10-MWe OTEC plant module. This can be compared to the 1.4-m diameter required for the ammonia flow of the CC system (George and Richards, 1980). A conceptual design study of a 100-MWe (net) open-cycle plant by Westinghouse (Coleman and Rogers, 1981) called for a turbine tip diameter of 44 m (144 ft), a power plant diameter of 113 m (373 ft), and a height of 98 m (320 ft).

A second difference between open- and closed-cycle OTEC systems is in the elimination in the OC system of the very large areas of metal surface required for heat transfer in the CC system. In a nominal closed-cycle power plant, 6.2 m² (67 ft²) heat exchanger surface area will be required per kilowatt of net power delivered, or approximately 60,000 square meters for a 10-MWe power output. Elimination of this metal surface and the pressure vessels and pumps of the ammonia system results in a major reduction in the complexity of the heat transfer equipment compared to the OC direct contact exchangers. The need for biofouling control equipment is also eliminated. Since heat transfer coefficients are higher for the OC system, the water flow needs may also be lower. A full discussion of OC systems is presented in Chapter 5.

3.3.2.2 Conclusions Regarding Power Plant Type

An assessment of the relative merits of the CC versus the OC concept for total system engineering performance and costs cannot be made without further research and development on the OC system; however, the large volume requirements of the OC plant raise serious questions about its suitability for grazing OTEC plantships. The choice will involve evaluation of the dependence of power plant cost on plant size, implications of the OC size and performance on total platform size and system cost, and possible advantages in fresh-water production as well as electric power. It must also be recognized that OC may have special advantages for small tropical island installations, which could offer an attractive way to begin commercialization of OTEC technology.

3.3.3 *Energy Delivery Options*

The immediate energy output of the OTEC process is mechanical power produced by the vapor turbine. Part of this power may be used at the OTEC platform to operate pumps or other operating equipment, with some benefits in net power output (Anderson and Anderson, Jr., 1966); however, it will be necessary to convert most of the OTEC mechanical energy into electrical energy before it can be used effectively. For this purpose, turbogenerators will be installed in the OTEC loop for direct production of electrical power.

Two general methods are available for using OTEC electric power to relieve dependence on petroleum fuels. The net power may be used at the OTEC platform to produce a fuel that can be stored on board and then at intervals shipped to distribution points to supply transportation needs, or at suitable near-shore sites electric power transmitted to shore can operate fuel production plants or replace power otherwise produced by burning fossil fuels in conventional power plants.

3.3.3.1 OTEC Plantships

For on-board production of fuel to be feasible and economically attractive, several conditions must be met and resolved by systems engineering methods:

1. An efficient means of converting electrical energy to fuel energy must be employed. The high efficiency of electrochemical processes qualifies them for this role.
2. The size and cost of the energy conversion equipment must be small enough to allow the total OTEC plant investment to be financed at an annual cost that will permit production and delivery of product fuel at a price competitive with alternatives. A comparison of sizes and costs of commercial ammonia and methanol plants indicates that such plants could be accommodated on OTEC platforms at a reasonable cost. Economic viability will depend on costs of petroleum-based vehicle fuels and cost benefits associated with favorable environmental and economic security features of OTEC.

3.   Operation of the plantship must meet safety and environmental standards.
      These requirements will be met by the design of OTEC plantships in compli-
ance with the rules of the American Bureau of Shipping and other international
organizations that govern ship construction and operation.

3.3.3.2 OTEC Plants that Transmit Electric Power to On-Shore Users

This option offers the highest efficiency in conversion of ocean thermal energy to
delivered electric power. OTEC electric plants may be mounted on floating
platforms stationed near shore or be shelf mounted or shore based. To be cost
effective, floating power plants will need to generate power of 100 MWe (net) or
more. The successful mooring of the Mini-OTEC and OTEC-1 platforms in water
over 1000 m deep and commercial power plant experience with underwater power
cables indicate the feasibility of both aspects; however, the development of a new
mooring technology will be required for large OTEC plants, and new power cable
designs that will withstand the tensions and twisting associated with platform
motions will be necessary. Considerable progress was made in engineering investigations of these problems, and satisfactory solutions appeared to be available
when work was terminated after 1980.
      The mooring and power-transmission problems can be avoided if the OTEC
plant is shelf or shore mounted; however, this option imposes a requirement for a
cold-water pipe several kilometers long that must be entrenched or buried in the
shallow water zone. Thus, costs are high.
      Much effort has been expended to investigate tradeoffs among the different
OTEC electric power options. Decisions are difficult because of the site-specific
nature of the evaluation, which increases in difficulty as the cold-water pipe
increases in diameter; however, the number of sites with high ΔT near enough to
shore to allow economical OTEC construction, and with large enough demand to
allow cost-effective production of OTEC electric power, is limited so land-based
or near-shore OTEC plants would not provide a large enough resource to make a
significant contribution to world fuel needs. Even so, the potential of such plants to
offer an entry for OTEC commercialization should not be overlooked. Many
tropical islands might achieve a self-sufficient economy by the installation of a
small OTEC plant (1–10 MWe) that could supply electric power, fresh water, air
conditioning, fuel, and food via mariculture. An active effort to investigate these
possibilities is in progress in Hawaii with the support of the Hawaiian state
government, the Pacific International Council for Technology (PICTR), several
commercial enterprises, and the U.S. DOE. Innovative technology to reduce plant
investment per unit of output to reasonable levels is required.
      Shore-based OTEC plants will be favored for tropical island sites with a low
(~10 MWe) electric power demand, and may be economically attractive if the
demand is reasonably constant. A market for fresh water and/or cold water for
mariculture at the site will improve the cost effectiveness. Open-cycle OTEC may
be favored for such plants.
      Siting studies of OTEC plants of 40 to 100 MWe, designed for transmission
of electric power to an on-shore utility grid, could find either shore-based or off-

shore floating designs to be optimum, depending on the local conditions and detailed engineering evaluation of the site characteristics. Commercialization of OTEC power plants in this range would be constrained by the limited number of electric power utilities that could accommodate power additions of this magnitude.

Grazing OTEC plantships can operate anywhere on the tropical oceans and will be concentrated in regions where the ΔT exceeds 22°C. Large plants (100–400 MWe) will be necessary to achieve commercial fuel production at competitive cost. These plants have a potential capability for supplying a significant fraction of world fuel requirements. Only such plants will be able to achieve the full OTEC alternative energy potential.

Further research and development is required to evaluate the potential role of OC in alternative energy production. Such plants could have advantages for small on-shore or shelf-mounted installations, where the combination of electric power, fresh water, air conditioning, fuel production, and mariculture could produce attractive, self-sufficient island economies.

3.3.3.3 Conclusions Regarding Energy Transfer

The primary means of OTEC energy transfer will be via liquid fuel produced on OTEC plantships grazing in tropical oceans and shipped to world ports. Siting and ΔT restrictions will limit the output of shelf-mounted and off-shore moored plants to a small fraction of the grazing plantship potential. Small, shelf-mounted plants can play an important role in introducing OTEC commercialization.

3.3.4 *OTEC Plantship System Design Alternatives*

Complete OTEC systems can be installed on land or on floating platforms; however, the previous discussion shows that the system objectives will be met only by the latter option.

For the final floating-platform system to be technically and economically attractive, it must have the following characteristics:

1. Size large enough to accommodate an OTEC power plant of 100- to 400-MWe (net) output, with associated fuel, plant, and supporting facilities for equipment and personnel;
2. Design to optimize arrangement and performance of all system components to minimize total system costs (this includes selection of an optimum power level for minimum cost product delivery);
3. Structural strength of the platform–CWP combination to meet American Bureau of Shipping (ABS) and related standards for normal operation and to withstand maximum wave, current, and wind loads predicted for 100-year storms;
4. Components designed to withstand corrosion and fatigue for the 30+ year life of the system or, for some components, designed convenient replacement (the need for the materials of the water ducting system to be inert with regard to the heat exchangers must be recognized);
5. Technology and facilities for construction and deployment of the platform–CWP design at low risk and at reasonable cost and time scale.

The design of OTEC platforms can be based on the marine engineering technology that has been developed by the off-shore oil industry and on the established principles involved in shipbuilding; however, the technology for building and deploying the large cold-water pipe and for attaching it to the platform in a manner that will ensure reliable operation and survival of the 100-year storm has not been demonstrated. In addition to conventional ship and barge platforms for OTEC, other types are possible that could minimize platform motions and reduce the stresses on the CWP, with possible systems benefits or disadvantages in size, constructability, risk, and cost.

The evaluation of optimum platform-CWP configurations for OTEC proceeded along two paths:

1. Identification of an optimum design for grazing OTEC plantships, and
2. Investigation of designs suitable for production of OTEC electrical power for transmission to onshore utilities.

3.3.4.1 OTEC Plantship Design

In 1974, work focused on the use of OTEC to produce an energy product started at JHU/APL under MARAD/Dept. of Commerce sponsorship. This activity led to a requirement for a floating OTEC system with integrated power production, water distribution system, and an energy-product production plant to minimize total system cost. Since power output would peak sharply for operations in regions of maximum ocean surface temperature. the studies showed that the plantships should be designed to be maneuverable so that they could stay in regions of maximum surface temperature as seasonal changes occurred. The need for a mooring system was also avoided. Since only a 0.5-knot cruising speed was necessary, systems engineering studies showed that a low-cost concrete barge platform with CWP suspended from the platform by a flexible mounting would be optimum for this method of operation. Successful commercial construction at low cost and deployment of a concrete barge, of the approximate dimensions suitable for a 40-MWe OTEC demonstration, supported the selection of this design for the plantship mode of operation (Mast, 1975; Magura and Mast, 1979). Later work under DOE support showed that the JHU/APL configuration could also be adapted well to moored operation (George and Richards, 1980).

3.3.4.2 Off-Shore Moored Platforms

During the period from 1977 to 1980, significant effort was devoted to systematic investigation of the marine engineering aspects of potential OTEC platform–CWP combinations, with support by the Energy Research and Development Administration (ERDA, later DOE). By direction, the efforts were restricted to evaluation of off-shore OTEC plants designed to produce electric power for transmission by underwater cable to on-shore utility grids. For this application, system designs were sought that would minimize the problems of station-keeping and power transmission. In addition to conventional ship and barge platforms, other types were considered that could minimize platform motions and reduce the stresses on the CWP, with possible systems benefits or disadvantages in size, constructability, risk, and cost. Generic studies were supported of OTEC configurations based on semi-submersible, spar, sphere, cylinder (or disk), and submerged platforms, as well as barge-ship types. All incorporated suspended CWPs (Waid, 1978; Scott and Rogalski, 1978; Basar et al., 1978). Schematic designs illustrating the concepts are shown in Fig. 1-11. Studies of shore- or shelf-mounted plants were conducted by Brewer (1979) and Brewer et al. (1979).

   Structural requirements and the apparent relative costs of OTEC power plants designed for off-shore power generation at levels ranging from 50 to 500 MWe were evaluated for sites in Hawaii; Key West, Florida; Puerto Rico; and south of New Orleans in the Gulf of Mexico. A DOE requirement was imposed that stated all options of a given power output should use the same power plant, and that a single CWP sized for an internal water flow velocity of 6 ft/s would be used. All investigations selected a 25-MWe power module as a basic unit which could be repeated for higher power levels. A figure of merit was developed for each design that included weighting factors for technical feasibility, the state of the technology required for construction of the design, the estimated risk, and cost. The Waid study found the types designated as circular barge, submersible, spar with detached modules, and ship to be essentially equivalent for off-shore OTEC power production. Waid noted that if the grazing operation had been included, only the ship would have qualified. The Scott–Rogalski study gave the highest rating to the ship-barge. Spar and semi-submersible ranked second for power ranges from 100 to 350 MWe, while sphere and spar had second ranking for powers in the range 350 to 500 MWe. As in the Waid study, only the ship-barge options would have qualified for grazing operation. Basar et al. (1978) considered a wide range of surface and submerged options in which the CWP was either pinned to the platform or separately supported from a buoy. They concluded that surface-supported configurations should be eliminated because stresses in the pinned CWP would require unacceptably high wall thickness and weights. Submerged designs could alleviate this problem. Their study gave highest ranking to a novel tubular truss designed by Rosenblatt & Sons, Inc. The assumption that the CWP would be pinned to the platform made the ranking invalid for configurations using a flexible joint. The work of Waid and of Scott and Rogalski showed that CWPs supported by a flexible joint at the platform would be feasible at diameters of at least 10 m and could be made with either concrete or fiber-reinforced plastic. (The subject is discussed in detail in Chapter 4.)

3.3.4.3 Conclusions Regarding Platform–CWP Design Options
The design investigations indicate that a ship-barge configuration with a flexible CWP suspended from the platform by a flexible joint will be optimum for the grazing mode of OTEC operation. This configuration will also be a candidate for moored operation; however, problems of power transmission from a bobbing ship-barge suggest that other alternatives could have overall system advantages for this role. A DOE decision in 1980 to concentrate OTEC research and development effort on shore-mounted or shelf-mounted systems has left unresolved the question of optimum design for moored or dynamically positioned OTEC power plants.

Manufacturing procedures and facilities developed for construction of large pipes of concrete or fiber-reinforced plastic (FRP) are applicable to OTEC CWPs. New equipment will be necessary to meet the specific needs. Further design effort supported by systems trade-off studies is needed to define the optimum CWP and supporting framework. A demonstration at sea of a scale version is necessary to satisfy potential investors that this phase of the construction will not involve unacceptable risk.

3.4 SUMMARY OF RESULTS OF SYSTEMS ENGINEERING EVALUATION

The survey in this chapter shows that OTEC offers an important option for using solar energy to produce fuels in large enough amounts to replace a significant fraction of petroleum fuels in transportation. To achieve this goal, OTEC must be incorporated in plantships. These will use OTEC power generated on board to produce hydrogen via water electrolysis, which will then be converted to a liquid fuel in an on-board plant. The fuel will be stored on the plantship and then shipped at roughly monthly intervals to world ports. The plantships will be designed for grazing operation on the high seas in the tropics so that they can make maximum use of the vast resource of water at temperature over 22°C, which is available in the surface waters of the tropical oceans.

Examination of design alternatives and applicable technology indicates that priority should be given to the following options:

Development goals:

1. Immediate: Demonstrate the commercial promise of OTEC in reduced-size on-land and floating plants. Ultimate: Supply liquid fuel via OTEC at a cost low enough for OTEC fuel to become a significant alternative to gasoline and diesel fuel for transportation.
2. Siting: Tropical oceans and selected islands with annual average ΔT exceeding 22°C.
3. Product: Methanol, ammonia, and liquid hydrogen at OTEC plantships. Electric power, fresh water, mariculture, air conditioning, and fuel at tropical islands.
4. Engineering design: Shore-based 1- to 10-MWe closed- or open-cycle plants for support of tropical island economies; 100- to 400-MWe floating plantships for fuel production.

Research and development priorities to establish performance and minimize risk:

1. Update systems engineering for minimum cost of small shore-based OTEC multiproduct plants and large OTEC/fuel plantships;
2. Select a preferred commercial method for construction and deployment of CWP for floating plants;
3. Engineer and demonstrate optimum CWP–platform flexible joint;

4. Scale up for OTEC and demonstrate electrolyzer modular units of 2- to 5-MWe capacity;

5. Construct and operate an OTEC demonstration plantship of 40–80 MWe (net) power with ammonia or methanol product to assure minimum risk and provide firm cost estimates for commercial OTEC plant construction and operation; and

6. Design and evaluate economic potential of small, shore-based OTEC plants for selected sites.

REFERENCES

Anderson, J. H., and J. H. Anderson, Jr., 1966. "Sea solar power." *Mechanical Engineering*, 41.

Avery, W. H., D. Richards, and G. L. Dugger, 1985. "Hydrogen generation by OTEC electrolysis, and economical energy transfer to world markets via ammonia and methanol." *Int. J. of Hydrogen Energy*, **10**, 727.

Basar, N. S., J. C. Daidola, and N. M. Maniar, 1978. "OTEC ocean systems evaluation." *Proc. 5th Ocean Thermal Energy Conversion Conf.*, Miami Beach, Fla., Feb., **2**, IV-15.

Bretschneider, C. L., 1975. "Operational sea state and design wave criteria, a generalized study for OTEC." *Proc. 3rd Workshop Ocean Thermal Energy Conversion, OTEC*, Houston, May, 87.

——, 1977, "Operational sea state and design wave criteria: State-of-the-art of available data for USA coasts equatorial latitudes." *Proc. 4th Annual Conf. on Ocean Thermal Energy Conversion*, New Orleans, Mar., **4**, 61.

——, 1978. "Operational sea state and design wave criteria for potential OTEC sites." *Proc. 5th Ocean Thermal Energy Conversion Conf.*, Sept., **2**, IV-178.

——, 1979. "Appendix A 100-year design hurricane wave criteria for potential OTEC sites." *Proc. 6th OTEC Conf.*, Washington, D.C., June, **1**, 6.10-7.

Brewer, J. H., 1979. "Land-based OTEC plants–cold water pipe concepts." *Proc. 6th OTEC Conf.*, **1**, 5.10-1.

——, J. Minor, and R. Jacobs, 1979. *Feasibility design study. Land-based OTEC plants.* U.S. Department of Energy, C00-4931-0.

Churgin, J., and H. Iredale, 1979. "OTEC data base and data products: NOAA." *Proc. 6th OTEC Conf.*, Washington, D.C., June, **1**, 13.4-1.

Claude, G., 1930. "Power from the tropical seas." *Mechanical Engineering*, **52**, 1039.

——, and P. Boucherot, 1928. "Conferences faites grand amphitheatre de la Sorbonne." *Conf. on Ocean Energy*, Feb., 3.

Coleman, W. H., and J. D. Rogers, 1981. "Design and development of an open-cycle OTEC turbine with particular emphasis on mechanical design aspects of rotor blades." *Proc. 8th Ocean Energy Conf.*, Washington, D.C., **2**, 277.

Davidson, Jr., H., and T. E. Little, 1978. "OTEC platform station keeping analysis." *Proc. 5th Ocean Thermal Energy Conversion Conf.* **2**, IV-237.

Flavin, C., 1992. "Building a bridge to sustainable energy." in *State of the World 1992*, Worldwatch Inst., p. 27.

George, J. F., D. Richards, and L. L. Perini, 1979. *A baseline design of an OTEC pilot plantship.* Johns Hopkins University Applied Physics Laboratory, SR-78-3A.

——, and D. Richards, 1980. *Baseline designs of moored and grazing 40-MW OTEC pilot plants.* Johns Hopkins University Applied Physics Laboratory, SR-80-A.

Giannotti, J. G., K. A. Stambaugh, and W. K. Jawish III, 1981. "OTEC cold water pipe hydrodynamic loading and structural response: At-sea measurements, model experiments and analytical simulations." *Proc. Int. Symp. on Hydrodynamics in Ocean Engineering,* Norwegian Inst. of Technology, 963.

Goldman, G. G., and D. Pesante, 1979. "Preliminary results of a program to study OTEC oceanic environmental parameters at Punta Tuna, PR." *Proc. 6th OTEC Conf.,* Washington, D.C., June, **1,** 13.9.

Hill, F. K., 1979. *Satellite tropical ocean surveys of SST.* Johns Hopkins Univ. Applied Physics Lab., FKH 79-4.

——, and G. L. Dugger, 1979. "Use of satellite-derived sea surface temperatures for cruising OTEC plants." *Proc. 6th OTEC Conf.,* **2,** 13.8.

Kajikawa, T., 1982. "Status of R&D activities in Sunshine project." *Proc. Marine Technology Soc. Ocean Energy Workshop,* Washington, D.C., Sept., 35.

Magura, D. D., and R. F. Mast, 1979. "Ocean thermal energy conversion (OTEC) 10-MWe preliminary plantship design." *Proc. 11th Annual Offshore Technology Conf.,* **3,** 2059.

Marchand, P., 1980. "The French ocean energy program." *Proc. 7th Ocean Energy Conf.,* Washington, D.C., June, **1,2,** 2.7.

Marchetti, C., 1987. "The future of hydrogen—an analysis at world level with a special look at air transport." *Int. J. of Hydrogen Energy* **12**(2), 61.

Mast, R. F., 1975. "The ARCO LPG terminal vessel." *Conf. on Concrete Ships and Floating Structures.* Berkley, Cal., Sept., 1.

Munier, R. S., T. N. Lee, and S. Chiu, 1979. "Observations of water mass structure and variability north of St. Croix, U.S. Virgin Islands for OTEC assessment." *Proc. 6th OTEC Conf.,* Washington, D.C., June, **2,** 13.10.

Renner, N., 1989. "Rethinking transportation." in *State of the World 1989,* Worldwatch Inst., p. 101.

Ridgway, S. L., 1977. "The mist flow OTEC plant." *Proc. 4th Annual Conf. on Ocean Thermal Energy Conversion,* New Orleans, VIII-37.

Scott, R. J., and W. W. Rogalski, 1978. "Considerations in selection of OTEC platform size and configuration." *Proc. 5th Ocean Thermal Energy Conversion Conf.,* Miami Beach., Fla., Sept., **2,** IV-77.

Shea, C. P., 1988. "Renewable energy: Today's contribution, tomorrow's promise." *Worldwatch Paper 81,* Worldwatch Inst., Jan. p. 5.

Statistical Abstract of the United States, 1990. U.S. Dept. of Commerce Bureau of the Census, Table 957.

Uehara, H., 1982. "Research and development on ocean thermal energy conversion in Japan." *Proc. 17th IECEC Conf.,* American Inst. of Chemical Engineers, **4.**

Veziroglu, T. N., 1987. "Hydrogen technology for energy needs of human settlements." *Int. J. of Hydrogen Energy* **12**(2), 99.

Waid, R. L., 1978. "OTEC platform design optimization." *Proc. 5th Ocean Thermal Energy Conversion Conf.,* Miami Beach, Fla., **2,** IV-1.

Wolff, P. M., 1978. "Temperature difference resource." *Proc. 5th Ocean Thermal Energy Conversion Conf.,* Miami Beach, Fla., **III,** 11.

Wolff, W. A., W. E. Hubert, and P .M. Wolff, 1979. "OTEC world thermal resource." *Proc. 6th OTEC Conf.,* Washington, D. C., **II,** 13-5.

Zener, C., and J. Fetkovitch, 1975. "Foam solar sea power plant." *Science,* **189,** 294.

4

CLOSED-CYCLE OTEC SYSTEMS

4.1 OTEC POWER PLANT

4.1.1 *Thermodynamics*

The Rankine closed cycle is a process in which beat is used to evaporate a fluid at constant pressure in a "boiler" or evaporator, from which the vapor enters a piston engine or turbine and expands doing work. The vapor exhaust then enters a vessel where heat is transferred from the vapor to a cooling fluid, causing the vapor to condense to a liquid, which is pumped back to the evaporator to complete the cycle. A layout of the plantship shown in Fig. 1-2.

The basic cycle comprises four steps, as shown in the pressure–volume (p–V) diagram of Fig. 4-1.[1]

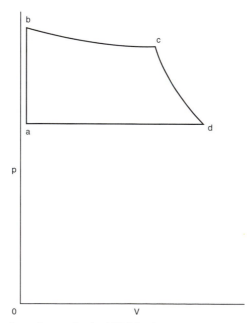

FIG. 4-1. Pressure–volume diagram for the OTEC Rankine cycle.

1. Starting at point a, heat is added to the working fluid in the boiler until the temperature reaches the boiling point at the design pressure, represented by point b.
2. With further heat addition, the liquid vaporizes at constant temperature and pressure, increasing in volume to point c.
3. The high-pressure vapor enters the piston or turbine and expands adiabatically to point d.
4. The low-pressure vapor enters the condenser and, with heat removal at constant pressure, is cooled and liquefied, returning to its original volume at point a.

The work done by the cycle is the area enclosed by the points a,b,c,d,a. This is equal to $H_c - H_d$, where H is the enthalpy of the fluid at the indicated point. The heat transferred in the process is $H_c - H_a$. Thus the efficiency, defined as the ratio of work to heat used, is:

$$\text{efficiency}(\eta) = \frac{H_c - H_d}{H_c - H_a}. \tag{4.1.1}$$

Carnot showed that if the heat-engine cycle was conducted so that equilibrium conditions were maintained in the process, that the efficiency was determined solely by the ratio of the temperatures of the working fluid in the evaporator and the condenser.

$$\eta = \frac{T_E - T_C}{T_E} \tag{4.1.2}$$

The maximum Carnot efficiency can be attained only for a cycle in which thermal equilibrium exists in each phase of the process; however, for power to be generated a temperature difference must exist between the working fluid in the evaporator and the warm-water heat source, and between the working fluid in the condenser and the cold-water heat sink. The rate of heat transfer increases as the temperature differences are increased, but the pressure difference between the working fluid in the evaporator and in the condenser is reduced, leading to an optimum distribution of temperatures in the total system for maximum power production. A theoretical analysis of the ideal power production process is presented in Wu (1987).

Values of the enthalpy and entropy of fluids that could be considered for use in OTEC power plants are available in the technical literature; therefore, the efficiency may be calculated for any Rankine cycle for which the state variables are known at the points of change in the cycle. The actual values of the variables will differ from the ideal values because of heat losses in the boiler, pressure losses in the lines and process equipment, and friction and inefficiencies in the prime mover. Estimation of the performance of a new design must be based on the use of manufacturers' specifications for components such as pumps, turbines, and demisters, and calculated values for temperatures and pressures derived from heat transfer rates and pressure losses associated with the flow properties, geometry, and construction. Thermodynamic relationships, combined with measured properties dependent on temperature and pressure, determine the power that can be generated for any particular plant design for which heat transfer coefficients and component efficiencies are known. An elegant analysis of the maximum net OTEC power that could be produced by an actual design was published by Owens (1979). Detailed discussion is deferred to Section 4.1.4. The following discussion follows Owens' exposition. The OTEC power cycle is shown schematically in Fig. 4-2 (Dugger et al., 1983). The net power of the OTEC power plant is the net power of the thermal cycle minus the power required for pumping the warm- and cold-water flows to the evaporator and condenser:

$$P_N = P_G - P_{CS} - P_{WS} \tag{4.1.3}$$

$$P_{CS} = \dot{m}_{CS} Z_{CS}, \tag{4.1.4}$$

where $\dot{m}_{CS}$ is the mass flow rate of the cold seawater and Z_{cs} is the sum of three terms: losses in pressure in water flow through the heat exchanger and associated ducting, (Δp_{HXC}), a pressure loss due to friction of the flow through the cold-water pipe (Δp_{CWP}), and a loss caused by the difference in density between the water in the CWP and the surrounding warm seawater (Z_{dw}):

$$Z_{CS} = \frac{(\Delta p_{HXC} + \Delta p_{CWP} + Z_{dw})}{\rho}. \tag{4.1.5}$$

The temperature range available for the Rankine cycle is restricted to about half the difference in temperature between warm and cold water by the requirements that a temperature difference must exist to enable heat to be transferred from the warm water to the working fluid in the evaporator, and another must exist for heat transfer from the working fluid to cold water in the condenser. Thus, the temperatures and flow variables change as shown in Fig. 4-2.

The principal mechanical components of the closed-cycle heat engine are the two heat exchangers (evaporator and condenser), the turbine generator, the water and working-fluid (ammonia) pumps, the demister (which is required for removal of liquid droplets from the vapor before it enters the turbine), the ducts that conduct

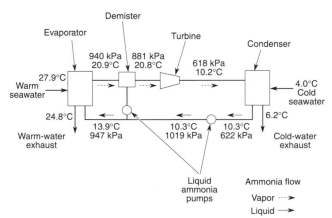

Fig. 4-2. Typical process parameters for the 40-MWe OTEC power system (Dugger et al., 1983).

warm and cold water to the heat exchangers, and the ducts that carry working-fluid vapor and liquid around the cycle. Only the heat exchangers and cold-water pipe have required development of new technology for application to OTEC.

4.1.2 *Heat Exchangers*

4.1.2.1 General Considerations

Heat exchangers for closed-cycle Rankine heat engines transfer heat between the working fluid and the heating or cooling fluid (water) through a separating surface. The fluids may move in parallel, in cross flow, or in opposite directions. Many structural configurations have been developed to satisfy the requirements for particular applications, which involve overall heat transfer coefficient versus pressure drop, thermal efficiency, durability, size, weight, cost, and availability. Because of the low temperature gradient, OTEC heat exchangers must have a heat transfer area per kilowatt of power generated roughly 10 times that of heat exchangers for conventional steam plants. Thus, instead of being a small part of the total plant cost, OTEC heat exchangers could become a major part if costs per unit surface area were to remain the same as for steam power plants. A goal of OTEC development has been to achieve heat exchanger designs that will have a small cost per kilowatt and minimal needs for supporting structures (for example, ducting and platform space) so that a low cost per kilowatt of the total OTEC installation may be achieved. The goal can be attained through engineering design that recognizes the following facts:

1. OTEC operating conditions are similar to those of refrigeration systems rather than conventional power plants. The low temperatures and pressures of OTEC permit large reductions in the strength of heat exchanger materials relative to conventional power plants. Low-cost aluminum alloys can replace titanium, copper–nickel, or stainless steel.
2. The less stringent OTEC operating conditions simplify and ease structural, supporting systems, and installation requirements.
3. Because the OTEC power system can be standardized for floating plants, and demands for large numbers of identical units are projected, cost reductions associated with scaling and multiple production of identical units will be achieved. This is not as feasible for land-based power systems, which are designed to meet specific user site requirements and are purchased as individual systems by competitive bidding procedures. As a general rule, production of identical units will reduce costs by a factor of 0.85 to 0.95 for each doubling of the number produced. Thus, the total cost of the 32 evaporators of a 160-MWe plant composed of 5-MWe units would be 19 to 27 times the cost of the first unit; that is, the average unit cost would be 0.60 to 0.84 times the cost of the first unit. As the unit size is increased, scaling costs vary with size to the 0.6 to 0.8 power. This implies that increasing the size of a power module from 5 to 25 MW would increase the cost by a factor of 2.5 to 3.5, rather than by 5. In combination, these factors will reduce power system unit costs by 50% or more between the demonstration unit and full production.

4.1.2.2   Heat Exchanger Types

The design of heat exchangers to meet industrial requirements for efficiency, durability, ease of manufacture, packaging, systems integration reliability, and cost has led to an extensive technology devoted just to this subject. Because of the complexities of the fluid flow and heat transfer processes, which depend on the geometry and state variables, rigorous analysis is generally superseded by empirical relationships valid only for the geometrical design for which they were developed. It is to be expected that the special requirements of OTEC will be met by heat exchangers with operating characteristics that differ from those of conventional designs. Particular objectives are to make major gains in overall heat transfer coefficients in ways that will reduce heat exchanger costs per kilowatt of net power generated. This has led to exploration of a variety of potential heat exchanger types with features designed to be optimal for OTEC applications. The improved designs will be briefly described here. Analytical methods for predicting heat exchanger performance and experimental results of tests of promising design improvements are compared in Section 4.1.3.

Shell-and-Tube Heat Exchangers.   The most widely used type of heat exchanger for industrial evaporator and condenser applications is the shell-and-tube exchanger. A schematic design is shown in Fig. 4-3 (Dugger et al., 1981). The heat transfer surface is provided by parallel tubes, typically about 25 mm (1 in.) in diameter, which are sealed at the ends into headers or "tube sheets." The tube sheets separate the middle section of the cylindrical vessel from the two end compartments. For OTEC applications, water flows through the tubes and the working fluid flows across the tube bank in the middle section. In one mode of conventional

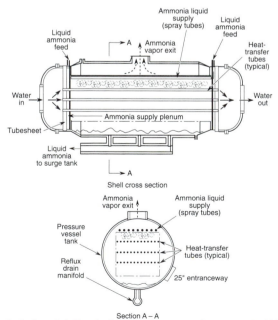

Fig. 4-3.   Schematic design of shell-and-tube spray evaporator (Dugger et al., 1981).

operation as an evaporator, the working medium enters the central compartment of the vessel as a liquid and is vaporized at the outside surfaces of the tubes, which are heated internally by warm seawater flowing through the tubes from the header at one end of the heat exchanger vessel to the header at the other end. The vapor produced by boiling at the surface of the tubes bubbles upward through the spaces between the tubes and exits from the top. In an alternative design, the liquid working fluid is sprayed over the tubes from the top and vaporizes as it flows or drips down. Other arrangements may also be used.

In the shell-and-tube condenser, the working medium enters as vapor at the top of the middle section, condenses on the tubes cooled by internal flow of cold fluid, and flows out the bottom. The heat exchangers may be mounted horizontally or vertically.

The performance of conventional shell-and-tube exchangers is limited by the thermal resistance of the liquid laminar boundary layer that forms on the surface of smooth tubes on both the water and working fluid sides. In the pool-boiling types, smooth tubes offer few sites to initiate bubble formation in the working fluid, which causes superheating of the film before boiling can begin. In the condenser, condensation of vapor on the tube surface can cause a thick liquid film to form on the surface (flooding), reducing heat transfer.

On the water side the heat transfer coefficient is nearly proportional to water velocity but pressure loss increases as velocity squared. Therefore a balance must be sought between improved heat transfer rates and increased pumping power losses, as velocity through the tubes is increased.

New designs that address these problems by changing (enhancing) the surface geometry have demonstrated great improvements in the overall heat transfer coefficients of experimental shell-and-tube heat exchangers for OTEC. This improvement may be accompanied by an increase in fluid friction with a consequent increase in pressure losses. Greater costs per tube will also be incurred. Both of the latter factors must be included in an assessment of benefits of "enhancements."

Fluted and Corrugated Tube Designs.   Gregorig (1954) noted that liquid flowing down the surface of a vertical tube with longitudinal ridges would be drawn into the troughs between the ridges by surface tension. This would cause the liquid film over the ridges to be thinned, with a possible large increase in the heat transfer coefficient for evaporation or condensation (Fig. 4-4). Application of this concept to the design of OTEC heat exchangers was proposed by Zener and Lavi (1973), who also investigated the geometrical requirements for maximizing heat transfer. The concept was further developed by Lavi (1974), Rothfus and Lavi (1978), and Webb (1978) and applied to the design of OTEC heat exchanger modules by Bakstad and Pearson (1979) and Kajikawa et al. (1979).

Nucleating Surfaces.   The tendency for a working fluid in contact with a smooth hot surface to become superheated before boiling commences can be reduced by coating the surface to provide a high concentration of nucleation sites. Effective coatings include a thin porous aluminum surface devised by Linde (Czikk et al., 1978) and a matrix of small wires investigated by the Electrotechnical

Laboratory in Japan (Kajikawa et al., 1980). Heat transfer coefficients on the water side are relatively low because practical flow velocities of from 1 to 2 m/s allow a thick laminar boundary layer to form on smooth surfaces, which impedes heat transfer. Various enhancement methods to thin or energize the boundary layer have been devised, which include surface roughening, extended surfaces, enhancement devices on the surface, and swirl-inducing techniques. A sketch of proposed methods is shown in Fig. 4-5 (Lorenz and Yung, 1980).

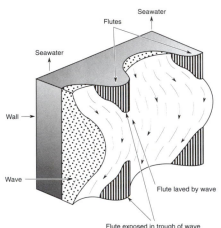

FIG. 4-4.   Fluid flow over a vertical Gregorig surface (Gregorig, 1954).

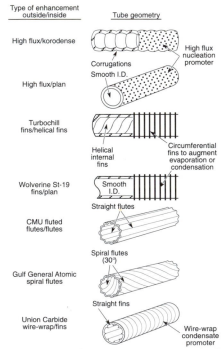

FIG. 4-5.   Types of surface treatments to enhance evaporator heat transfer (Lorenz and Yung, 1980).

An excellent discussion and analysis of the design approaches, as well as ratings of their effectiveness relative to the magnitude of induced pressure losses, are given by Bergles and Jensen (1977). The analysis indicates that overall heat transfer could be improved by water-side enhancement with the best configurations to give a 30% reduction of the heat exchanger surface area without increasing pumping power.

Plate Heat Exchangers. The large volume requirements and difficult packaging of shell-and-tube heat exchangers for OTEC applications have led to investigation of alternatives that would be more compact and would offer advantages in performance and economy. In the plate type of heat exchanger, the water and working fluid flow in alternate channels separated by parallel plates (see Fig. 1-8). Appropriate manifolds are used to route the fluids into the proper channels. Surface treatments to enhance heat transfer akin to those used in shell-and-tube exchangers are also applicable to plate designs. Gains in heat transfer coefficients of 100 to 200%, compared with conventional shell-and-tube designs, are possible with this type of heat exchanger. Although plate types of heat exchangers are simple in principle and efficient in space utilization (compact), difficulties arise in practice in providing reliable, low-cost methods of fabricating and sealing the channels to prevent leakage and avoid corrosion. Some of the volume saved in the heat exchange section may be lost in the need for additional manifolding and ducting volume. Also, the flat plate construction requires sufficient plate thickness or internal bracing to prevent distortion or collapse under the pressure differentials between the parallel passages and between the internal pressures and the external environment. Pressure differentials induced by startup or shutdown or loss of working fluid must be considered in the design. This may require significantly more metal thickness than is needed for the tubing in shell-and-tube or folded-tube exchangers. Finally, for sites where the use of chlorine for biofouling control is considered undesirable, development of practical mechanical methods of biofouling removal may not be possible.

Folded-Tube Heat Exchangers. In this type of heat exchanger, shown schematically in Fig. 4-6 (George and Richards, 1980), the working fluid flows inside long tubes, folded for compactness to form vertical elements, which are immersed in seawater flowing under a gravity head in suitable channels in the OTEC platform. The overall heat transfer coefficient for this type of exchanger is about the same as that of conventional shell-and-tube types. Advantages of the design are that it eliminates the need for special pressure vessels for the heat exchangers, reduces materials and fabrication costs, and permits the water ducting to be incorporated as an integral part of the structural design of the platform. The relatively low waterside heat transfer makes the heat transfer coefficient less sensitive to biofouling than in high-performance heat exchangers. Design studies indicate that despite the lower heat exchanger performance, costs per kilowatt of net power of OTEC systems using these heat exchangers can be lower than for alternative types. A major disadvantage of the folded-tube design is its unconventionality and the lack of manufacturing and operational experience with heat exchangers of this type. This induces an impression of higher risk. No information

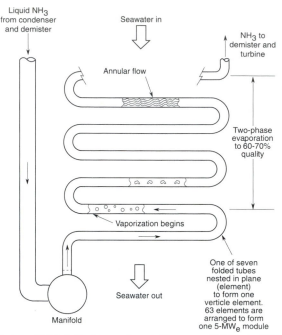

FIG. 4-6. Schematic design of folded-tube heat exchanger (George and Richards, 1980).

is available on potential performance gains with the use of tubing geometries that would enhance heat transfer, for example, fluted or twisted configurations, which are effective in more conventional heat exchanger designs.

4.1.2.3 Working-Fluid Selection

The Rankine closed cycle can employ as a working medium any fluid with an appropriate vapor pressure at the temperature of the hot source and physical and chemical properties suitable for the total power system design. Most of the working fluids developed for air-conditioning systems are potential OTEC candidates.

Desirable Characteristics of Working Fluids. The major factors are:

1. Vapor pressure in the range of 700 to 1400 kPa (100 to 200 psi) at 27°C (80°F).
2. Low volume flow of working medium per kilowatt of power produced.
3. High heat transfer coefficient, that is, low thermal resistance to heat transfer from the bulk vapor to the heat exchanger surface through the liquid film.
4. Chemical stability and compatibility with materials and structures of the power cycle, including heat exchangers, turbine, seals, and lubricants.
5. Safety.
6. Environmental acceptability.
7. Low cost.

Evaluation of Candidate Fluids. The thermodynamic properties of four working fluids that are potential candidates for OTEC use are summarized in Table 4-1.

Table 4-1. Properties of heat engine working fluids of interest for OTEC (15.56°C [60°F])

Property[a]	Ammonia	Propane	Butane	R-22
Formula	NH_3	C_3H_8	C_4H_{10}	CHF_2Cl
Molecular weight (M)	17.03	44.09	58.12	123.46
Density(1) (kg/m³)	616.73	508.60	583.73	1229.78
Density(v) (kg/m³)	5.82	16.19	4.61	33.89
Vapor pressure(sat) (kPa)	741.88	741.40	179.40	801.93
Heat of vaporization (kJ/kg)	1214.63	350.63	374.64	193.51
Specific heat(l) (kJ/kg K)	4.68	2.56	2.37	1.22
Specific heat(v) (kJ/kg K)	2.92	2.01	1.76	0.78
Viscosity(l) (Pa s)	1.596×10^{-4}	1.199×10^{-4}	1.782×10^{-4}	2.121×10^{-4}
Viscosity(v) (Pa s)	1.100×10^{-5}	8.681×10^{-6}	7.606×10^{-6}	1.277×10^{-5}
Thermal conductivity(l) (W/m K)	0.50343	0.101551	0.1211	0.092382
Thermal conductivity(v) (W/m K)	0.02595	0.018684	0.015397	0.01038
B (kPa M kJ/kg)	1.535×10^7	1.146×10^7	3.906×10^6	1.916×10^7
$\kappa^3 \rho^2 \lambda / \mu$	7.796×10^2	1.678×10^2	1.889×10^2	1.816×10^2

[a]l, saturated liquid; v, vapor.

In the list of desirable working fluid properties, item 2 is inversely proportional to the product of pressure p_v, enthalpy of vaporization, H_{lg}, and molecular weight M. A maximum value of this product, B, corresponds to the minimum volume requirement for the turbine and associated ducts and valves:

$$B = p_v \ \Delta H_{lg} \ M. \tag{4.1.6}$$

Item 3 depends on the heat transfer process. For vapor condensation on horizontal or vertical tubes of diameter D, the heat transfer coefficient h may be expressed by the Nusselt relationship (Perry and Green, 1984):

$$h = \left(\frac{k^3 \rho^2 H_{lg}}{D\mu \ \Delta T} \right)^{1/4}, \tag{4.1.7}$$

where k is the thermal conductivity, ρ the liquid density, and μ the dynamic viscosity. For a given temperature and tube diameter,

$$h = \text{const}(k^3 \rho^2 H_{lg} / \mu)^{1/4} = \text{const} \times \Phi. \tag{4.1.8}$$

A maximum value of Φ is associated with minimum heat exchanger area. The dependence, however, is weak since the waterside heat transfer coefficient is usually significantly lower than the value on the working fluid side and has a larger effect on the value of the overall heat transfer coefficient, U. An extensive tabulation of the pertinent parameters is presented in Marchand (1980), which shows that ammonia is the best choice with regard to the selection criteria listed.

Materials Compatibility. Practical use of the working fluid requires that it be compatible with materials of construction and handling. Ammonia is not compatible with copper alloys; therefore, the selection of ammonia for OTEC precludes the use of copper not only for heat exchanger components exposed to the working fluid but also for piping, valves, and other structures commonly employed in commercial equipment. Ammonia does not dissolve grease and oil, which may be used for lubrication of turbines or valves. This means that if any vapors are generated by these lubricants, they will tend to deposit and remain on the cool condenser heat transfer surfaces with eventual degradation of the condenser performance. Hydrocarbons and Freons do not have these drawbacks; however, the other advantages of ammonia for OTEC have led to its adoption by most designers. The disadvantages can be avoided by appropriate system design.

Safety. This item involves consideration of toxicity and fire and explosion hazards associated with the use of the working fluid. Although the relative values differ among the fluids, all of those considered for OTEC use are widely used commercially and are approved for public use in compliance with the safety standards that have been established.

Because of the large amounts of working fluids that would be involved in OTEC, special consideration is required of the hazards that would arise with catastrophic accidents. From this standpoint, the hydrocarbon fluids would be the most hazardous because any significant leak could form a cloud of denser-than-air gas that could produce fires and explosions involving the entire amount of working fluid and destruction of the OTEC vessel or installation. Several hundred tons or more of flammable hydrocarbons could be discharged in a catastrophic accident involving a large OTEC plant. The safety regulations of the Interstate Commerce Commission classify ammonia as a nonexplosive material for shipping and handling. A large leak of ammonia would produce a gaseous mixture that would be lethal after a few minutes of exposure wherever ammonia concentrations exceeded one part per thousand. Ammonia is highly soluble in water, and if it were rapidly mixed with seawater in large amounts, could lead to large heat evolution, boiling, and vigorous steam production. Concentrations of dissolved ammonia that would be lethal to marine life could be found to an appreciable distance (a kilometer or more), depending on ocean currents and mixing rates. The fluorinated hydrocarbons are nontoxic and nonflammable but are denser than air. A large spill would create an asphyxiating cloud, perhaps several hundred meters in diameter.

In view of the widespread safe distribution of hydrocarbons and ammonia by barge and tanker in quantities of millions of tons per year, as well as the large use of halogenated hydrocarbons as working fluids in commercial refrigeration, the hazards of their use for OTEC appear to be small enough to be disregarded in the choice of optimal working fluids.

Environmental Considerations. Widespread commercial use of OTEC plants and plantships would inevitably lead to some release of working fluid into the atmosphere and ocean environments. For hydrocarbons and ammonia, the quantities emitted would be insignificant in comparison to those associated with industrial uses of hydrocarbons for transportation and heating, and ammonia uses for fertilizers. Effects of the fluorine–chlorine–carbon compounds (Freons) are more questionable because these compounds entering the atmosphere can affect the ozone layer and in seawater can enter the food chain, becoming progressively concentrated and eventually reaching possibly toxic concentrations (for example, plankton are eaten by small fish that are eaten by larger fish that may be eventually consumed by humans). Insufficient information is available to allow any estimate of these potential long-term effects; however, the initial impact would be small, and time would be available to allow the collection of data and to permit informed assessments to be made.

Working Fluid Cost. Depending on the power plant design and provisions for storage, the working fluid cost would be about $5/kW for ammonia and $35/kW for Freon R-22. Thus, working fluid cost is not a significant consideration in working fluid selection. Total assessments of all of these factors by the OTEC community have led to a consensus on the selection of ammonia as the preferred working fluid for OTEC.

4.1.3 *Heat Exchanger Design for OTEC*

4.1.3.1 Basic Heat Exchanger Design Factors

The flow of heat from a heat source at temperature T_1, through a dividing wall to a heat sink at temperature T_2 is governed by the general expression:

$$\dot{Q} = UA(T_1 - T_2). \qquad (4.1.9)$$

Since heat is extracted as water flows over the heat exchanger surfaces, the water temperature at the heat exchanger exit is lower than at the inlet. Thus, the rate of heat transfer from water to ammonia varies with position in the heat exchanger, and appropriate average values must be used for these quantities. If U does not depend on ΔT or flow position in the heat exchanger,

$$\dot{Q} = \Delta T_m \ UA = \frac{UA(\Delta T_i - \Delta T_o)}{\ln(\Delta T_i / \Delta T_o)}, \qquad (4.1.10)$$

where ΔT_i is the temperature difference between water and working fluid at HX inlet, ΔT_o the temperature difference between water and working fluid at HX outlet, T_m the log mean temperature difference, and U the overall heat transfer coefficient.

Equation (4.1.10) is valid for parallel flow, cross flow, and counter flow if the temperature of the working fluid is constant. This will be approximately true for OTEC evaporators and condensers. Accurate calculations require finite difference methods. For OTEC conditions, this refinement is generally not needed. The value of the coefficient, U, is dependent on the thermodynamic properties of the fluids, flow velocities, fraction of each fluid vaporized at a particular point in the heat exchanger pressure, composition and construction of the separator material that forms the bounding surface between the fluids, character of the surfaces on both sides of the separator (roughness, presence or absence of scale, biofouling, etc.), and detailed design and dimensions of the heat exchanger.

As shown in Fig. 1-7, if a quantity of heat $\dot{Q}$ flows from water to ammonia, it must pass first from the water interior to the separating surface. This requires a temperature difference $T_w - T_{sb}$, the magnitude of which depends on the character of the flow at that point, which in turn depends on the local geometry of the flow passage. The heat then passes through any buildup of corrosion and biofouling on the water side of the surface, where a further temperature drop, $T_{sb} - T_{mo}$, occurs. Similarly, the heat passes through the other resistances until it is finally absorbed in the bulk ammonia flow. We then have per unit area,

$$T_w - T_a = (T_w - T_{sb}) + (T_{sb} - T_{mo})$$
$$+ (T_{mo} - T_{mi}) + (T_{mi} - T_{sa}) + (T_{sa} - T_a). \qquad (4.1.11)$$

Also, since

$$\dot{Q} = h_w(T_w - T_{sb}) = h_{sb}(T_{sb} - T_{mo})$$
$$= k/x(T_{mo} - T_{mi}) = h_{sa}(T_{mi} - T_{sa})$$
$$= h_a(T_{sa} - T_a) = U(T_w - T_a), \qquad (4.1.12)$$

it follows that

$$U = 1/h_w + 1/h_{sb} + x/k + 1/h_{sa} + 1/h_a. \qquad (4.1.13)$$

Of the terms in Eq. (4.1.13), $1/h_w$ and $1/h_a$ depend on fluid-dynamic parameters applicable to the specific heat exchanger design. The term $1/h_{sb}$ is time dependent and varies with local environment and biofouling countermeasures; x/k can be accurately calculated from the physical properties and dimensions of the separating surfaces; and $1/h_{sa}$ will be unimportant if the working fluid does not become contaminated with materials that can deposit on the surface, and the wall material and working fluid are noninteractive. With proper design, this term will be negligible. Because the physical processes are complex and geometry dependent, it is generally not possible with actual heat exchangers to predict accurate values of $1/h_a$ and $1/h_w$ from basic principles. Many formulas have, however, been derived for idealized flow situations and correlated with experimental data, which may be used to provide design and performance predictions for particular heat exchanger configurations. These are discussed briefly in the following sections, supplemented by references to the extensive technical literature.

4.1.3.2 Shell-and-Tube Heat Exchangers

Performance prediction requires analysis of the fluid dynamics of water flow in tubes and of two-phase evaporative and condensing cross flow over the external surfaces of nested tubes.

Water-Side Heat Transfer and Pressure Drop.   Heat transfer rates for single-phase fluid flow inside long tubes in which turbulent flow is established can be correlated in terms of the dimensionless quantities—the Nusselt number, Nu; the Reynolds number, Re; and the Prandtl number, Pr (ASHRAE, 1985):

$$Nu = \text{const} \times Re^m \, Pr^n. \qquad (4.1.14)$$

This is equivalent to

$$\frac{hD}{\kappa} = \text{const} \times \left[\frac{\rho v D}{\nu} \right]^m \left(\frac{\mu C_p}{\kappa} \right)^n. \qquad (4.1.15)$$

The values of the constant , m, and n depend on the heat loading, tube friction, and Reynolds number range of operation. For heat exchanger conditions typical of OTEC operation, the Sieder–Tate relationship is generally used (Holman, 1981):

$$hD/\kappa = 0.027\,\text{Re}^{0.8}\text{Pr}^{0.333}(\mu/\mu_W)^{0.14}. \qquad (4.1.16)$$

Usually the final term is omitted and the Sieder–Tate equation is given as

$$hD/\kappa = 0.023\,\text{Re}^{0.8}\,\text{Pr}^{0.4}. \qquad (4.1.17)$$

Experiments at the Argonne National Laboratory under OTEC conditions show that these equations underpredict the measured values, but that formulas developed by Sleicher–Rouse and Petukhov–Popov give agreement with the measured values within 5% over the entire range of OTEC operating conditions (Panchal et al., 1981; Lorenz et al., 1982). The Sleicher–Rouse correlation is

$$hD/\kappa = 5 + 0.015\,\text{Re}^a\,\text{Pr}^b, \qquad (4.1.18)$$

where

$$a = 0.88 - 0.24/(4 + \text{Pr})$$
$$b = 0.33 - 0.5^{-0.6\text{Pr}}.$$

The pressure drop (loss of hydraulic head) due to friction in fluid flow through a straight circular pipe is given by the Fanning equation (Perry and Green, 1984):

$$\Delta p = \text{head loss} = 4f'(L/D)v^2/2g, \qquad (4.1.19)$$

where f' is defined by an equation of Moody quoted by Owens (1983):

$$f = 4f' = 0.0055[1 + (20000\epsilon/D + 10^6/\text{Re}^{1/3})]. \qquad (4.1.20)$$

The quantity ϵ is a measure of pipe roughness. For drawn tubing, ϵ has a value of approximately 0.0015 mm. For commercial steel pipe, the value is about 0.046 mm. A plot of friction factor versus Reynolds number and surface roughness based on Eq. (4.1.20) is shown in Fig. 4-7 (Perry and Green, 1984).

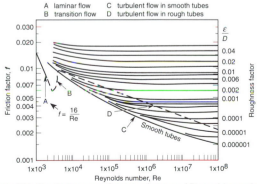

FIG. 4-7. Dependence of friction factor on surface roughness and Reynolds number (Perry and Green, 1984).

For an OTEC shell-and-tube heat exchanger with drawn tubes of 25 mm diameter and 15.2 m length, if the evaporator water temperature is 26.7°C (80°F), the condenser temperature is 4.44°C (40°F), and the water flow velocity is 0.76, 1.5, 2, or 3 m/s (2.5, 4.9, 6.6, or 9.8 ft/s), then Eqs. (4.1.18) and (4.1.20) predict the values h_w, Z_w (head loss), and pumping power (as a percentage of gross OTEC power) shown in Table 4-2. (Pumping power is based on an estimated water flow requirement of 2.35 kg/s per kilowatt of OTEC power output and pump efficiency of 75%. Pumping powers are based on the parameter values for optimal power listed in Table 4-21.

Table 4-2. Water-side heat transfer and pressure drop for flow in smooth tubes

Water temp. (°C)	Water vel. (m/s)	Re	Pr	h (W/m)	Head loss (m)	Pumping power (% net power)	f
4.4	0.76	1.26×10^4	11.2	4.5×10^3	0.52	2.59	0.0073
	1.5	2.48×10^4	11.2	8.1×10^3	1.69	8.48	0.0061
	2	3.31×10^4	11.2	1.0×10^4	2.80	14.06	0.0057
	3	4.96×10^4	11.2	1.5×10^4	5.73	28.81	0.0052
26.7	0.76	2.26×10^4	5.8	6.1×10^3	0.44	2.23	0.0063
	1.5	4.46×10^4	5.8	1.1×10^4	1.47	7.38	0.0053
	2	5.94×10^4	5.8	1.4×10^4	2.45	12.30	0.0050
	3	8.91×10^4	5.8	2.1×10^4	5.04	25.35	0.0046

Table 4-2 shows that improving heat transfer by increasing water velocity will be accompanied by sharp increases in water pumping power for conventional heat exchanger designs. Note that ducting pressure losses and the head requirement imposed by the higher density of the cold seawater are not included in the table.

Working Fluid Heat Transfer: Pool Boiling. Transfer of heat from a submerged surface to a liquid at a temperature above the boiling point will cause bubble formation if the temperature and heating rate are high enough to surpass convective heat transfer to the body of the fluid and if nuclei are available that can initiate bubble formation at the local fluid temperature. Bubble formation causes a sharp increase in heat transfer so that the coefficient increases with heating rate; however, with further heating, the evolved vapor begins to interfere with the access of liquid to the surface so that the heat transfer rate decreases, eventually reaching a maximum, as vapor bubbles interfere with heat transfer to the surface. With higher heat loading the heat transfer rate drops rapidly. Therefore, pool-boiling heat transfer coefficients are highly dependent on surface characteristics, working fluid purity, and geometrical factors that affect convection patterns, as well as temperature differentials and fluid thermodynamic properties. Several equations have been derived that can be useful in estimating the effects of changes in scale or operating conditions on the performance of particular designs (Rohsenow, 1952; Holman, 1991; Perry and Green, 1984). A thorough discussion is given by Holman.

Many methods have been studied for enhancing pool-boiling heat transfer by alteration of the surface characteristics to promote bubble formation. Porous coatings formed by spraying or sintering powders or fibers, as well as surface roughening to create nucleation sites, have been found effective. Order of magnitude improvements in the shell-side heat transfer coefficient in comparison with smooth tubes have been achieved (Berenson, 1962; Czikk et al., 1975, 1977, 1978; Sabin and Poppendiek, 1977, 1978, 1979; Kajikawa et al., 1980, 1983; Mailen, 1980). The benefits with surfaces coated with piled fibers were investigated in detail by Kajikawa et al. (1983). Twenty-six kinds of surface structures were used in the experiments, and an equation was developed relating boiling heat transfer performance to the physical characteristics of the fiber sintered material. Model tests in the Argonne facility with a pool-boiling heat exchanger using the Linde Hi Flux™ coating indicate that the ammonia-side heat transfer coefficient for OTEC operating conditions will be in the range of 25.5 to 28.4 kW/m² K (Thomas et al., 1980). Mochida et al. (1983) obtained similar values with their enhanced surfaces by using Freon-type working fluids (Flon). Enhanced tubes of this type were employed in the pool boiler of the 100-kW demonstration plant at Nauru, which operated successfully over an 8-month period.

Spray Evaporator. Spray evaporators are designed to distribute a thin film of the liquid working fluid over all of the tubes so that vaporization takes place from the film surface or via nucleate boiling. The tube banks are mounted horizontally to facilitate uniform distribution of the working fluid from nozzles mounted above the tube banks. Liquid is injected at a rate that will prevent dry-out as the film progresses down the tubes. Unvaporized fluid drips down to a collecting pool at the bottom of the tube bank and is recirculated to the spray nozzles. This process is illustrated in Fig. 4-8 (a–c) (Lorenz and Yung, 1978). Subcooled spray falling on the top tube of a vertical bank forms a liquid film (a,b). As the fluid flows around the tube, heat is conducted into the film, changing the temperature profiles, as shown in (c). When the temperature wave reaches the surface, surface warming begins and is accompanied by evaporation. With further flow around the tube, a steady state is reached in which film heating balances film cooling by evaporation. Depending on mass flow, this condition may be reached before the liquid drips off the first tube. The liquid then remains at the equilibrium temperature as it drips from tube to tube to the bottom of the tube bank. For a given surface temperature, the vapor pressure in the evaporator is then a measure of the ΔT between the tube wall and the film surface. Nucleate boiling may also occur, varying in intensity with the number of nucleation sites and the degree of super-heating required to initiate boiling. The dependence of heat transfer on Reynolds number, Prandtl number, heat flux, and flow rate for potential OTEC working fluids and operating conditions has been studied by many investigators (for example, Lorenz and Yung, 1978; Owens 1978). Lorenz and Yung (1978) proposed equations to predict heat transfer rates in spray evaporators.

Owens (1978), in experiments with ammonia evaporation from a 51-mm-diameter horizontal stainless steel tube and by using other data for tube banks, showed that the following correlations are valid for OTEC, within ±10% (see Fig. 4-9).

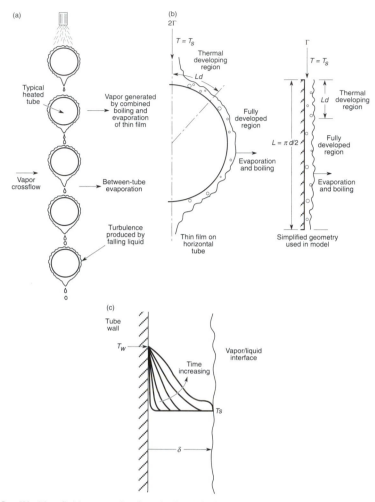

Fig. 4-8. Working fluid evaporation from horizontal tubes in a spray evaporator (Lorenz and Yung, 1978).

Turbulent flow, nonboiling:

$$h\left(\frac{\mu^2}{g\rho^2 k^3}\right)^{1/3} = 0.185(H/D)^{0.1}\mathrm{Pr}^{1/2} \tag{4.1.21}$$

Turbulent flow, boiling:

$$h\left(\frac{\mu^2}{g\rho^2 k^3}\right)^{1/3} = 0.0175(H/D)^{0.1}(\dot{Q}/A)^{1/4}\mathrm{Pr}^{1/2} \tag{4.1.22}$$

where H is the distance between the top and bottom of two adjacent tubes in a column.

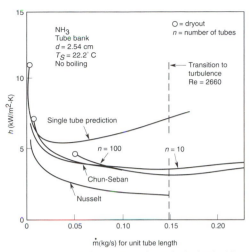

FIG. 4-9. Predicted heat transfer in NH₃ flow over horizontal tube banks (Owens, 1979).

Predictions by Owens (1978) for ammonia and other potential OTEC working fluids are shown in Fig. 4-10. The Owens values are about 10% above those predicted by Lorenz and Yung. Either of the sets of equations appears satisfactory for predicting working-fluid-side heat transfer coefficients for smooth horizontal tubes. Major improvements are possible in heat transfer coefficients with surface enhancements that promote nucleate boiling or thinning of the flowing film. For thin film evaporation of ammonia, Sabin and Poppendiek (1978) compared the conductances of films on smooth tubes with those on tubes having five types of enhanced surfaces: (1) porous nickel plating, (2) Linde Hi Flux™, (3) fine turned and burnished grooves, (4) diamond knurl, and (5) straight knurl. As shown in Fig. 4-11 the diamond knurled surface gave the best results, slightly better than the Linde surface. The improvement is between two and three times, depending on weight flow, compared with the smooth tube. Czikk et al. (1978) measured the film conductances in a test bundle of 13 Hi Flux–coated tubes. Heat transfer values range from 28.4 to 34 kW/m² K, and the heat flux varied from 9.5 to 30 kW/m². The performance was essentially the same as for the pool-boiling condition. A test of the horizontal spray evaporator with Hi Flux–coated tubes at Argonne National Laboratory (ANL) gave a value of 26.0 kW/m² K (Panchal et al., 1981).

Condensation. In the horizontal shell-and-tube condenser, vapor enters the shell side and condenses as it impinges on the tube surfaces, forming a liquid film that drips from tube to tube in the bank, finally flowing to the collecting pool at the bottom. The process is described in detail by Holman (1981). Nusselt showed that for laminar flow over the tubes, the average heat transfer coefficient was given by the following equation:

$$h = 0.725 \frac{[\rho_l(\rho_l - \rho_v)gh_{fg}k^3]^{1/4}}{[\mu nd(T_g - T_w)]^{1/4}},$$

(4.1.23)

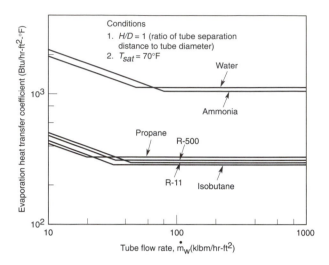

FIG. 4-10. Predicted heat transfer in spray evaporators with various working fluids (Owens, 1978).

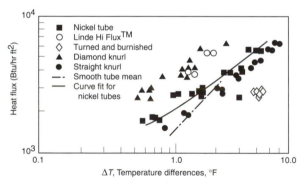

FIG. 4-11. Heat conductance in ammonia films on various surfaces (Sabin and Poppendiek, 1978).

where n is the number of tubes in a vertical bank. The coefficient will increase by about 20% if ripples form in the film, which is generally the case. Equation (4.1.23) may be expressed as a function of the Reynolds number:

$$h = 1.514 \, \text{Re}^{-1/3} \frac{[k^3 \rho_1 (\rho_1 - \rho_v) g]^{1/3}}{(\mu^2)^{1/3}}$$
$$= 1.514 \Phi^{1/3} \, \text{Re}^{-1/3}. \tag{4.1.24}$$

(Note that Φ has the dimensions of heat transfer cubed.)

As the Reynolds number increases, a flow velocity is reached at which the Eq. (4.1.24) must be replaced by an empirical relation proposed by Kirkbride (1934). Above this Reynolds number, the condensing film becomes turbulent:

$$h = 0.077 \Phi^{1/3} \text{Re}^{0.4}. \tag{4.1.25}$$

These equations predict a transition Reynolds number of 1400 for an ammonia saturation temperature of 50°F. For a heat exchanger with forty 25-mm-diameter tubes in a vertical bank and a wall-to-surface temperature difference of 0.55°C, h will have a value of 6 kW/m^2 K. This corresponds to a film Reynolds number of 240. Thus turbulence will not occur unless the film thickness, that is, heat loading, causes the ΔT to become greater than about 3°C. Heat transfer enhancement methods that are effective for evaporators because they reduce the average film thickness can also improve the performance of shell-and-tube condensers, particularly at high heat loadings. Tests of horizontal tube condensation under OTEC conditions on an enhanced wire-wrapped surface of Linde design have been reported by Thomas et al. (1979). The observed value of h for the nominal operating condition was 30 kW/m^2 K, roughly four times the predicted smooth tube value. Sabin and Poppendieck (1979) reported that shallow spiral grooves also produce a many-fold improvement in the condensation heat transfer coefficient on horizontal tubes. It should be noted that the enhancement ratio is based on the smooth tube area.

Vertical Shell-and-Tube Condensers. Condensation of vapor on smooth vertical tubes produces a liquid film that increases in thickness, with a lowering of the heat transfer coefficient, as the liquid flows to the bottom of the vessel. If, however, the tube surface is altered to provide grooves or channels, surface tension tends to draw the liquid into the grooves and allows high heat transfer through the thinned regions of the film. The advantages of this design were first discussed by Gregorig (1954). A theoretical study of Zener and Lavi (1973) indicated that major gains were possible with such surfaces. They proposed the use of vertical condensers as a way to make major reductions in OTEC heat exchanger size. Their analysis yielded a method for defining the optimal shape for rounded longitudinally grooved surfaces. Heat exchanger concepts based on this analysis are shown in Fig. 4-12; however, an extensive further review of the literature showed that the actual flow was too complicated to be described simply and that further experimental data were needed before complete analytical design methods could be developed (Rothfus and Lavi, 1978). Experiments with vertical fluted tubes at Carnegie–Mellon

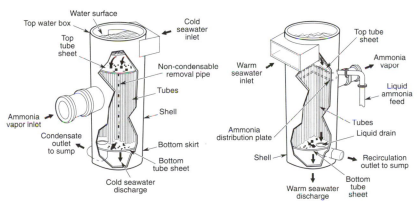

Fig. 4-12. Schematic of CMU vertical heat exchangers. Condenser: left; evaporator: right (Rothfus and Lavi, 1978).

University confirmed the predicted improvement and led to the design and construction of a fluted evaporator and condenser for tests at ANL. The flutes were a constant 0.66-mm radius machined on the outer surface of a 25.4-mm-diameter tube. The area ratio compared with the smooth 25.4-mm-diameter tube was 1.50 (Rothfus and Neumann, 1977). Tests were conducted with both ammonia and Freon R-11. Heat transfer enhancement factors of 3 to 5 were demonstrated. Extensive studies of longitudinally fluted tube condensers were done by Uehara et al. (1978), Suematsu (1974), and Miyoshi et al. (1977) using water, R-11, and R-113 as working fluids. Tube diameters of 25.4 and 51 mm, flute heights of 0.15 to 1.07 mm, and flute pitches of 1.0 to 3.6 mm were studied. Smooth tubes were also included. The data were correlated by the relationship:

$$\text{Nu}/[(d/P)^{0.8}(P/l)^{0.6}] = 1.8(\text{Ga Pr} L/H)^{0.4} \; , \qquad (4.\,1.26)$$

where Nu is the Nusselt number, d the flute depth, P the flute pitch, Ga the Galileo number $= gl^3/v^3$, l the effective length of fluted tube, $H = C_p(T_{sat} - T_w)/L$, L the latent heat of vaporization, and v the viscosity.

For the Freons, the constant had a value of 0.94 for smooth tubes, 1.8 for the tube with flute height 0.15 mm and 2.00-mm pitch, and 2.6 for the remaining tubes. Additional experiments by Kawano et al. (1983) with a series of fluted tube designs have shown that the equation in the figure correlates all of the experimental results within ±20%, including the data of Uehara et al. (Fig. 4-13).

	Tube	P(mm)	d(mm)	l(mm)	Fluid
◇	Fluted tube-1	2	0.15	510	R-11
▽	Fluted tube-2	1	0.3	510	
△	Fluted tube-3	1	0.5	446	
□	Fluted tube-4	1.5	0.5	466	
▲	Fluted tube-3	1	0.5	945	R-114
○	Fluted tube-3	1	0.5	1370	R-22
▲	Fluted tube-3 (Uehara)	1	0.5	1561	R-113

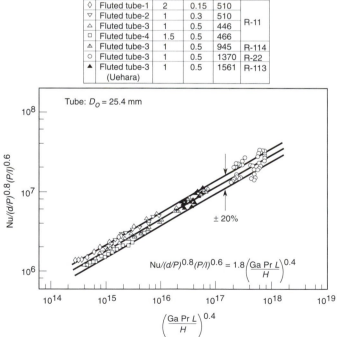

FIG. 4-13.   Heat transfer coefficients for downward flow over fluted tubes (Uehara, 1978).

The Nusselt number for the fluted tubes was found to be 1.5 to 7 times the value for smooth tubes. A cross-sectional view of the fluted tubes is shown in Fig. 4-14 (Kawano et al., 1983). The experimental data show the highest heat transfer coefficient for the fluted tube with 1-mm groove pitch and 0.5-mm groove depth (tube 3). This tube design was adopted for the proposed 100-kW demonstration plant. Although vertical flutes are effective in increasing heat transfer, with long tubes the liquid film floods the surface, causing the heat transfer coefficient to approach smooth tube values. The difficulty may be alleviated by providing gutters or separators at intervals to drain off the film. In the demonstration plant, separators were provided at 1.2-m intervals. Various methods have been devised for preventing buildup of thick films. A wire-wrapped configuration developed by Linde was tested at ANL (Thomas et al., 1979). Studies have also been made by Webb (1978), Combs (1979), and Kajikawa et al. (1979), among others. Takazawa et al. (1983) showed that spirally fluted tubes with drainage gutters, as shown in Fig. 4-15, could be particularly effective. The heat transfer coefficient of the tube that gave the best performance is compared with the smooth tube in Fig. 4-16. This shows an enhancement ratio of 4 for a heat duty of 8.9 kW/m^2, which is the design value for the 100-kW demonstration plant.

4.1.3.3 Plate Heat Exchangers

Plate heat exchangers have long been used for forced-convection heat transfer between two liquids. Their use for OTEC was explored by Alfa-Laval as part of the European program begun in 1970 to explore industrial uses of the oceans (Lachmann, 1978). Heat exchangers made from thin plates separated by gaskets designed to channel the water and working fluid between alternate corrugated

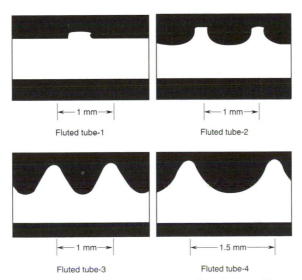

FIG. 4-14. Profiles of tube surfaces used in fluted tube condensation tests (Kawano et al., 1983).

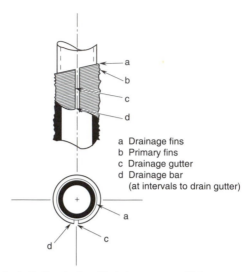

a Drainage fins
b Primary fins
c Drainage gutter
d Drainage bar
   (at intervals to drain gutter)

Fig. 4-15. Design of spirally fluted tube with drainage gutters (Takazawa et al., 1983).

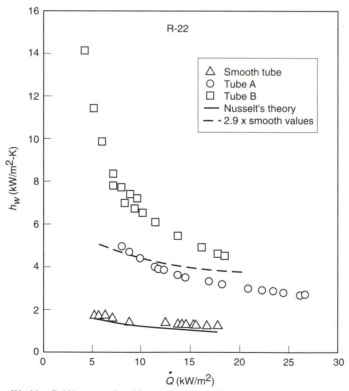

Fig. 4-16. Working fluid heat transfer with smooth and spirally grooved tubes (Takazawa et al., 1983).

plates had been developed by Alfa-Laval for a variety of applications, but new designs were required for two-phase flow (Berndt and Connell, 1978). A schematic view of the design is shown in Fig. 4-17. A plate heat exchanger supplied by Alfa-Laval was used in the Mini-OTEC demonstration (Owens and Trimble, 1980), and was also tested in the ANL program (Panchal et al., 1981, 1984). Tests of a commercial plate heat exchanger made by the Tranter Company were also conducted in the ANL program (Panchal et al., 1983).

The corrugations on the heat exchanger plates act as Gregorig surfaces to enhance heat transfer and also increase the heat transfer area by 20 to 30% relative to the projected area. They also provide structural rigidity and serve to distribute the fluid flow uniformly. The plate surfaces may also be coated with thin layers of fibers or powders to enhance nucleate boiling. In combination, these features can permit major reductions to be made in the size of OTEC heat exchangers; however, the heat transfer advantages of the plate heat exchanger tend to be accompanied by significant pressure drops, which reduce the system benefits. The plate concept may be extended to include thin plates that may be formed and combined by welding or adhesives to carry water and working fluid in parallel channels, with heat transfer through a dividing wall. Such designs may offer great advantages in volume reduction and constructibility. Anderson and associates explored the potential

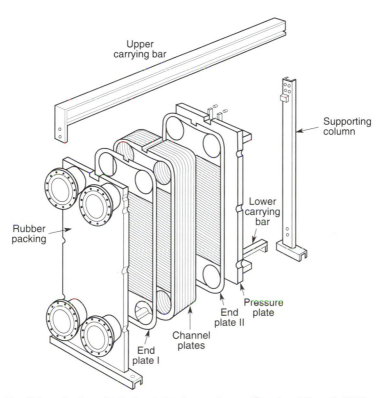

Fig. 4-17. Schematic view of Alfa-Laval plate heat exchanger (Berndt and Connell, 1978).

benefits of several variations of this approach (Anderson and Anderson, Jr., 1977; Anderson, 1978; Anderson and Pribis, 1979; Anderson et al., 1980). A drawing of a recent test configuration, designed to be suitable for scale-up to OTEC sizes, is shown in Fig. 4-18 (Anderson et al.,1980).

A modification of the plate concept that combines compact design and ease of fabrication with high performance, and avoids sealing and crevice corrosion problems, has been proposed by the Trane Company (Ashworth and Slebodnick, 1980; Foust, 1981). The concept is illustrated in Fig. 4-19. Aluminum billets are extruded to form flat plate sections with internal water passages, as shown. Longitudinally corrugated sections are rolled from thin aluminum sheet to form ammonia flow channels. The heat exchanger is then constructed by aligning the plates alternately and brazing the assembly. A diagonal cut enables two ammonia sections to be joined at right angles so that ammonia can enter the channels horizontally at the top, then turn to flow vertically, until it reaches the bottom where another diagonal cut returns the flow to the horizontal direction. A cutaway view of the assembly is shown in Fig. 6-25. Water flows downward under gravity from pools above the heat exchangers. This simplifies ducting and makes the water channels easily accessible for inspection and cleaning. A heat exchanger of this type was tested at the 1-MWt level at ANL (Panchal et al., 1981).

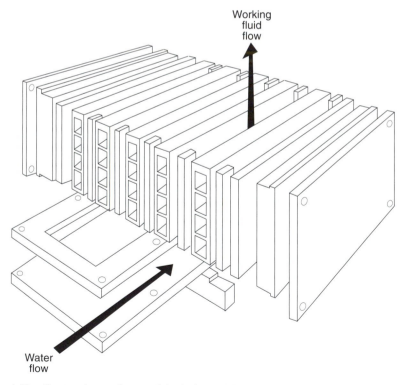

FIG. 4-18. Compact heat exchanger of the Andersons' design (Anderson et al., 1980).

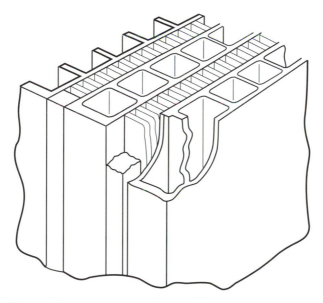

Fɪɢ. 4-19.   Diagram of Trane heat exchanger (Foust, 1981).

4.1.3.4   Folded-Tube Heat Exchangers

The folded-tube heat exchanger design differs from the other OTEC concepts discussed in having the working fluid contained in long tubes, folded and mounted in parallel banks that are immersed in seawater flowing downward under gravity from overhead pools. The rectangular channels that direct the water flow are an integral element of the OTEC platform structure. Because this heat exchanger is unique, the description to follow is somewhat more detailed than those for the other types, for which an extensive literature exists.

The concept is illustrated schematically in Fig. 4-6, which shows the working fluid flow for the evaporator. A drawing of a 2.5-MWe assembled condenser module is presented in Fig. 4-20 (George and Richards, 1980). In the design shown, 7.6-cm-diameter aluminum tubes are fabricated with U bends connecting straight sections at approximately 7.6-m intervals to form single "folded" tubes 213 m long. The U bend adds 0.9 m to the length of each tube, so there are 25 horizontal passes in the 213-m length. Seven tubes are then nested to form a vertical tube bank 17 m deep. Sixty-five banks are combined to form a 5.0-MWe (net) evaporator or condenser module.

The concept originated in the application of systems engineering principles to achieve an integrated total OTEC system of minimal cost at a given net power level. Emphasis was placed on optimizing tradeoffs between heat exchanger performance and total system cost reduction through simplicity of system design, minimal packaging volume, use of low-cost materials (for example, roll-welded aluminum tubing at $1.5/kg versus extruded aluminum tubing at $2.75/kg versus titanium

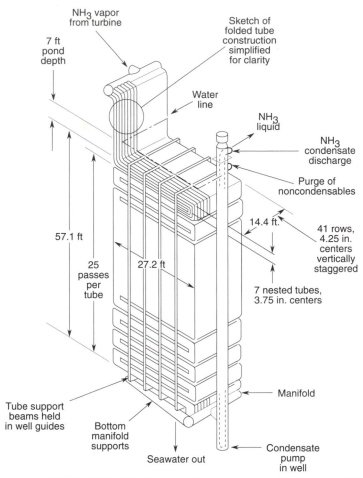

FIG. 4-20. Design of JHU/APL 2.5-MWe folded-tube heat exchanger module (George and Richards, 1980).

tubing at $10/kg), ease of manufacturing through use of standard procedures, and compatibility with system requirements and components for minimal platform costs. The performance of this type of exchanger was first investigated by Olsen and Pandolfini (1975) and further reported by George et al. (1979), Pandolfini et al. (1980), and Pandolfini et al. (1981). In the evaporator, liquid ammonia enters at the bottom of the tubes at a flow velocity of from 0.2 to 0.6 m/s, slightly subcooled. Nucleate boiling begins at about the end of the first pass. The vapor produced accelerates the flow, producing successive hydraulic regimes as flow proceeds along the tube, which are characterized by the fraction of the flow volume occupied by vapor. The flow regimes depend not only on axial distance but also on heat loading and have picturesque descriptions, such as bubbly, churning, stratified, annular, annular spray, and dry-out. Various semiempirical equations have been proposed that correlate experimental data for internal two-phase flow for specific designs, working fluids, and operational conditions. Since the heat transfer and flow resistance progressively change as the fluid moves forward, the use of average values derived from inlet and outlet measurements can be misleading and should be used with caution, particularly in comparisons of overall heat transfer behavior from one type of heat exchanger to another. Data on water-side heat transfer and internal two-phase flow ammonia heat transfer, directly applicable to OTEC heat exchanger conditions, were obtained in experimental programs conducted at JHU/APL (Pandolfini et al., 1978, 1980). These were followed by tests at the first on-land OTEC installation at Kea-Hole Point, Hawaii (Pandolfini et al., 1977). Data on overall heat transfer were measured later at the 1-MWt test facility at ANL. In the initial tests, water-side and ammonia-side heat transfer rates and pressure drops were measured separately.

Water-Side Heat Transfer. The assembly shown schematically in Fig. 4-21 was used for the water-side measurements. Instrumented, electrically heated tubes, as shown in Fig. 4-22, were placed at the indicated positions for heat transfer measurements, and pressure sensors measuring Δp were placed above and below the tube bank. Heat transfer rates varied with radial position on the tubes and were slightly different for the upper and lower tubes, compared with tubes internal to the bank. The data were correlated by the Zukauskas formula:

$$\mathrm{Nu} = 0.40 \, \mathrm{Re}^{0.6} \, \mathrm{Pr}^{0.36}. \tag{4.1.27}$$

The measured values were roughly 25% higher than those predicted by the formula (Pandolfini et al., 1981; Afgan and Schlunder, 1974).

Water-Side Pressure Drop. The pressure drop in flow across a staggered bank of tubes is given by the following relationship:

$$\Delta p = \mathrm{Eu} \, \rho v_m^2 (2N - 1) / g, \tag{4.1.28}$$

where Eu is the Euler number ~ 0.1 for the folded-tube bank, V_m the water velocity at minimal clearance between tubes, and N the number of tubes in one vertical column.

Measured values of the variation of pressure loss with flow velocity are shown in Fig. 4-23. The graph indicates a small decrease in Euler number with Reynolds number.

Ammonia-Side Heat Transfer. Heat transfer coefficients for internal two-phase flow of ammonia were measured by using the apparatus shown schematically in Fig. 4-24 (Keirsey et al., 1979). By adjusting the heat input to the steam preheater, ammonia at any desired quality (mass percentage of vapor to total mass) could be introduced to the electrically heated test section, from subcooled to 20% quality. Wall surface temperature measurements and ammonia pressures were then used to derive heat transfer coefficients. In the all-liquid flow regimes, heat transfer rates agreed reasonably well with the Dittus–Boelter correlation:

$$h_{lA} = 0.023(k/d) \, \mathrm{Re}^{0.8} \, \mathrm{Pr}^{0.4}, \tag{4.1.29}$$

where h_{lA} is the liquid ammonia heat transfer coefficient.

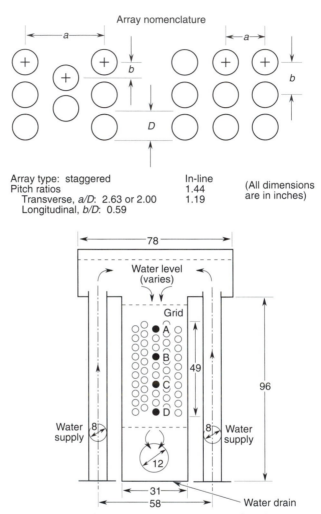

FIG. 4-21. Drawing of apparatus for measuring water-side heat transfer in folded-tube HX (Pandolfini et al., 1977).

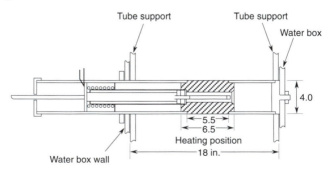

All dimensions are in inches

FIG. 4-22. Instrumented tube used in water-side heat transfer measurements (Pandolfini et al., 1977).

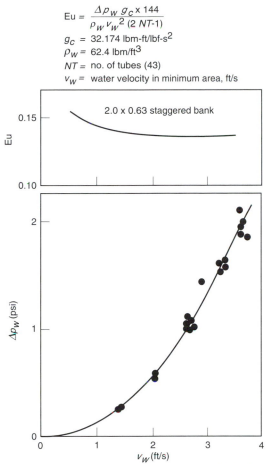

$$Eu = \frac{\Delta p_W \, g_c \times 144}{\rho_W v_W^2 \, (2 \, NT\text{-}1)}$$

g_c = 32.174 lbm-ft/lbf-s^2

ρ_W = 62.4 lbm/ft^3

NT = no. of tubes (43)

v_W = water velocity in minimum area, ft/s

2.0 x 0.63 staggered bank

FIG. 4-23. Variation of pressure loss with water velocity in folded-tube heat exchanger tests (Pandolfini et al., 1981).

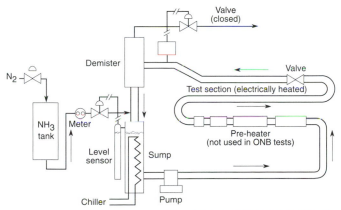

FIG. 4-24. Diagram of apparatus for measuring heat transfer in internal two-phase flow of ammonia (Keirsey et al., 1979).

In the two-phase flow regimes, the Chaddock–Brunemann relation gave good agreement:

$$h/h_{lA} = 1.91[\text{Bo}\exp(4 + 1.5X_{tt}^{-2/3})]^{0.6}, \qquad (4.1.30)$$

where h is the two-phase flow heat transfer coefficient, Bo the boiling number = $(\dot{Q}/A)\rho_1 v_1 H$, and X_{tt} the Martinelli parameter

$$[(1 - x)/x]^{0.9}(\rho_g/\rho_l)^{0.5}(\mu_l/\mu_g)^{0.1}.$$

The experimental results for the two-phase flow regime are presented in Fig. 4-25.

Ammonia-Side Pressure Drop. The pressure drop in two-phase flow may be estimated by use of a correlation proposed by Chisholm (Collier, 1972):

$$\Delta p = (\phi_f)^2(1 - x)^{1.75}\Delta p_l, \qquad (4.1.31)$$

where x is the vapor mass quality and Δp_l is the all-liquid pressure drop given by the Fanning equation for the same conditions:

$$\phi_f^2 = 1 + CX_{tt}^{-1} + X_{tt}^{-2}, \qquad (4.1.32)$$

where X_{tt} is the Martinelli parameter.

The experimental results showed that the appropriate value for the constant C was 0.9. A comparison of predicted versus experimental quality is shown in Fig. 4-26 (Pandolfini et al., 1981).

4.1.4 *Model Tests of OTEC Heat Exchangers*

With support from the U.S. Department of Energy, a heat exchanger test facility was constructed at ANL in 1978. The specific purpose was to allow proposed OTEC heat exchanger designs to be tested at a large enough scale to give valid design data

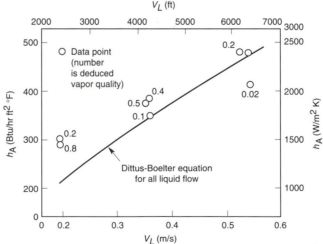

FIG. 4-25. Variation of liquid ammonia heat transfer with flow velocity in internal flow (Pandolfini et al., 1980).

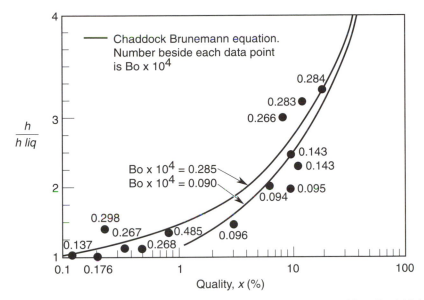

Fig. 4-26. Ratio of two-phase flow to all liquid heat transfer in ammonia internal flow (Pandolfini et al., 1981).

for demonstration-size OTEC power units. A nominal heat loading of 1 MWt was selected for the tests, and participants in the OTEC program were encouraged to fabricate innovative heat exchanger designs for comparative evaluations. Particular care was taken to ensure accuracy in the measurement of heat inputs and outputs and pressure drops across the exchangers so that accurate values of overall heat transfer and pumping losses could be obtained. It was not feasible to measure the water and working fluid heat transfer coefficients directly. These values were inferred, via the Wilson plot procedure, that is, from the measured change in the overall heat transfer coefficient at constant heating rate when the flow rate of one fluid was varied while the other fluid was held constant. The facility design, instrumentation, and methods of data analysis are described in Sather et al. (1978) and Lewis and Sather (1978). The careful design of the equipment and operating procedures is attested by the accuracy of the data, which were reproducible within a few percent. A schematic of the test loop is shown in Fig. 4-27.

4.1.4.1 Linde Flooded-Bundle Evaporator

The Union Carbide flooded-bundle test evaporator, of conventional horizontal shell-and-tube design, employed a proprietary surface coating called Hi Flux on the shell side of 3. 8-cm-OD titanium tubes. Tests were conducted during November and December 1978 and were supplemented by additional experiments during March to May 1979. Details of the tests are given in Lewis and Sather (1978). A schematic of the heat exchanger is presented in Fig. 4-28. Nominal design parameters are listed in Table 4-3.

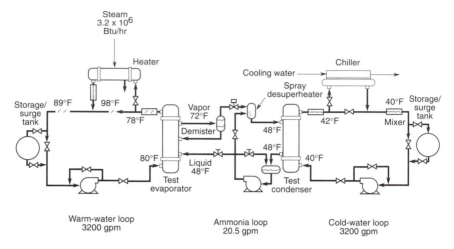

FIG. 4-27.   Schematic of ANL test facility for 1-MWt model heat exchangers (Lewis and Sather, 1978).

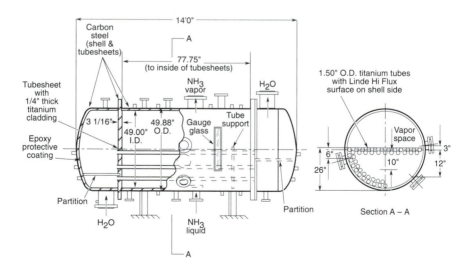

FIG. 4-28.   Diagram of Linde flooded-bundle evaporator (Lewis and Sather, 1978).

Table 4-3. Nominal design parameters for the Linde flooded-bundle evaporator

Parameter	English units	SI units
Heat duty	3.2 MBtu/h	0.94 MWt
Ammonia inlet temperature	72°F	22.22°C
Water inlet temperature	80°F	26.67°C
Water outlet temperature	78°F	25.56°C
Number of tubes	279	279
Hi Flux-coated tube length	75 in.	1.90 m
Tube outside diameter	1.5 in.	0.139 m
Tube inside diameter	1.43 in.	0.133 m
Effective outside heat transfer area[a]	685 ft²	63.63 m²
Mean water velocity in tubes	6.8 ft/s	2.07 m/s
Overall water-side pressure drop	2.75 psia	18.96 kPa
Shell inside diameter	49 in.	1.24 m
Shell-side design pressure	215 psig	1482 kPa
Tube-side design pressure	100 psig	689 kPa
"Clean" overall heat transfer coef.[b]	768 Btu/ft² h °F	4365 W/m² °C

[a]Based on actual coated tube length of 75 in.; the total tube length between tube sheets is 2.75 in. longer.
[a]Based on outside tube surface area. Assumes no tube-side or shell-side fouling.

The tests fully confirmed the predicted improvement in the overall heat transfer coefficient through the use of the Hi Flux surface coating. A summary of the test results in normal operation is given in Table 4-4, showing a value of $U = 4336$ W/m² °C and pressure drop of 18.8 kPa for the nominal water velocity of 2.1 m/s. The variation of water-side heat transfer with water velocity is shown in Fig. 4-29. The observed data are correlated by the following equation:

$$h = 114.6 v_{WE}^{0.8}. \qquad (4.1.33)$$

As in other experiments, the measured values are about 20% above the predictions of the Sieder–Tate and Dittus–Boelter correlations. An unexpected observation during the tests was that the surface tended to become deactivated if liquid ammonia was allowed to stand in contact with the surface under nonboiling conditions. The effectiveness of the surface was restored in operation, but approximately 100 h of operation was required for the full effectiveness to be regained.

Table 4-4. Values of U_0 as a function of water flow rate for the Linde flooded-bundle evaporator

Flow rate		U_0		Δp	
gpm	kg/s	Btu/ft	W/m² °C	psi	kPa
1492	97	514	2917	0.57	3.93
2404	156	665	3775	1.55	10.68
3171	205	764	4336	2.72	18.75
4008	259	840	4768	4.32	29.78
4997	323	909	5159	6.61	45.57

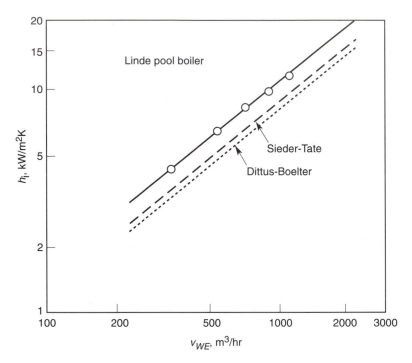

Fig. 4-29. Variation of water-side heat transfer with flow rate in Linde flooded-bundle evaporator (Lewis and Sather, 1978).

4.1.4.2   Linde Sprayed-Bundle Evaporator

The sprayed-bundle evaporator employed horizontal 3.8-cm-OD titanium tubes, smooth on the inside and treated on the outside with Linde Hi Flux coating. A diagram of the test unit is shown in Fig. 4-30. Details of the tests, which were conducted during October and November 1978, are presented in Hillis et al. (1979). The nominal parameters for the tests are listed in Table 4-5.

The heat transfer performance of this evaporator was almost the same as that of the Linde flooded-tube version. The value of U was measured to be 4313 W/m² °C at the nominal conditions. For the nominal conditions, the value of h_w was found to be 2600 W/m² °C from a Wilson plot. With this value, the dependence of the water-side heat transfer coefficient on water flow was accurately represented by the following equation:

$$h = 107.2 v_{WE}^{0.8}. \qquad (4.1.34)$$

For the spray evaporator setup, the pressure drop was 27.6 kPa (including ducting losses), a significantly higher pumping power requirement than for the flooded-bundle design. As with that design, the enhanced surfaces were deactivated by being allowed to remain in contact with nonboiling liquid ammonia. For this

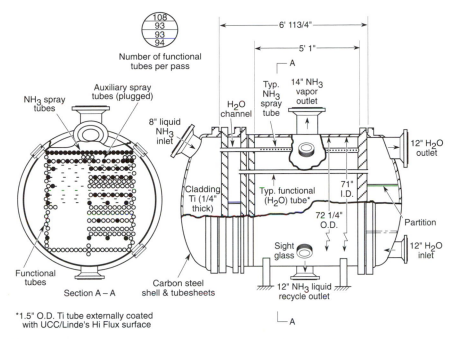

FIG. 4-30.   Diagram of Linde sprayed-bundle evaporator (Hillis et al., 1979).

Table 4-5. Nominal design parameters for the Linde sprayed-bundle evaporator

Parameter	English units	SI units
Heat duty	3.2 MBtu/h	0.94 MWt
Ammonia inlet temperature	72°F	22.22°C
Water inlet temperature	80°F	26.67°C
Water outlet temperature	78°F	25.56°C
Number of tubes ·	388	388
Hi Flux-coated tube length	55 in.	1.40 m
Tube outside diameter	1.5 in.	0.038 m
Tube inside diameter	1.43 in.	0.036 m
Effective outside heat transfer area[a]	698 ft^2	64.7 m^2
Mean water velocity in tubes	6.6 ft/s	2.01 m/s
Water-side pressure drop[b]	2.1 psi	14.5 kPa
Shell inside diameter	71 in.	1.803 m
Shell-side design pressure	215 psig	1482 kPa
Tube-side design pressure	100 psig	690 kPa
"Clean" overall heat transfer coef.[c]	740 Btu/ft^2 h °F	4200 W/m^2 °C

[a]Based on actual coated tube length of 55 in.; the total tube length between tube sheets is 6 in. longer.
[b]Does not include tube and nozzle inlet and exit losses.
[c]Based on outside tube surface area. Assumes no tube-side or shell-side fouling.

situation, the surfaces could be reactivated by drying them out by maintaining warm-water flow; however, if the surfaces were allowed to dry out during operation, the dry surface areas could not be rewetted simply by increasing the ammonia flow rate. The surface could be rewetted only by "flooding" the surface or subcooling it. A summary of the test results is given in Table 4-6.

Table 4-6. Values of U_0 as a function of water flow rate for the Linde sprayed-bundle evaporator

Flow rate		U_0	
gpm	kg/s	Btu/ft^2 h ^{0}F	W/m^2 °C
1502	95	500	2837
2384	150	637	3615
3177	200	760	4313
4012	253	835	4738
5004	315	905	5135

4.1.4.3   Linde Enhanced-Tube Condenser

The Union Carbide test condenser, of conventional horizontal shell-and-tube design, employed 3.8-cm-OD aluminum tubes fluted on the inside to promote water-side heat transfer and wire-wrapped on the ammonia side to promote drainage of the condensing ammonia film. Details of the test are presented in Yung et al. (1979). The nominal test conditions are listed in Table 4-7. A diagram of the condenser is shown in Fig. 4-31, and details of the tube design are given in Fig. 4-32.

Table 4-7. Nominal design parameters for the Linde enhanced-tube condensers

Parameter	English units	SI units
Heat duty	3.2 mBtu/h	0.94 MWt
Ammonia condensing temperature	48°F	8.9°C
Water inlet temperature	40°F	4.4°C
Water outlet temperature	42°F	5.6°C
Number of tubes	147	147
Externally enhanced tube length	154.5 in.	3.92 m
Tube outside diameter (excluding wrapped wire)	1.50 in.	0.0381 m
Tube inside diameter	1.37 in.	0.0348 m
Effective outside heat transfer area[a]	743 ft^2	69.1 m^2
Mean water velocity in tubes	4.7 ft/s	1.43 m/s
Overall water-side pressure drop	2.0 psia[b]	13.8 kPa
Shell inside diameter	29.25 in.	0.743 m
Shell-side design pressure	215 psig	1482 kPa
Tube-side design pressure	100 psig	690 kPa
"Clean" overall heat transfer coef.[c]	740 Btu/ft^2 h °F	4427 W/m^2 °C

[a]Based on outside diameter without wrapped wire and on actual enhanced tube length of 154.5 in., the total tube length between the tube sheets is 6.4 in. longer.
[b]Based on communication with Union Carbide, Linde Division, Tonawanda, New York.
[c]Based on outside tube surface area. Assumes no tube-side or shell-side fouling.

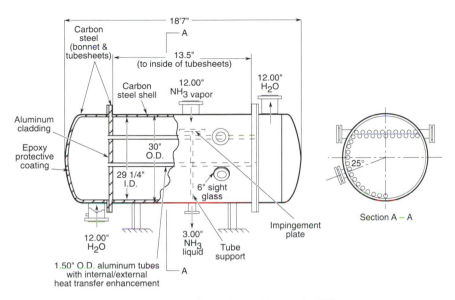

FIG. 4-31. Diagram of Linde enhanced-tube condenser (Yung et al., 1979).

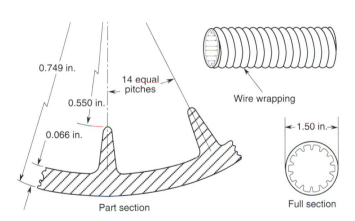

FIG. 4-32. Details of Linde condenser tube design (Yung et al., 1979).

The performance of the enhanced-tube condenser fully confirmed the expectations. For the nominal conditions, the measured values of U and the pressure drop were 4642 W/m^2 K and 12.4 kPa. The data are presented in Table 4-8.

The derived values of the water-side heat transfer coefficient, based on a Wilson plot, are shown in Fig. 4-33. The values are accurately represented by the following equation:

$$h_{sw} = 192.7\dot{m}_{swc}^{0.66}.$$

(4.1.35)

Table 4-8. Heat transfer and pressure drop for Linde enhanced-tube condenser
(Yung et al., 1979)

Water flow		U_0		Δp	
gpm	kg/s	Btu/h ft² °F	W/m² K	psi	kPa
1516	98.0			0.6	4.1
1540	99.6	546	3098		
1930	124.8	625	3547		
2398	155.1			1.1	7.6
2410	155.9	702	3983		
3190	206.3	818	4642		
3198	206.9			1.8	12.4
3970	256.8	921	5226		
3993	258.3			2.7	18.6
4965	321.2			3.9	26.9
5010	324.1	1014	5754		

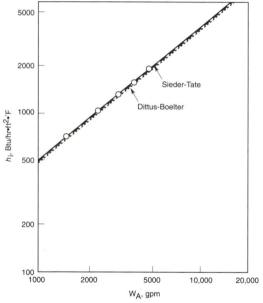

Fig. 4-33. Water-side heat transfer coefficient versus water flow rate in Linde condenser (Yung et al., 1979).

The measured values are from 1.5 to 1.75 times the Sieder–Tate predictions for smooth tubes of the same nominal diameter. Since the actual wetted area was 1.8 times the smooth-tube area, reasonable agreement would result from substituting the actual wetted perimeter of the fluted tube for the smooth-tube value in the Sieder–Tate equation. The experimental program also included tests that showed that the ammonia-side heat transfer was not affected by the heat duty, within the range from 13,620 to 22,700 W/m² °C. This implies that the average thickness of the condensing film is effectively maintained at a constant level by the wire-wrapping technique.

4.1.4.4 Carnegie–Mellon Vertical Fluted-Tube Evaporator

The Carnegie–Mellon evaporator design was based on analyses of liquid flow on vertical tubes by Zener and Lavi (1973) and Lavi (1974). Fluted aluminum tubes of nominal 2.32 cm ID and 3.09 cm OD were arranged vertically in the evaporator test vessel is shown in Fig. 4-34. Details of the tube design and a sketch of the liquid ammonia "applicator," which allowed liquid ammonia in a plenum chamber to be applied as a thin film at the top of the tubes, are shown in Fig. 4-35 (Lorenz et al., 1979). The nominal design parameters are listed in Table 4-9.

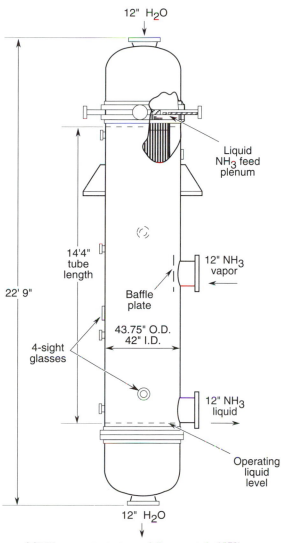

FIG. 4-34. Diagram of CMU evaporator test vessel (Lorenz et al., 1979).

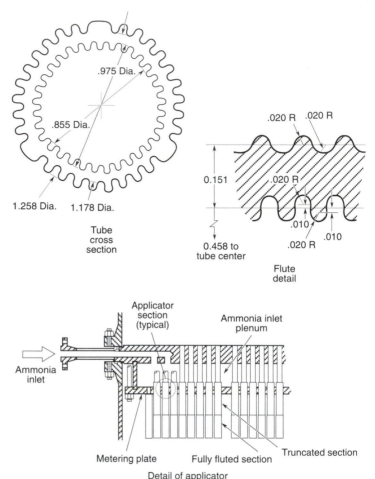

FIG. 4-35. Details of CMU evaporator design (Lorenz et al., 1979).

For the nominal operating conditions, the overall heat transfer coefficient, U, was 4680 W/m² K, and the pressure drop was 22.1 kPa. The value of U is based on surface areas computed as 2π times the mean inside and outside diameters times the total tube length. If the actual area were used, the U value would be 2264 W/m² K. The value of h_a computed from a Wilson plot was 10,214 W/m² K. This is 2.5 times the smooth-surface value that would be predicted by the Chun–Seban correlation. Since the enhanced area due to fluting was 1.85 times the smooth-tube area, it is apparent that the flutes produced the beneficial behavior predicted by Zener and Lavi (1973). The water-side heat transfer coefficient was 1.95 times the smooth-tube value computed by using the Sieder–Tate equation. This value is in close agreement with the enhancement in internal area by a factor of 2.07 due to the water-side fluting. Figure 4-36 and Tables 4-10 and 4-11 summarize the test results.

Table 4-9. Nominal design parameters for the CMU vertical fluted-tube evaporator

Parameter	English units	SI units
Heat duty	3.2 MBtu/h	0.94 MWt
Ammonia evaporating temperature	72°F	22.22°C
Water inlet temperature	80°F	26.67°C
Water outlet temperature	78°F	25.56°C
Number of tubes	240	240
Tube length	172 in.	4.4 m
Tube outside diameter (to midpoint of flutes)	1.218 in.	0.031 m
Tube inside diameter (to midpoint of flutes)	0.915 in.	0.023 m
Outside heat transfer area[a]	1097 ft²	102.0 m²
Mean water velocity in tubes	6.5 ft/s	1.98 m/s
Overall water-side pressure drop	2.7 psia	18.6 kPa
Shell inside diameter	42.25 in.	1.07 m
Shell-side design pressure	215 psig	1482 kPa
Tube-side design pressure	100 psig	690 kPa

[a]Based on outside tube diameter to midpoint of flutes

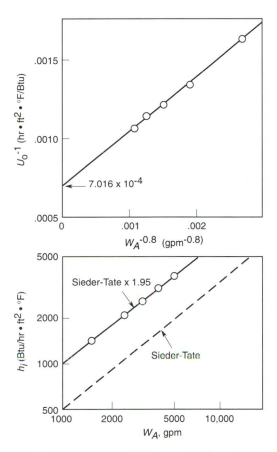

FIG. 4-36. Heat transfer results in tests of the CMU evaporator (Lorenz et al., 1979).

Table 4-10. Values of U_0 as a function of water flow rate for the CMU fluted-tube evaporator

Water flow velocity[a]		U_0	
gpm	kg/s	Btu/ft² h °F	W/m² °C
1510	95	610	3461
2420	152	750	4256
3200	202	825	4681
4050	255	885	5022
4940	311	945	5362

[a]The gpm values can be converted to tube-side velocity v (ft/s) by the relationship $v = 0.002033w$.

Table 4-11. Values of h_i as a function of water flow rate for the CMU fluted-tube evaporator

Water flow rate		h_i	
gpm	kg/s	Btu/ft² h °F	W/m² °C
1510	95	1431	8120
2420	152	2047	11616
3200	202	2610	14810
4050	255	3150	17875
4940	311	3692	20950

The tests showed that the results were not reproducible if the tubes were allowed to dry out, presumably because of difficulty in re-establishing surface wetting. It also appeared that the applicator did not produce uniform films at the tops of the tubes, which caused scatter in the data.

The Carnegie–Mellon condenser, shown in Fig. 4-37, was of the same general design as the Carnegie–Mellon evaporator. It employed fluted aluminum tubes, of 3.09 cm average OD and 2.33 cm average ID. Details of the tube design are shown in Fig. 4-38 (Lewis and Sather, 1979). The nominal design parameters are given in Table 4-12.

The performance of the condenser was in accord with the predictions of the bench-scale tests conducted by the Carnegie–Mellon group. At the nominal conditions, U was 5918 W/m² K, and the water-side pressure drop was 23.0 kPa. As with the evaporator, the U value is based on the projected area of the tubes. If the actual surface areas are used for the calculation, the value of U is 2917 W/m² K for the nominal test conditions. The ammonia heat transfer coefficient deduced from Wilson plots at heat duties of 2.3, 3.2, and 4.0 × 10⁶ kJ/h shows an increase in value from 18,556 to 21,166 W/m² K. This result shows that the film thickness under the test conditions did not reach the critical loading values. The Sieder–Tate equation would predict a value of U of 2463 W/m² K for a smooth tube of this diameter. The experimental result is in accord with other observations that give about 20% higher overall heat transfer values for smooth tubes than the Sieder–Tate equation.

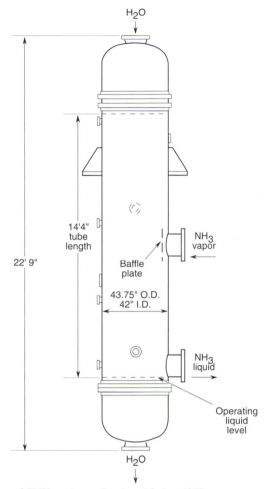

FIG. 4-37. Diagram of CMU condenser (Lewis and Sather, 1979).

4.1.4.5 Alfa-Laval and Tranter Plate Heat Exchangers

The Alfa-Laval heat exchanger used in the ANL tests was a refurbished version of the unit constructed earlier as the evaporator for the OTEC-1 at-sea test. The number of titanium plates was reduced from 260 in OTEC-1 to 52 to 136 plates for the ANL 1-MWt test series. The Tranter heat exchanger was a commercially available small heat exchanger closely similar in design to the Alfa-Laval heat exchanger, which was used to measure the performance benefits of applying the Linde Hi Flux coating to the plate surface. A sketch of the Alfa-Laval plate heat exchanger design is shown in Fig. 4-39. Gaskets attached to alternate sides of the plates are used to direct the flow into alternate channels between the plates. The physical characteristics of the plates used in the Alfa-Laval and Tranter plate heat exchangers are listed in Table 4-13.

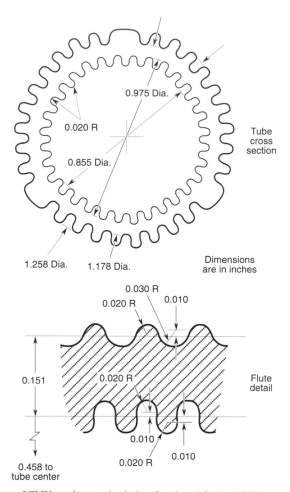

Fig. 4-38. Diagram of CMU condenser tube design (Lewis and Sather, 1979).

A series of experiments was made with the Alfa-Laval plate heat exchanger at several values of heat flux to measure the effects of flow rates of water and ammonia on overall heat transfer and pressure (Panchal et al., 1983). Three types of corrugation (chevrons) were tested. Both ammonia and Freon 22 were used as working fluids. The data for these tests are shown in Tables 4-14 and 4-I5.

The results of the evaluation of the Hi Flux coating on the performance of the Tranter heat exchanger are shown in Table 4-16.

4.1.4.6   Japanese Plate Heat Exchangers (Saga University)

Extensive tests of plate heat exchangers were conducted by Uehara and associates at Saga University (1978, 1983). Both ammonia and R-22 were used as working fluids. The fluid distributions are indicated in Fig. 4-40. Specifications for the plates are given in Table 4-17.

Table 4-12. Nominal design parameters for the CMU vertical fluted-tube condenser

Parameter	English units	SI units
Heat duty	3.2 MBtu/h	0.94 MWt
Ammonia inlet temperature	48°F	8.89°C
Water inlet temperature	40°F	4.44°C
Water outlet temperature	42°F	5.56°C
Number of tubes	240	240
Externally enhanced tube length	172 in.	4.37 m
Tube outside diameter (to midpoint of flutes)	1.218 in.	0.0309 m
Tube inside diameter (to midpoint of flutes)	0.915 in.	0.0232 m
Effective outside heat transfer area[a]	109 ft²	10.13 m²
Mean water velocity in tubes	6.5 ft/s	1.98 m/s
Overall water-side pressure drop	2.95 psi	20.3 kPa
Shell inside diameter	42.25 in.	1.073 m
Shell-side design pressure	215 psig	1482 kPa
Tube-side design pressure	100 psig	690 kPa
"Clean" overall heat transfer coefficient[b]	1045 btu/ft² h °F	5930 W/m² °C

[a]Based on outside tube diameter to midpoint of flutes.
[b]Based on mean outside tube surface area (1097 ft²). Assumes no tube-side or shell-side fouling.

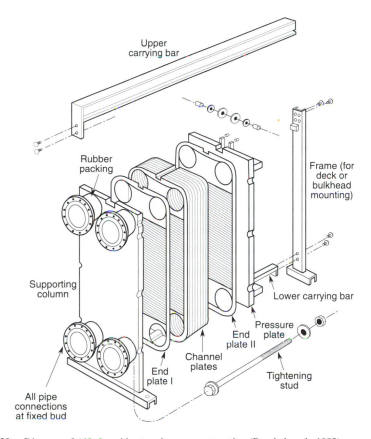

Fig. 4-39. Diagram of Alfa-Laval heat exchanger construction (Panchal et al., 1983).

Table 4-13. Characteristics of Alfa-Laval AX30-HA heat exchanger plates

Property	English units	SI units
Dimensions		
Plate thickness	0.024 in.	0.00061 m
Overall height	88 in.	2.235 m
Overall width	35.5 in.	0.902 m
Manifold diameter	11.25 in.	0.286 m
Plate separation	0.154 in.	0.00391 m
Water-side flow parameters		
Length	72 in.	1.829 m
Width	32.4 in.	0.823 m
Heat transfer area	2330 in.^{2}	59.2 m^2
Channel flow area, cross section	4.85 in.^{2}	0.0031 m^2
Weight, including gasket	12 lb	5.44 kg

Table 4-14. Effects of chevron angles on the overall performance of the Alfa-Laval plate heat exchanger

Run no.	Heat flux (kW/m^2)	H$_2$O flow rate (kg/s/channel)	NH$_3$ flow rate (kg/s/channel)	Water-side pressure drop (kPa) M[a]	H[a]	L[a]	Overall heat transfer coefficient (kW/m^2 K) M[a]	H[a]	L[a]
1	6.98	1.55[b]	0.023	44.1	102.0	13.1	3.29	3.47	2.56
2	6.96	2.26	0.035	92.4	203.4	25.5	3.55	3.68	2.84
3	7.04	0.97	0.047	15.2	40.7	3.4	3.17	3.41	2.34
4	10.00	1.03	0.510	17.2	46.2	4.1	3.37	3.73	2.44
5	9.96	1.67	0.039	47.6	116.5	14.5	3.63	3.92	2.71
6	11.94	1.55	0.039	42.1	99.3	13.1	3.75	3.84	2.67
7	11.95	2.25	0.058	89.6	202.0	26.2	4.15	4.34	3.00
8	11.99	0.91[c]	0.045	16.6	35.5	4.14	3.41	3.59	2.37

[a]M = mixed plates; H = high-angle plates; L = low-angle plates.
[b]Water flow rate for mixed plates was 1.60 kg/s/channel.
[c]Water flow rate for mixed plates was 0.97 kg/s/channel.

Table 4-15. Comparison of the thermohydraulic performance of the Alfa-Laval heat exchanger with ammonia and Freon 22 as working fluids (per channel)

Run no.	Heat flux (kW/m^2)	H$_2$O flow rate (kg/s)	Working-fluid flow rate (kg/s) A[a]	F[a]	Exit quality A[a]	F[a]	Water fluid-side pressure drop (kPa) A[a]	F[a]	Overall heat transfer coeff. (kW/m^2 K) A[a]	F[a]
1	7.09	1.59	0.023	0.143	0.73	0.74	6.0	11.1	3.29	2.57
2	6.98	2.90	0.023	0.141	0.73	0.73	6.1	10.8	3.58	2.58
3	6.99	2.57	0.31	0.165	0.63	0.63	6.5	11.9	3.56	2.61
4	7.08	2.26	0.035	0.214	0.49	0.50	7.1	13.3	3.55	2.63
5	9.91	1.92	0.027	0.152	0.91	0.98	5.7	10.2	3.51	1.46
6	9.6	1.66	0.039	0.229	0.62	0.64	6.8	15.0	3.63	2.94
7	9.80	1.81	0.066	0.374	0.37	0.39	8.8	22.0	3.83	2.87
8	9.88	1.43	0.074	0.424	0.33	0.35	9.7	24.8	3.80	2.86

[a]A = ammonia; F = Freon 22.

Table 4-16. Comparison of the thermal performance of the Tranter plate heat exchanger with smooth and with Linde Hi Flux surface

	Smooth surface	With Hi Flux surface		
		Same heat flux	Same approach temperature[a]	Same mean temp. difference[b]
Heat duty (Btu/h)	68,950	68,030	117,380	132,400
Water flow rate (gpm)	40.3	40.3	40.7	41.3
Ammonia flow rate (gpm)	0.88	0.84	1.48	1.69
Exit quality	0.51	0.52	0.50	0.50
Approach temperature (°F)	9.85	5.94	9.90	11.04
Mean temp. difference (°F)	8.13	4.25	7.01	7.83
Overall heat transf. coeff. (Btu/h ft^2 °F)	615	1205	1265	1280
Ammonia-side heat transfer coeff.[c] (Btu/h ft^2 °F)	1010	5185	6515	6930

[a]Approach temperature = inlet H_2O temp. − mean NH_3 saturation temp.
[b]Mean temperature difference = mean H_2O temp. − mean NH_3 saturation temp.
[c]Based on an estimated water-side heat transfer coefficient of 2700 Btu/h ft^2 °F.

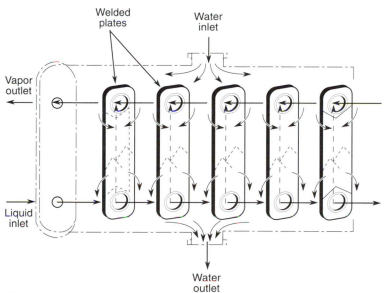

FIG. 4-40.   Fluid flow in Saga University plate heat exchangers (Uehara et al., 1983).

Table 4-17. Specifications of plates for the Saga University heat exchangers

Item	S plate	IP plate	P plate	No. 1 plate	No. 2 plate
Length (mm)	1255	1255	1450	1255	1450
Width (mm)	415	415	235	415	235
Thickness (mm)	0.8	0.8	1.0	0.8	1.0
Material	titanium	titanium	titanium	titanium	titanium
Number	60	50	100	54	168
Total surface area (m^2)	20.8	8.16	21.95	21.6	40.7

Results of the Uehara tests parallel those of Panchal et al. (1983). The best plate configurations enhance overall heat transfer by a factor of about 4 in comparison with the smooth plate. Plots of overall heat transfer coefficient versus water flow velocity are shown in Fig. 4-41 for the evaporator and Fig. 4-42 for the condenser. Pressure loss (hydraulic head) as dependent on water velocity is shown in Figs. 4-43 and 4-44.

4.1.4.7   Folded-Tube Heat Exchanger (Johns Hopkins University APL)

The JHU/APL folded-tube heat exchanger constructed for the ANL test had three 7.62-cm-OD (3 in.) folded tubes, each folded into 43 passes and mounted in a vertical plane. Ammonia flowed inside the tubes, and water flowed externally from

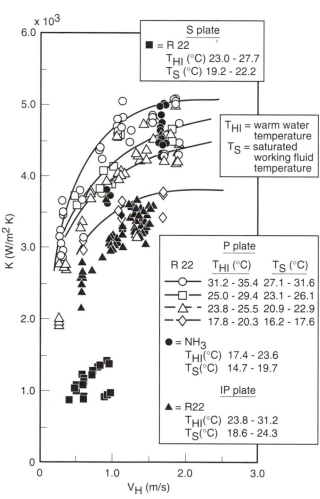

Fig. 4-41. Heat transfer versus water flow for Saga University plate evaporators (Uehara et al., 1978, 1983).

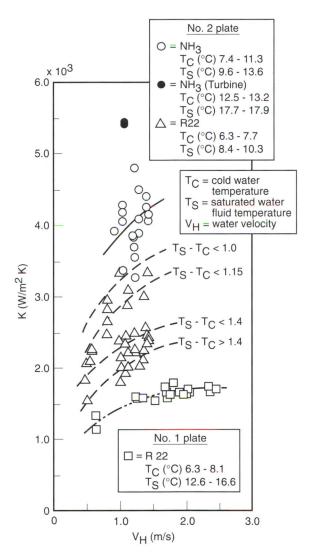

FIG. 4-42. Heat transfer versus water flow for Saga University plate condensers (Uehara et al., 1978, 1983).

a head pond above the tubes. The equipment was designed to permit swirlers to be inserted in the three inlet tubes of the evaporator to test whether they would improve the heat transfer at the low flow velocity of the first pass. A diagram of the ANL test article is shown in Fig. 4-45 (Yung et al., 1981). The heat exchanger was tested both as an evaporator and as a condenser. Operation was unusually stable and reproducible. The performances of both the evaporator and condenser exceeded predictions based on published correlations and earlier smaller-scale experiments. This re-

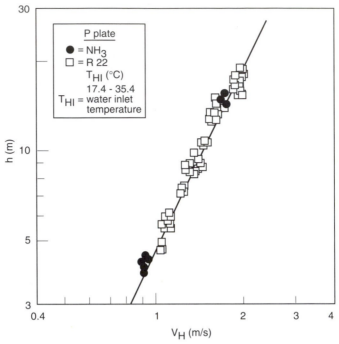

FIG. 4-43. Pressure loss (hydraulic head) versus water flow velocity for Saga University plate evaporator (Uehara et al., 1983).

quired development of modified analytical techniques to correlate the results. Full details of the analysis are presented in Pandolfini et al. (1981). The nominal test conditions for the folded-tube evaporator are shown in Table 4-18.

The measured values of U versus flow rates of ammonia and water in evaporator operation are shown in Fig. 4-46. All of the data can be correlated in terms of the Reynolds numbers of the water and ammonia by the following equation:

$$U^* = 0.0243(\mathrm{Re}_W \, \mathrm{Re}_A)^{0.442}, \qquad (4.1.36)$$

where U^* is defined by the following relationship:

$$
\begin{aligned}
1/U^* &= (1/U) - \text{thermal resistance of tube wall} \\
&= 1/U - 0.000127 .
\end{aligned}
\qquad (4.1.37)
$$

The pressure drop of ammonia in the evaporator, as dependent on the ammonia flow rate and heat loading, is well represented by the following equation:

$$\Delta p = 0.472(1 + 0.87\dot{m}_A)(1 + 1.42\dot{Q}), \qquad (4.1.38)$$

where P is the pressure (psi); $\dot{m}_A$, the ammonia flow rate (lb/s); and $\dot{Q}$, the heat loading (10^6 Btu/h).

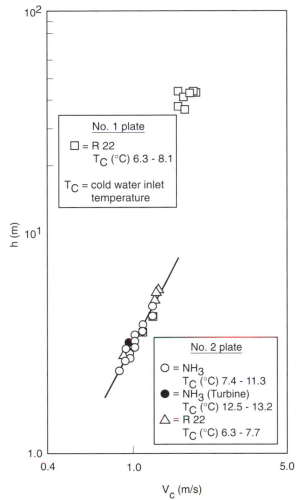

Fig. 4-44. Pressure loss (hydraulic head) versus water flow velocity for Saga University plate condenser (Uehara et al., 1983).

Accurate measurements of the water-side pressure drop were not possible because of foaming of the water in the head pond caused by a rust preventative added to the water. However, the data were consistent with predictions for an Euler coefficient of 0.11, that is, a pressure drop of 6.55 kPa for the nominal water velocity of 0.76 m/s at the point of closest spacing between tubes in the staggered-tube configuration.

A series of tests with the swirlers showed no significant effect on performance. The analysis of the data showed that the evaporator net power output could be increased 33% above the nominal values by adjusting the ammonia and water flow velocities to optimize the tradeoff with pumping power. The resulting value of U was 2724 W/m^2 °C. The performance of the evaporator at nominal conditions is summarized in Table 4-19. The condenser test conditions are shown in Table 4-20.

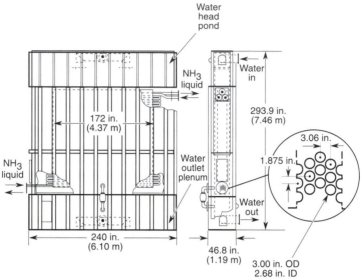

Fig. 4-45.   Diagram of test module for JHU/APL folded-tube heat exchanger (Yung et al., 1981).

Table 4-18. JHU/APL evaporator test conditions

Parameter	Nominal value (Eng. units)	Range	Nominal value (SI units)	Range
Heat duty	3.2 Mbtu/h	2.8–4.7	0.94 MWt	0.82–1.47
Water flow rate	3200 gpm	1500–4800	201.6 kg/s	95–302
Water inlet temp.	80°F	78–85	26.67°C	6.0–28.3
NH$_3$ feed rate[a]	36 gpm	16–72	2.27 kg/s	1.00–4.54
NH$_3$ side				
Pressure[b]	133 psia	131–138	917 kPa	903–951
Saturation temp.[b]	72°F	71–74	22.22°C	21.9–22.8
NH$_3$ inlet subcooling	10°F	2–20	5.55°C	1.1–11.1
NH$_3$ exit quality	55%	22–85	55%	25–85

[a]Total ammonia feed rate for all three tubes.
[b]Arithmetic average of inlet and outlet values.

Heat transfer rates for the condenser are shown in Fig. 4-47. Analysis of the data showed that the stratified flow model of Nusselt accurately predicted the heat transfer rates over the expected OTEC operating range. For this model (Collier, 1972):

$$h_a = F \frac{\rho_1(\rho_1 - \rho_v)gi_{fg}k^3]^{1/4}}{[d_i\mu(T_{vi} - T_{li})]}, \qquad (4.1.39)$$

where F is the ratio of gas "wetted" circumference to total surface.

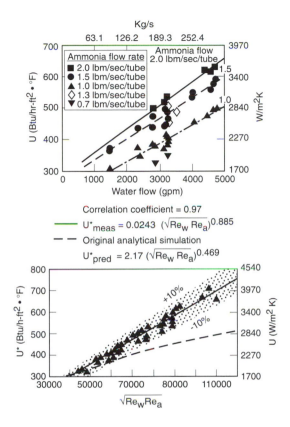

Fig. 4-46. Heat transfer versus flow rates of water and ammonia in ANL tests of the JHU/APL folded-tube evaporator (Yung et al., 1981).

Table 4-19. Performance of JHU/APL evaporator at nominal conditions
(Yung et al., 1981)

Parameter	Predicted	Measured[a]	
		With swirlers	**Without swirlers**
U (Btu/h ft^2 °F)	360	390 ± 10	410 ± 20
Ammonia-side Δp (psi)	5.6	5.1 ± 0.3	5.0 ± 0.2
Water-side Δp (psi)	0.95	0.95	0.95

Water flow = 3200 gpm (v = 2.54 ft/s); ammonia flow = 1.0 lbm/s/tube; heat duty = 3.2 × 10^6 Btu/h; exit quality = 58%.

[a]Average ammonia inlet subcooling. Standard deviation for 7 runs with swirlers = 9.0 ± 2.4°F, and for 4 runs without swirlers = 7.3 ± 2.4°F.

Table 4-20. JHU/APL condenser test conditions (Yung et al., 1981)

Parameter	Nominal value	Range
Heat duty (Btu/h)	3.4×10^6	2.5×10^6–5.0×10^6
Water flow rate (gpm)	3200	1500–4600
Water inlet temperature (°F)	40	40–42
Rate of condensation (lb/h)	6500	4000–9500
NH$_3$ side		
Pressure (psia)[a]	86	84–116
Saturation temperature (°F)[a]	48	47–64
NH$_3$ inlet superheating (°F)	4	0–8

[a]Arithmetic average of inlet and outlet values.

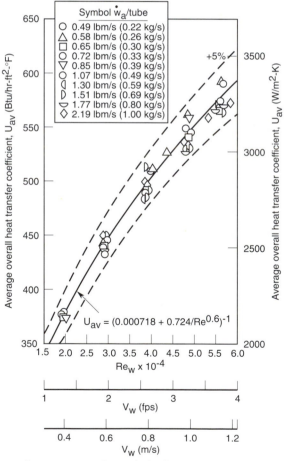

FIG. 4-47. Heat transfer versus water flow velocity in ANL tests of the JHU/APL folded-tube condenser (Yung et al., 1981).

In the stratified flow regime, vapor condenses at the tube wall and flows down the sides, forming a channel of liquid ammonia at the bottom that increases in depth as the vapor progressively condenses toward the end of the tube. As vapor flow rates increase, the liquid flow at the bottom is thrown more and more to the sides and top, eventually producing the condition called annular flow. The test results show that stratified flow analysis applies to flow rates up to about 0.39 kg/s in the 0.068-m-ID folded tube, after which a combination of equations for annular flow and stratified flow gives better agreement with the measured results. (The nominal OTEC value of ammonia flow for the ANL tests was 0.32 kg/s. Algorithms for finding h_a were developed by Pandolfini et al., 1981).

A comparison of the observed overall heat transfer coefficients with the predictions of stratified and annular flow models is shown in Fig. 4-48.

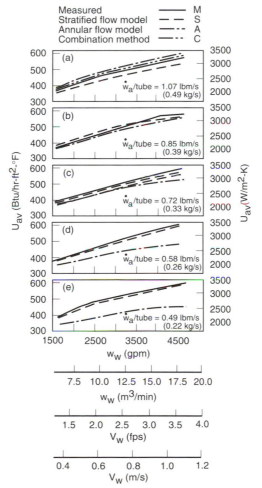

FIG. 4-48. Heat transfer in two-phase flow of ammonia in tests of the JHU/APL folded-tube condenser (Pandolfini et al., 1983).

4.1.4.8   Rocketdyne Axial-Fluted Heat Exchanger

The Rocketdyne "Variflux" heat exchanger employed concentric vertical aluminum tubes with water flow in the inner tube and ammonia flow upward in the
annulus between inner and outer tubes. Both sides of the inner tube had longitudinal
fins (Wright et al., 1978). The design of the concentric tube assembly is shown in
Fig. 4-49. Limited tests at ANL as an evaporator gave an overall U of 6500
W/m² °C with ammonia-side and water-side coefficients of 17,200 and 20,700 W/
m² °C, respectively. The water-side pressure drop was 10.3 kPa. The water velocity
was not specified. The high value of U and low pressure drop made the performance
of this heat exchanger comparable with that of the Trane heat exchanger (Panchal
et al., 1981).

4.1.4.9   Rosenblad Plate-and-Shell Evaporator

The Rosenblad evaporator used dimpled plates assembled so that they could be
incorporated in a shell. It could be operated as a falling film evaporator, pool boiler,
or condenser. In the falling film mode at nominal conditions, the overall heat
transfer coefficient was measured as 2468 W/m² °C. At the nominal conditions
chosen for the ANL tests, the ammonia-side coefficient was 3547 W/m² °C and the
water-side coefficient was 16,172 W/m² °C. The water-side pressure drop was
173.7 kPa, indicating a need for operation of the heat exchanger at water flow rates
much lower than the nominal values chosen for the ANL tests, if intended for OTEC
use (Panchal et al., 1981).

4.1.5 *OTEC Power Plant*

4.1.5.1   OTEC Power System Performance

The thermal and hydraulic characteristics of the heat exchangers described in
Section 4.1.3 may be used for quantitative prediction of the power that can be

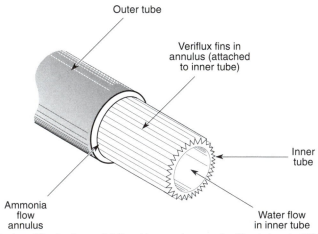

FIG. 4-49.   Design of Rocketdyne axial-fluted heat exchanger tube (Panchal et al., 1981).

produced by OTEC plants of various types. The analysis follows the exposition of Owens (1983), to which the reader is referred for details. Owens derived a closed-form solution for the OTEC thermal cycle that permits the maximum net power to be expressed in terms of the optimum saturation temperature difference, which is shown to be dependent for a particular design only on the seawater temperature difference and constants associated with vapor and condensate flow. The net power per unit of total heat exchanger area is then shown to be primarily dependent only on the heat transfer coefficients as dependent on seawater flow velocities in the evaporator and condenser, and on the associated pressure losses, which can be defined in terms of the length to hydraulic diameter ratios. With the Owens analysis a computer program can be used to predict the power output of OTEC plants designed to employ heat exchangers of the advanced types tested in the ANL program. The optimum selections of flow conditions are then displayed as a ratio of net power output to maximum theoretical output.

A summary of the model heat exchanger test conditions, the heat transfer coefficients, and pressure drops measured by ANL, for which complete data have been reported by ANL, is given in Table 4-21. The data are used to predict the OTEC power plant performance, using the Owens analysis procedure, as shown in the table. Data from performance tests of heat exchangers for a 100-kW demonstration plant designed by Japanese investigators for the island of Nauru are also included (Panchal et al., 1981; Hillis et al., 1979; Pandolfini et al., 1981; Mochida et al., 1983; Kawano et al., 1983).

The predictions of system performance based on the results of the ANL HX tests at the nominal conditions are presented in Table 4-21 in the columns with "nom" in the heading. For each system design a second column is also displayed, with "opt" in the heading. This shows the heat loading and water flow that yield the optimal net power output per unit of heat exchanger area as estimated by use of the analysis procedure of Owens (1983). A second column, labeled "opt2," is displayed for the folded-tube HX design to show the power system performance when water flows are varied to minimize CWP size. The implications of the analysis with regard to OTEC system design are large but have not been investigated because of the termination of program support. The results, however, demonstrate that major improvements in power plant performance for OTEC, compared with systems based on conventional heat exchanger designs, are possible with enhancement techniques, improved hardware concepts, and optimization of operating conditions. The data also show the significant impact of pressure losses in water ducting on the net power output of the systems.

The ANL tests and the Japanese investigations, along with research and development by other organizations, show that OTEC heat exchangers are practical to construct that will have a heat transfer area requirement of between 4 and 9 m^2 of heat exchanger surface per megawatt of net power developed by the power system. For a particular design, minimum surface area per kilowatt will imply minimum heat exchanger cost, but area per kilowatt of net power is not a good basis for comparing the merits of heat exchangers of different designs. Power plant cost per kilowatt of delivered power is a better measure; however, the only valid basis

Table 4-21. OTEC power plant performance comparison

	APL HX			ANL-SP3		CMU		Trane		Japan		Alfa-Laval	
	(nom)	(opt 1)	(opt 2)	(nom)	(opt)	(nom)	(opt)	(nom)	(opt)	(nom)	(opt)	(nom)	(opt)
Heat loading (kWt)	937	1591	570	937	1332	938	2019	938	1615	4091	4664	938	938
Water flow area (evaporator) (m²)	0.26	0.26	0.26	0.10	0.10	0.26	0.26	0.089	0.089	0.20	0.20	0.31	0.31
Water flow area (condenser) (m²)	0.26	0.26	0.26	0.14	0.14	0.26	0.26	0.089	0.089	0.20	0.20	0.31	0.31
Mass flow, warm water (kg/s)	202	202	75	202	202	202	202	202	202	409	409	202	202
Mass flow, cold water (kg/s)	202	202	30	202	202	202	202	202	202	392	392	202	202
Working fluid (WF)	NH_3	NH_3	NH_3	NH_3	NH_3	NH_3	NH_3	NH_3	NH_3	R-22	R-22	NH_3	NH_3
Mass flow, WF (kg/s)	0.76	1.28	0.46	0.76	1.09	0.77	1.66	0.78	1.35	21.47	23.50	0.77	1.81
Δp (sat) (kPa)	402	271	268	378	283	446	276	412	279	320	320	440	266
ΔT (sat) (K)	15.83	10.61	9.64	14.80	10.98	17.42	10.59	15.94	10.59	10.66	9.73	17.63	10.66
ΔT (lmtd) (evaporator)	3.27	5.55	2.75	3.52	5.01	2.26	4.86	2.51	4.32	3.82	4.35	2.48	5.02
ΔT (lmtd) (condenser)	3.11	5.27	6.35	3.90	5.69	2.50	5.39	3.73	6.42	4.52	4.77	2.09	4.95
ΔT in WW flow through evaporator (K)	1.16	1.97	1.90	1.16	1.65	1.16	2.50	1.16	2.00	2.50	2.85	1.16	2.75
ΔT in CW flow through condenser (K)	1.13	1.91	4.61	1.13	1.60	1.13	2.43	1.13	1.94	2.43	2.76	1.13	2.67
U_0 (evaporator) (W/m² K)	1469	1469	1464	2844	2844	2709	2709	4030	4030	2593	2593	1956	2461
U_0 (condenser) (W/m² K)	1503	1509	644	2678	2678	3423	3423	2811	2811	1886	1886	2406	2406
Net OTEC power (kWe)	25.69	32.31	12.86	19.65	23.24	23.92	38.97	21.49	28.33	80.27	91.77	17.52	37.68
Cycle net power/max cycle power	0.825	1.000	1.000	0.891	1.000	0.683	1.000	0.817	1.000	0.994	1.000	0.663	1.000
Net power/total heat exchanger area (kWe/m²)	0.092	0.116	0.046	0.147	0.173	0.173	0.282	0.195	0.257	0.197	0.225	0.115	0.247
CW flow for 40-MWe (net) power (m³/s)	7.86	6.25	2.33	10.28	8.69	8.44	5.18	9.39	7.13	4.88	4.27	11.53	5.36
CWP diam. for 40-MWe (net) power (m)	14.15	12.61	7.71	16.18	14.87	14.66	11.49	15.47	13.47	11.15	10.43	17.13	11.68

for heat exchanger optimization is the power cost for a total OTEC system, suitably amortized over the life of the plant. In this selection the significant impact of the layout and volume requirements of the heat exchanger and its associated ducting on the total OTEC system cost must be considered. These factors are discussed in the next section.

4.1.5.2   OTEC Power System Space Requirements

It was realized in the early design studies that the selection of shell-and-tube heat exchangers for OTEC commercial plants would make the heat exchangers a major volume element in the total OTEC plant installation. Thus, it was important to investigate alternative designs that would reduce heat exchanger size per kilowatt of net power produced. This consideration focused research and development attention on so-called "compact" heat exchanger designs.

Estimates of the layout dimensions of 10-MWe (net) OTEC power systems employing various heat exchanger types were made in DOE-sponsored programs designed to identify the requirements for demonstration of OTEC commercial capabilities. The programs were conducted in response to Public Law 96-320, which called for the demonstration of 100 MWe of OTEC power by 1985. Power system conceptual design studies were supported at Westinghouse (1978), Lockheed (1981), and General Electric (1983), and a preliminary design of a power system for baseline, barge-mounted, complete OTEC systems was prepared by JHU/APL (George et al., 1979; George and Richards, 1980). Also, a preliminary design of a shore-based OTEC power plant for a site at Kahe Point on the island of Oahu, Hawaii, was prepared by Ocean Thermal Corporation (OTC) (1984) as the second phase of a DOE cost-sharing plan to build a demonstration plant in cooperation with the Hawaiian Electric Corporation.

The approximate volume requirements of the three 10-MWe (net) power system modules designed in these programs, which incorporated different types of compact heat exchangers, are compared with a module employing a conventional shell-and-tube type exchanger in Table 4-22. The arrangement of the water ducts significantly affects the space assigned to the machinery part of the modules. The comparison shows that the compact designs will reduce HX volume requirements by 55 to 70% in comparison with the shell-and-tube values, with a reduction of from 45 to 60% in total module space requirements. Proposed layouts for the four power system types are illustrated in Figs. 6-35 and 6-36.

4.1.5.3   Other Power System Components

The principal characteristics of the major subsystems of the baseline 40-MWe OTEC demonstration vessel are listed in Table 4-23.

Turboelectric Generators.   The technology for production of vapor turbines is highly developed and embodied in installations throughout industry and the public utilities. Turboelectric steam generators with outputs exceeding 50 MWe in individual units are commercially available. The technology is described in standard references, for example, *Marks' Standard Handbook for Mechanical Engineers*, 8th Ed., Baumeister, Avallone, and Baumeister, eds. Large turbines for

Table 4-22. Dimensions of alternate types of OTEC 10-MWe heat exchanger modules

HX type[a]	Designer	Evaporator or condenser				
		Length (m)	Width (m)	Depth (m)	Area (m²)	Volume (m³)
S&T	OTC	21.0	21.9	15.9	460	7300
PHE	LMSD	10.8	13.3	14.6	145	2070
Trane	Trane	23.8	9.5	14.6	225	3290
Folded tube	JHU/APL	16.5	9.1	18.0	150	2700

[a]S&T, Ocean Thermal Corp. (1984). PHE, Lockheed (1979). Trane, Ashworth and Slabodnick (1980). Folded tube, George and Richards (1980).

Table 4-23. OTEC power plant subsystems [40 MWe (net), 10 MWe modules]

Subsystem or component	Approx. dimen. (m)			Wt (mt)	Power (kW)
	L	W	Z		
HX (8)	33.0	9.1	18.0	1163	
Turbo-generator	10.7	4.9	3.7	102	13000
WW pump	3.8	Diam.	12.2	45	
Motor & drive	3.7	2.6	6.0	20	1030
CW pump	3.5	Diam.	9.1	39	
Motor & drive	3.6	2.6	6.0	25	1440

Z, vertical; WW, warm water; CW, cold water.

use with ammonia vapor are not in use because there has not been a demand for large sizes. The basic design information and manufacturing technology for steam turbines, however, are directly applicable to scale-up of ammonia turbine designs to sizes of OTEC interest. The conceptual and preliminary design studies by American and Japanese teams showed that several commercially available turbine designs employing ammonia and other potential working fluids would be suitable for OTEC. Firm quotes for the manufacture of ammonia-powered multi-megawatt turbine–generator sets, with delivery within 12 to 18 months from receipt of an order, were obtained in the DOE-sponsored OTEC conceptual and preliminary design studies. Engineering designs for turbine–generator assemblies of 13-MWe gross power were received from six organizations in 1984 in response to a request for bids for OTC's planned installation at Kahe Point. The Ocean Thermal Corporation preliminary design program chose the Toshiba design, which is an axial, three-stage, double-flow type.

Liquid Ammonia Pumps. Liquid ammonia pumps with capacities in the range from 20 to 200 kg/s are required for recycling condensate from the condenser to the evaporator, for returning liquid from the demister or evaporator sump to the evaporator, and for transferring ammonia to and from storage and purification areas. Ammonia pumps are commercially available that are suitable for these purposes.

Electrical Power Generation and Distribution. In an OTEC system based on the modular concept, the electrical system must be designed to combine multiple, parallel generator outputs, for example, 10 MW (net), with appropriate signal processing of individual unit voltages and phases to satisfy the total power requirements. This includes power to operate water and ammonia pumps, power for system operation and maintenance including personnel needs, and net power output and transfer to the consumer. Provision must be made for transition to standby diesel-generated power for essential services during shutdown or start-up operations and for transitions to and from operation with one or more units temporarily shut down. The design must satisfy all requirements and specifications of IEEE Standard No. 45, USCG Electrical Engineering Regulations, CG-259, ABS rules, and appropriate standards and specifications of NEMA, IEEE, NFPA, and UL. Commercial equipment and industrial procedures are available for these purposes. The OTEC requirements can be met by straightforward application of electrical and marine engineering practice.

A discussion of the electrical systems for the 40-MWe baseline moored OTEC plant is given Section 6.5. The layout is representative of the designs prepared in the other studies.

Biofouling Protection. Experiments with warm and cold flowing seawater, conducted at the Natural Energy Laboratory of Hawaii (NELH) over a period of 5 years, showed that negligible biofouling of heat exchanger surfaces occurred in operating conditions typical of OTEC condenser operation. Biofouling did build up slowly in the evaporator operating regime; however, fouling was completely prevented by the injection of chlorine at a concentration of 50 to 70 parts per billion (ppb) for 1 hour per day (24-h average of 2–3 ppb) (Larsen-Basse, 1983). This concentration is well below the level established by the Environmental Protection Agency (EPA) as acceptable for continuous chlorine injection in power plant operation at coastal sites. An average chlorine concentration of 2 ppb is 1% of the minimum level typically specified for protection of public drinking water.

At an average chlorine concentration of 2 ppb, the amount needed to prevent biofouling of the heat exchangers in a 10-MWe (net) power module would be between 5 and 10 kg of chlorine per day. This could be supplied by shipment of bottled liquid chlorine to the OTEC plant at monthly intervals, or a small electrolytic chlorine generator could be installed. In either case, the cost would be minimal.

Since chlorination is economical and convenient to use, it is the preferred method of biofouling control; however, ultrasonic radiation (Pandolfini et al., 1977), circulating Amertap rubber sponges, and MANN brushes have also been shown to be effective biocontrol methods (Draley, 1981). Slurries of silica particles of various sizes and concentrations circulated past the heat exchanger surface were of limited effectiveness (Draley, 1981).

4.2   WATER DUCTING

The system design studies summarized in Table 4-21 show that a 10-MWe (net) OTEC power module operating with a temperature difference of 22. 8 K will require a flow of warm and cold water at rates which range from 26 to 75 m³/s, dependent on the design parameters. If the water velocity is 2.5 m/s, which provides a reasonable compromise between drag loss and duct area, water ducts with area between 1 and 3 m²/MW of net power delivered will be required.

The size of OTEC water ducts makes the engineering a major task for OTEC plants of commercially attractive sizes. The cold-water pipe (CWP) poses unique engineering problems; however, the successful development of off-shore oil drilling installations and platforms has provided analytical techniques, facilities, and operational data that are directly applicable to OTEC cold-water pipe design. This background, along with experience gained in scale tests and in French, American, and Japanese at-sea programs, gives confidence that the necessary technical base exists for successful construction, deployment, and operation of CWPs large enough to meet the demands of OPEC plants of commercial interest. Preliminary estimates project a nominal cost, including deployment, of the CWP for floating platforms that is 10–20% of the total plant investment.

For shore-based plants, which require a CWP 2000 m or more in length secured at intervals to an uneven sea bottom, the estimated cost of the CWP is a major fraction of the total cost.

4.2.1   *Cold-Water Pipes for Floating OTEC Installations*

4.2.1.1   General Requirements

Commercial floating OTEC plants will be designed to generate net power in the range of 100 to 350 MW. This implies that cold-water pipes with diameters of 20 m or more might be required, if a single CWP were to be used for the largest OTEC plants. Studies, however, of alternative plant design layouts, in conjunction with manufacturing and deployment assessments, show that the modular approach found optimum for the power system should be extended to include the water inlet and outlet systems (Richards et al., 1980). This leads to design of OTEC systems with 60- to 80-MWe (net) modules as basic units, which can then be combined for the creation of larger plant sizes. With this design doctrine CWP diameters can be restricted to 10 to 12 m. The following discussions assume that CWPs will be limited to these diameters.

4.2.1.2   Structural Requirements

The static and dynamic behavior of cold-water pipes can be analyzed by application of formulas and techniques developed for cylindrical shells under tension. The analysis is complex because the loads are distributed along the length of the pipe, depend on materials and methods of construction, and vary with time. The principal dynamic driving force is provided by the platform motions in the seaway that are transmitted to the CWP through its attachment to the OTEC platform. Additional loads are imposed by underwater currents and shedding of vortices from the pipe. Loads imposed during pipe deployment must also be carefully analyzed, and designs and strategies adopted that prevent pipe damage in this critical phase of the operation.

The following discussion is based on analytical procedures and experimental data derived in DOE-sponsored programs conducted during the period 1975 to 1980. As shown in Fig. 4-50, many alternative concepts have been proposed (Anderson, 1974b, 1975; Bell-Aerospace, 1979; McGuiness and Griffin, 1979). The programs have included parametric examinations of CWP requirements for floating OTEC plants of power ranging from 100 to 1000 MWe (Little, 1976) and included analytic and experimental studies to provide CWP design information for barge-mounted 40-MWe (net) OTEC demonstration vessels (George et al., 1979; McGuiness, 1982; McHale and Bender, 1979). Design studies of cold-water pipes for 10- and 40-MWe shore-based plants for sites in Hawaii and Puerto Rico were made by Brewer (1979). The studies have been refined and extended as part of conceptual and preliminary designs of a 40-MWe (nom) shore-based OTEC plant for Hawaii by OTC (1982). Design studies of shore-based CWPs for small plants in the 1- to 10-MW range have also been reported by Stevens (1982, 1986).

Nominal parameters for design of the 40-MWe demonstration vessel CWP were:

Net OTEC power	40 MWe
Gross OTEC power	50–60 MWe
CWP pipe diameter	9.1 m
CWP pipe length	90–1000 m

Pipe pivoted near the center of gravity of the platform

For the demonstration program the CWP diameter was selected to represent the maximum consistent with the use of existing construction techniques, and facilities that could reasonably be expanded. The water flow requirement would be met by a significantly smaller diameter, as shown in Table 4-21 (APL HX, Opt 2).

A preliminary determination of the structural requirements for an OTEC CWP can be made from an estimate of the static loads. This estimate can be used to select suitable materials and methods of construction and to indicate design options for dynamic analysis and test. The final design options are then selected based on an evaluation of construction and deployment feasibility and cost.

4.2.1.3   Static Loads

Evaluation of the static loads provides a basis for preliminary selection of materials of construction and definition of structural details. By superposition of loads an estimate may be made of the extreme values expected. Cylindrical vessels of steel, aluminum, concrete, and reinforced plastics are in commercial use in diameters exceeding 10 m. These materials and vessel construction techniques could be adapted for CWP design (Brewer et al. 1979). The discussion to follow of the static loads and CWP design implications is based mainly on the analysis by Little (1976).

The static forces acting on the CWP during OTEC operation are:

1. Bending moments at the joint between the CWP and the platform. For a 1000-m-long CWP of 10 m diameter, rigidly attached to the platform, the bending moment would exceed 3×10^7 kg m if the near-surface current attained the projected 100-year-storm value of 3.2 knots. The magnitude of the moment shows that a pivoted connection is necessary. Bending moments also result from subsurface currents that vary with depth. These moments may be relieved by designing the CWP to be flexible, either by use of compliant materials or by incorporating flexible joints between rigid pipe sections. A pivot between the CWP and the platform is assumed in the following discussion.
2. Longitudinal tension due to the weight of the pipe in seawater. This is lessened slightly during OTEC operation by the drag force of the upward flow of cold water through the pipe.
3. Collapse loads produced by the suction pressure in the pipe.
4. Ovalling loads caused by circumferential pressure differences.

Bending Loads.   The force exerted by horizontal water flow past the CWP at a given depth will consist of three components:

$$F_d' = C_d' A \rho_w v^2 / 2$$
$$F_d'' = C_d'' A \rho_w v'^2 / 2$$
$$F_l = C_l A \rho_w (v' - v)^2 / 2$$
$$F_d = F_d' + F_d'' \ ,$$

where F_d' is the drag force associated with normal flow around a cylinder; F_d'', the intermittent force produced by eddy shedding; F_l, the intermittent lateral lift force associated with changes in flow; v', the instantaneous flow velocity; and v, the average flow velocity.

Maximum values of the forces can be estimated by assigning values of the coefficients of:

$$C_d' = 0.3, \ C_d'' = 0.03, \ C_l = 0.4 \ .$$

The maximum resulting shear force at a particular depth is then:

$$F = \left(F_d^2 + F_l^2 \right)^{1/2}. \tag{4.2.1}$$

Since there are underwater currents that vary with depth, the values of flow velocity used in computing shear forces and moments must be found by vector addition at each depth of the flow velocity past the CWP due to horizontal platform motion and to current velocity. The maximum bending loads are then found by integration over the length of the pipe.

Longitudinal Forces.   Tension in a uniform pipe caused by its submerged weight is:

$$T_z = (\rho_p - \rho_w)(L_0 - L_z), \tag{4.2.2}$$

where T_z is the unit stress; ρ_p, the average density of CWP wall; ρ_w, the density of seawater; L_0, the design pipe length; and L_z, the pipe length to depth z.

Upward force caused by internal drag on the pipe walls by flow of water is given by:

$$F = 4f \frac{L}{D} \rho \frac{v_p^2}{2g},$$ (4.2.3)

where f = pipe friction coefficient = $4f'$, f' = Fanning coefficient = 0.004

$$\frac{1}{f^{0.5}} = -4 \log_{10} \left[\frac{\epsilon}{3.7D} + \frac{1.256}{Re^{0.5}} \right],$$ (4.2.4)

ϵ = pipe surface roughness factor (0.3 mm for smooth concrete, and 0.05 mm for steel), Re = 1.7×10^6, v_p = flow velocity in the CWP, D = internal diameter of the CWP, L = CWP length.

Collapse Loads.   The suction pressure difference across the CWP is the sum of the Δps caused by flow drag in the pipe, by the density difference between the cold water and warm surface water, and by the Δp associated with accelerating the cold water to the flow velocity in the pipe. Minor losses due to the inlet screens and other restrictions in the flow must also be considered. Pumping power must also be provided to overcome pressure losses in the heat exchangers and associated ducts.

$$\Delta p = f \frac{L}{D} \rho \frac{v_p^2}{2g} + \int_0^L \frac{\rho_i (1 - \rho_o)}{\rho_i} dy$$
$$+ \rho \frac{v_p^2}{2g} + \Delta p_{HX} + \Delta p_{misc},$$ (4.2.5)

where ρ_0 is the density of water outside the CWP (depends on depth), v_p the flow velocity of water in the CWP, y the depth, f the friction factor = $4f'$, L/D the length to diameter ratio of the CWP, and ρ_i the density of water inside the CWP.

Values of Δp for cold-water flow computed for values of L/D and v representative of power levels of 40, 60, and 80 MWe (net) are shown in Table 4-24. The values are based on cold-water flow of 3.4 m³/s per MWe (net), which was found to be optimum for the 40-MWe (net) baseline design (George and Richards, 1980).

Ovalling Loads.   Ovalling of the pipe can result from a circumferential difference in pressure between the inside and outside of the pipe, which varies with current direction and velocity. The effect is small for CWP designs of practical interest and may be neglected.

4.2.1.4   Structural Design Options

Design of the CWP involves selection of materials and a configuration that will have sufficient structural strength to resist the operating loads during the design life of the pipe. The CWP must be feasible to construct and deploy at reasonable cost.

Corrosion or deterioration of the CWP wall must not produce products that could cause corrosion or fouling of the heat exchangers. Design concepts that have been proposed are shown schematically in Fig. 4-50.

Properties of materials that have been considered suitable for OTEC cold-water pipe construction are summarized in Table 4-25 (Little, 1976; George et al., 1979).

Table 4-24. Collapsing loads on CWP [cold water flow 3.4 m³/s per MWe (net)]

	Power MWe (net)					
	40	40	60	60	80	80
V (m/s)	2.00	2.50	2.00	2.50	2.00	2.50
D (m)	9.30	8.32	11.40	10.19	13.16	11.77
L (m)	1000	1000	1000	1000	1000	1000
P (kPa)						
Friction	3.53	6.16	2.88	5.03	2.50	4.36
Density effect	7.65	7.65	7.65	7.65	7.65	7.65
Dynamic loss	1.03	1.60	1.03	1.60	1.03	1.60
Miscellaneous	1.00	1.00	1.00	1.00	1.00	1.00
P(total) (kPa)	13.20	16.40	12.60	15.30	12.20	14.60
P(total) (psi)	1.90	2.40	1.80	2.20	1.80	2.10
P(800 m) (kPa)	12.50	15.20	12.00	14.30	11.70	13.70

The primary structural requirement for the CWP is ability to withstand buckling, either through collapse due to the reduced internal pressure associated with pipe flow, or by bending from lateral drag forces. The latter effects are minor if the pipe is free to pivot at the attachment to the platform. The wall thickness required to prevent collapse can be estimated from the formula (Little, 1976):

$$\frac{t}{D} = \left[(p_o - p_i) \left(\frac{1 - v^2}{2E} \right) \right]^{0.333} , \qquad (4.2.6)$$

where t is the wall thickness, E the elastic modulus, D the pipe diameter, P_o the external water pressure, and p_i the internal water pressure.

Values of wall thickness computed by Eq. (4.2.6) for the structural materials shown in Table 4-25 are listed in Table 4-26 for module sizes of 40, 60, and 80 MWe (net) for flow velocity of 2.5 m/s, and for CWP lengths of 800 and 1000 m. Data are listed for a cold-water volume flow of 3.4 m³/s that was selected as optimum for the 40-MWe baseline OTEC demonstration plant (George and Richards, 1980).

Major reductions in the wall thickness and weight of steel, aluminum, and glass-reinforced plastic cold-water pipes can be achieved by the use of circumfer-

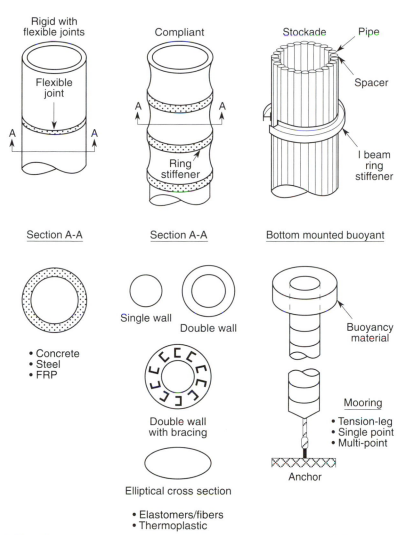

Fig. 4-50. Alternative concepts for cold-water pipe construction (McGuiness, 1979).

ential ring stiffeners. The minimum wall thickness of reinforced concrete CWPs is limited by the requirement for 4 to 6 cm of concrete covering of the steel reinforcing rods. Data for a range of materials, power levels, and pipe lengths are tabulated by Little (1976).

4.2.1.5 Dynamic Loads

Deflection by Currents and Ship Forward Motions.   A freely supported CWP attached to a moored platform subjected to steady ocean currents will assume

Table 4-25. Properties of CWP construction materials

Material	Density, ρ (kg/m³)	Yield strength, Sy (MN/m²)	Elastic modulus ($E \times 10^3$) (MN/m²)	Poisson's ratio, η
Steel	7850	138.0	207.0	0.30
Aluminum	2770	117.0	70.0	0.33
Prestressed concrete				
Concrete	2400	18.6	20.7	0.18
Tendons	2770	1034.0	207.0	0.30
Post-tensioned lightweight concrete				
Concrete[a]	1362	16.9	11.7	0.23
Tendons	7850	1850.0	199.0	0.30
Glass-reinforced				
plastic	1700	138.0	17.2	0.30

[a]Sy = 45% of compressive yield strength.

Table 4-26. Wall thickness needed to prevent CWP buckling (flow velocity = 2.5 m/s)

Power (MWe)	CWP length (m)	CWP diameter (m)	Δp (kPa)	Wall thickness (cm)			
				Steel	Al	Concrete	FRP[a]
40	800	8.32	16.4	2.76	3.94	7.31	6.30
	1000	7.13	16.1	2.35	3.35	6.23	5.37
60	800	8.74	15.0	2.82	4.02	7.45	6.42
	1000	8.74	14.1	2.76	3.93	7.30	6.29
80	800	10.09	13.6	3.15	4.49	8.33	7.18
	1000	10.09	14.4	3.21	4.57	8.49	7.32

[a]Fiberglass-reinforced plastic.

an angle representing a force balance between lateral moments imposed by the current profile and the restoring force due to the pipe weight. Values of the angle for a range of materials, pipe diameters and lengths were estimated by Little (1976) for a rigid pipe pivoted at the platform. The study showed that the forces on the pipe will be determined mainly by the dynamic interactions of natural vibrational modes of the pipe with wave-driven oscillations of the OTEC platform.

Analysis of Loads and Displacements. Wave and wind forces on the OTEC platform will cause time-varying displacements of the top of the cold-water pipe. These will induce lateral and longitudinal motions along the length of the pipe of a magnitude dependent on the natural vibrational modes of the pipe and the damping effects of the structure and of the water surrounding and flowing through the pipe. Additional loads will result from underwater currents and platform forward motions. Rigorous analyses of the effects of the forces on the pipe and platform produced by waves and currents, and of the time dependence of the resulting stresses and motions as functions of depth and time have been made by several investigators: Paulling (1970, 1979, 1980) Chang (1977), George and Blevins 1978), Giannotti et al. (1981), TRW (1982), and Vega and Nihaus (1985).

A detailed account of the theory and analytical procedures for a quasilinear frequency-domain analysis of the coupled platform and CWP system is presented in Paulling (1980). An alternative formulation of the frequency domain analysis is given by Chang (1977). A similar treatment for the time-domain analysis is given by TRW (1982). The discussion to follow is based on the analyses reported by Paulling (1979, 1980), George and Richards (1980), and Vega and Nihaus (1985). A thorough discussion of the analytical and experimental investigations is given in George and Richards (1979, 1980).

The analyses involve several fundamental assumptions:

1. The frequencies and amplitudes of the wave forces acting on the platform and CWP can be described by a spectral density function, $S_w(f, \theta)$, which expresses the relationship between the amplitudes of component waves and the frequency, f, and direction, θ, of each component. The function S_w is defined so that its integral equals the temporal mean square of the wave elevation Y_w.

$$\overline{Y_w^2} = \int_0^{2\pi} d\theta \int S_w(f, \theta) \, df \, . \tag{4.2.7}$$

For typical cases the waves will approach the platform from one or a few preferred directions. The function may then be considered as the superposition of unidirectional functions. The θ dependence is then eliminated.

2. If the response of the platform–CWP system to regular waves of unit amplitude may be expressed by a response transfer function whose amplitude is $R(f)$, then the spectral density of the response, $S_R(f)$, to a random seaway whose spectrum is $S_w(f)$ is given by:

$$S_R(f) = R(f)^2 S_w(f) \, . \tag{4.2.8}$$

The mean square value of the response is given by:

$$\overline{R^2} = \int_0^\infty S_R(f) \, df \tag{4.2.9}$$

$$R = (\overline{R^2})^{1/2} \, . \tag{4.2.10}$$

3. Many statistical properties of the response can be derived from the response function, in particular:
   a. The average of the peak values of response during a period of N cycles.
   b. The expected value of the largest single excursion during a large number, N, of cycles of the response.

4. The spectral density function of Bretschneider (1959) is a valid representation of $S_w(f)$. This formula embodies the results of a large number of sea state measurements in a form that may be easily used for computational purposes.

$$S_w(f) = 5\left(\frac{H_s}{4}\right)^2 \frac{1}{f_0}\left(\frac{f_0}{f}\right)^5 \exp\left[-\frac{5}{4}\left(\frac{f_0}{f}\right)^4\right] \, . \tag{4.2.11}$$

There are two independent parameters: H_s, the significant wave height of the random seaway, and f_0, the frequency of waves having peak energy of the seaway. Values of the significant wave height and mean frequency for potential OTEC sites have been calculated by Bretschneider (1978) from historical data on wind velocities developed in hurricanes. He has then used the data to derive equations relating wave height to wind velocity as dependent on geographical location.

5. The six degree-of-freedom motions of the combined system of platform and CWP can be modeled by a quasilinear frequency-domain analysis, which uses a finite-element model of the pipe structural dynamics and contemporary ship or stable platform theory for the platform motions.

With this rationale, the systems of equations are formulated and solved successively for a series of different wave frequencies, wave directions, and an assumed unit wave magnitude. For each node the resulting complex motion vector, x_i, may be considered a frequency-dependent transfer function for that node's motion. From the nodal motions the internal forces may be found for each element of the continuous pipe, or the joint forces for the segmented pipe. Having the transfer function for any response, such as platform motion, pipe nodal motion, element stress, the response of the total system to the random seaway may be computed.

Computer programs have been developed that solve the equations for systems with suspended CWPs attached to a floating platform by a flexible joint. Separate programs provide solutions as a function of pipe length, for motions and moments of flexible pipes of specified dimensions and physical properties (ROTEC), and for pipes made of rigid segments connected by flexible joints of given compliance (SEGPIP). The programs include internal and external water flow (Paulling, 1979).

4.2.1.6 CWP Requirements for 40-MWe Demonstration Plants

Dynamic Analysis. The equations and programs discussed have been used to estimate the expected stresses and displacements in the 9.1-m (30-ft) diameter CWP designed for the demonstration plants, either grazing at a mid-Atlantic site east of Brazil or moored at a site south of Punta Tuna, Puerto Rico.

The following significant wave heights, wave periods, and surface currents were selected as representative for design estimates:

		Atlantic site	Puerto Rico
Maximum operating	H_s (m)	5.50	5.50
condition	f_0 (s^{-1})	0.10	0.10
100-year-storm	H_s (m)	88.0	10.9
condition	f_0 (s^{-1})	0.083, 0.056	0.076
Surface currents	v_s (cm/s)	20.0	62.0

Two at-sea test programs, at one-sixth and one-fourth scale of the OTEC demonstration design, and two small-scale water tunnel programs have been conducted to measure the behavior of platform–CWP systems under conditions representative of expected OTEC operating regimes. By comparison of the observed behavior with the predicted results, the applicability of the CWP dynamic analysis routines for the prediction of forces and displacements that would be expected for full-scale OTEC operation under the extreme conditions of the hundred-year storm has been determined.

X-1 At-Sea Test of Steel Cold-Water Pipe.   The first at-sea experiments designed to provide information on the system dynamics of an OTEC platform–CWP combination were conducted during December 1979 and January 1980 by the Deep Oil Technology Corporation, with program direction by the Applied Physics Laboratory of The Johns Hopkins University (Blevins et al., 1980).

The tests were conducted in the ocean off Santa Catalina Island, California, and employed an available semisubmersible platform "Deep Oil X-1," developed as a drilling research platform for the off-shore oil industry. For the OTEC tests a steel pipe 1.5 m in diameter (5 ft) was constructed that could be assembled in 6-m (20-ft) sections with mating flanges to give a maximum length of 152.4 m (500 ft) ($L/D = 100$). The wall thickness was 0.47 cm (3/16 in.). A universal joint that could be inserted between any two pipe sections was provided. The pipe was supported at the platform by a universal joint with provision for locking the gimbals so that a fixed or flexible connection to the platform could be used.

Instrumentation for the test program included wave staffs and accelerometers for environmental documentation; heave, sway, and yaw accelerometers on the platform; and accelerometers and strain gauges spaced at optional intervals along the length of the cold-water pipe. Details of the instrumentation and test results are given in Blevins et al. (1980).

Thirty-nine files of data were recorded during the test period for eleven different platform–CWP configurations, ranging from the platform alone to the platform supporting 500 ft of pipe. Twelve of the data files were processed completely to yield a set of plots showing the power spectra of the pipe and the platform responses to the wave forces. The plots were then used to compare the observations with the predictions of four of the analytical programs discussed:

1. ROTEC-SEGPIP, a frequency-domain program developed by Paulling (1979, 1980).
2. MRDAP, a proprietary time-domain program developed by Deep Oil Technology, Inc.
3. NOAA/DOE code, a frequency-domain program developed by Hydronautics, Inc. (Barr et al., 1978).
4. HULPIPE, a time-domain program developed at TRW, Inc.

The most extensive comparisons of the test results with analytical predictions were done by JHU/APL using the ROTEC and SEGPIP programs. Detailed documentation is provided by Blevins et al. (1980). Good agreement was found for predictions of the frequencies of the spectral peaks, but the magnitudes of the peak values of stresses and accelerations were generally overestimated.

Superposition of predicted values for unidirectional seas was shown to give a satisfactory representation of results for a bidirectional sea. Two examples of bending stress comparisons are shown in Fig. 4-51.

General conclusions from the comparisons were:

1. The computer programs generally overpredict the pipe stresses and accelerations, thus tending to yield conservative design requirements.
2. Experimental data were inadequate for accurate estimates of water-added mass and drag for large pipes; however, inclusion of the missing data would have a minor impact on the estimates of forces and displacements and would not affect the major design conclusions.
3. Computer programs that do not include interactions between pipe and platform cannot provide accurate simulation.
4. Computer programs must consider directionality of the sea.

A thorough discussion of the results has also been published by Giannotti et al. (1981).

X-1 1/20-Scale-Model Test. Model tests of the X-1 OTEC configuration were conducted at 1/20 scale by Giannotti and associates to determine the usefulness of water tank testing for prediction of at-sea responses of OTEC systems (Giannotti et al., 1981). An accurate and reliable simulation method would be extremely valuable for assessing effects of design changes on OTEC dynamic response in sea environments. True simulation of scale effects is not possible because Froude number scaling, which simulates buoyancy effects, does not properly account for changes involving drag and viscous phenomena.

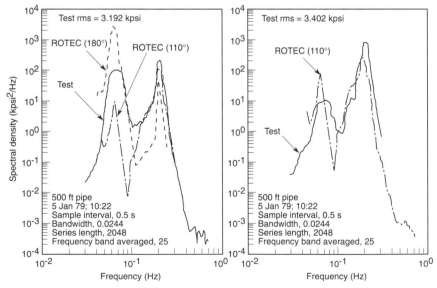

Fig. 4-51. Comparison of CWP dynamics measured in at-sea test, with analytical predictions of Paulling computer programs (O'Connor, 1981).

Fairly good agreement was found between the 1/20-scale-model behavior and full-scale X-1 test results; however, the predictions of full-scale motions and stresses from extrapolation of the 1/20-scale measurements were significantly less reliable than the predictions of full-scale behavior from the ROTEC and SEGPIP program, using as input parameters the physical properties of the barge and CWP and the measured environmental data.

At-Sea Test of Fiber-Reinforced Plastic (FRP) Cold-Water Pipe. With support by the National Oceanic and Atmospheric Administration (NOAA) an at-sea test was conducted in April and May 1983 of an 8-ft-diameter (2.44 m) FRP cold-water pipe, 400 ft long (122 m) (Kalvaitis et al., 1983). Design, construction, and deployment were conducted by the Hawaiian Dredging and Construction Company. The pipe was of sandwich construction with inner and outer face sheets 0.97 cm (0.38 in.) thick separated by a 3.3-cm (1.3-in.) layer of syntactic foam. The facing laminates were filament wound with filament strands at 60 degrees to the longitudinal axis and with unidirectional weft-roving parallel to the pipe axis to give an optimum balance between stiffness and strength properties.

The principal objectives of the test were to demonstrate the fabrication and deployment of a CWP of the FRP type and to document the responses of the pipe to the sea environment, which would provide design information for full-scale CWPs of this type. The system characteristic frequencies are shown in Fig. 4-52 (Kalvaitis et al., 1983).

The tests included: wave height measurements and current gauge measurements of the sea environment at the surface and at depth; measurement of platform accelerations at the center of gravity in heave, sway, and surge, and yaw, roll, and pitch; gimbal angles in roll and pitch; strain measurements at nine stations along the pipe and accelerations at one station; and differential pressures around the pipe at one station. An example of the average, RMS, and maximum values of the strain measured in a test with significant wave height of 1.31 m (4.3 ft) is shown in Fig. 4-53. The observed maximum strain of 1000 microstrains corresponds to a stress of 20.7 MPa (3000 psi) in a pipe wall with ultimate design stress of 393 MPa (57,000 psi).

Review and analysis of the test results and comparison of measured and predicted pipe and barge responses were conducted by Vega and Nihaus (1985) under NOAA support. General conclusions from the test and data analysis were:

1. The pipe response was dominated by the bending strains induced by barge motions coupling with the pipe, and was large enough to be important for design.
2. No evidence was found for vortex-excited vibrations, heave-induced axial strains, or torsional oscillations.
3. There was general agreement between the measured system responses and the predictions of the NOAA/ROTEC frequency-domain model. The predicted response of a barge as characterized by response amplitude operators agreed with the observed behavior within the limits of the measurement uncertainties. The response of the pipe as determined by the lengthwise distribution of standard distribution strain was underestimated by the model program. The

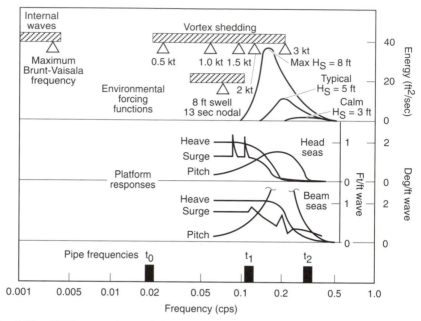

FIG. 4-52. OTEC system characteristic frequencies in at-sea operation (Kalvaitis et al., 1983).

high-end values from the range of predicted standard deviation strains were 10–20% less than the standard-deviation strains derived from measurements.

4. It was determined that the predictions were sensitive to the values assumed for the added mass and drag coefficients, indicating that a range of values should be selected for predictions. The values giving the highest response used in determining the standard deviation should be used to predict operational and extreme weather effects. The relationship between peak value and standard deviation should be used to estimate maximum stresses for a given number of cycles.

4.2.1.7   General Conclusions Regarding Suspended Cold-Water Pipes

The analytical and experimental studies discussed have established a firm engineering basis for the construction of OTEC cold-water pipes up to 10 m in diameter. Important points include:

1. Methods of construction have been demonstrated for cold-water pipes of 2.5 m (FRP) and 3 m (lightweight concrete) in diameter, which are directly applicable to fabrication of CWPs with diameters up to 10 m or more.
2. At-sea tests have demonstrated that analytical models can predict dynamic behavior and pipe stresses with sufficient accuracy to give high confidence of pipe survivability under extreme 100-year-storm conditions.
3. Lightweight concrete and FRP have been shown to be suitable CWP construction materials.
4. More accurate data are desirable on added mass and drag associated with lateral

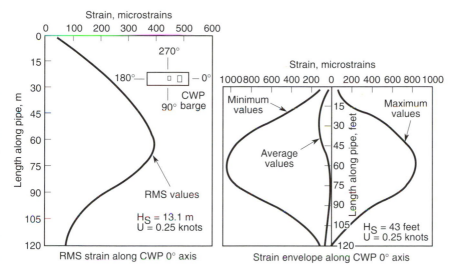

Fig. 4-53. Observed strains in at-sea test of FRP CWP (Kalvaitis et al., 1983).

motions of large-diameter pipes, and of effects of multidirectional seas on CWP dynamics.

4.2.2 Cold-Water Pipes for Shore-Based OTEC Plants

The design and construction of shore-based cold-water pipes involve several problems not encountered with suspended CWPs:

1. The design must be site specific.
2. The design must be based on detailed historical information at the site about ocean waves and currents for three (or more) ocean regimes:
   a. The near-shore regime associated with high waves and currents, requiring heavy structures and/or pipe emplacement below the sea floor;
   b. A shallow zone beyond the near shore where water depth gradually increases to 100 or more meters at the edge of the continental or island shelf;
   c. A final zone where depth increases precipitously to more than 1000 m.
   Total pipe length ranges from 2 km or more at the best sites to 5 km or more at possible locations. A typical profile is shown in Fig. 4-54 (Brewer, 1979).
3. Detailed bathymetry and morphological information about the sea bottom is required, whether the pipe is designed to float, with one or more catenary sections held in place by suitable anchors, or to be heavy and be anchored to the bottom or placed in a trench.

As with suspended pipes the technology developed in the off-shore oil industry for transferring oil and gas from emplaced rigs to collection points at considerable distances is applicable (Otteman et al., 1985).

Engineering information is also available from municipal sewage systems and electric power plants that discharge effluents through large underwater pipes extending to considerable depths and distances from shore.

Design studies of cold-water pipes, including a review of materials options and installation methods for 10- and 40-MWe shore-based plants for sites in Hawaii and Puerto Rico, were made by Brewer (1979). Facilities and materials for construction of pipes of concrete, steel, fiber-reinforced plastic (FRP), and elastomeric materials, were investigated. It was concluded that land-based plants of 10 to 40 MWe in size were feasible. The near-shore installation could be done with existing technology. Emplacement of the deep-water section of the pipe would require development of new techniques and facilities. This would require new development involving reasonable extension of demonstrated technology. The work led to recommendation of reinforced concrete as the lowest-cost material for construction and deployment at the Hawaii site where the sea bottom was relatively smooth. It was concluded that the irregular sea bottom at the Puerto Rico site would require a buoyant pipe using a catenary support anchored with piles driven into the sea floor. For this design FRP, although considerably more expensive than concrete, would be the preferred option. Concrete and FRP pipes in diameters exceeding 50 ft have been constructed for use as storage vessels and in off-shore platform structures and have been made with existing facilities. Pipe dimensions and materials for the two plant sizes and locations are listed in Table 4-27.

Similar studies have been conducted for sites at Punta Yeguas, Puerto Rico (40 MWe) (Moak et al., 1981); at Taiwan [8.7 MWe (net)] (Lin et al. 1983); at Tahiti (5 MWe) (Vilain et al., 1985); at Jamaica (1 MWe) (Berndt and Hogbom, 1983), and at the island of Bali (100 kW) (Bart, 1983). Design studies of CWPs for small

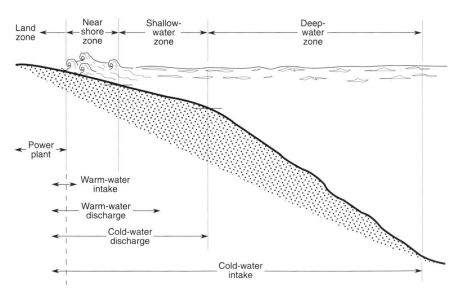

FIG. 4-54. Typical off-shore profile for island-based OTEC plant (Brewer, 1979).

Table 4-27. CWP dimensions for shore-based OTEC plants of 10.0 and 40.0 MWe (net) power

	Hawaii		Puerto Rico	
	10.0 MWe (net)	**40.0 MWe (net)**	**10.0 MWe (net)**	**40.0 MWe (net)**
ID (m)	6.80	7.50	7.61	10.05
Wall thickness (cm)	34.3	45.7	3.8–11[a]	3.8–11[a]
Length (m)	2440.0	2440.0	2440.0	2440.0
Material	concrete	concrete	FRP	FRP

[a]Thickness decreases with depth.

plants in the 0.07- to 2-5-MWe range have also been reported by Stevens (1982).

After review of previous work and consideration of factors affecting deployment costs, the French team involved in the design of a 5-MWe plant for the island of Tahiti selected an FRP pipe construction as the preferred design for a new deployment and emplacement method that would avoid the problems associated with the topography of the sea bottom at potential OTEC sites. The method, indicated in Fig. 4-55 (Vilain et al., 1985), involves two steps:

1. A series of anchoring fixtures are mounted on the sea bed at intervals along the pipe trajectory, to which cables supported by floats are attached. The anchors may be held in place by dead weights or by piles driven into the sea bottom. To each cable a sleeve (hawse) is attached at a depth determined by the pipe trajectory.
2. The flexible, slightly buoyant FRP pipe is assembled on shore and deployed by drawing it through the hawses. It is then attached to the end of the shelf-mounted pipe, which extends from the onshore plant through the surf zone to a suitable point near the edge of the shelf.

The most comprehensive studies of the design requirements for shore-based cold-water pipes were made in the preliminary design of a 50-MWe OTEC plant planned for operation at Kahe Point, Hawaii (Ocean Thermal Corp., 1984). That work led to a final pipe configuration in two segments with total length of 3667 m (12,030 ft). The near-shore segment would extend 935 m from the inlet of the OTEC plant, approximately 485 m off shore, to the edge of the shelf at 1410 m from shore. This section would be made of post-tensioned concrete and would be integral with two similar ducts carrying the mixed effluent from the plant to a discharge point at the edge of the shelf. The final section, extending from the shelf to a depth of 670 m, would be made of FRP segments with bellows joints at alternating intervals of 75 and 91 m along the 2560-m length of the deep-water section of the pipe. Details of the pipe construction and deployment are presented in Section 6.4 in connection with the complete plant design. It is worth noting that the final cost of the pipe, including deployment, was estimated to be 47% of the total system construction cost of $327 million.

From the preceding discussion, the follow conclusions regarding CWPs for

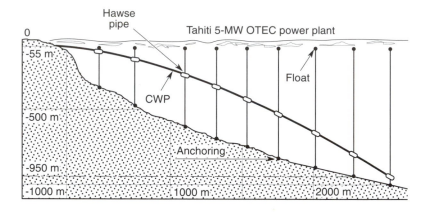

Fig. 4-55. Deployment method proposed for FRP CWP at Tahiti (Vilain et al., 1985).

shore-based installations can be made:

1. Construction and deployment of CWPs for shore-based OTEC plants of up to 50 MWe appear to be feasible, with some extension of existing technology for deployment of deep-water pipe sections.
2. The size and length of CWPs for shore-based OTEC plants, combined with requirements for heavy equipment operating in deep water for pipe deployment and emplacement, make the CWP a major cost concern.
3. Availability of flexible FRP pipes of up to 3 m in diameter will facilitate construction of OTEC plants of several megawatts net power. This would furnish accurate engineering design information for larger plant construction.

4.2.3 Warm-Water Ducting

Warm water is drawn from the mixed layer near the ocean surface. An inlet depth is selected that will give a suitable compromise among requirements for maximum water temperature, minimum pressure fluctuations due to waves, which decrease exponentially with depth, and compatibility with platform design restrictions. Mean depths ranging from an average of several meters below the surface (Anderson, 1976, 1986) to approximately 20 m are proposed (George and Richards, 1980). No ducting external to the platform is needed for floating and shelf-mounted OTEC plants. Shore-based plants draw warm water through one or more pipes that require special protection in the surf zone. The ducting arrangements are specific to a given site and plant design. An example is discussed Section 6.4.1.

The warm water for a heat exchanger module passes through screens that bar entrance of marine animals and debris and enters a warm-water pump that directs the flow to the selected module. The entrance area at the screen is designed to give a flow velocity of 0.5 m/s or less to prevent fish from being trapped against the screen. The ducts to and from the pumps are designed without sharp turns or contractions to minimize dynamic head losses. This aspect of OTEC design involves only standard marine engineering practice and needs no further comment.

Water passing through OTEC heat exchangers changes temperature by several degrees and will affect the temperature of the ambient ocean water into which it discharges. Since a density difference is associated with the temperature difference, both warm and cold discharges will be denser than the surrounding surface water. Thus, the warm exhaust water will tend to sink slowly and the cold water, relatively rapidly. If the flows are mixed before discharge to the ocean, the water will sink at an intermediate speed. In the absence of experimental data on the environmental impact of continuing exhaust flows from moored or fixed OTEC installations, it has been assumed that discharge pipes should be provided that would conduct the flow to a depth below the mixed layer where no interaction with surface waters could occur; however, the limited data obtained in the OTEC-1 tests showed that the mixed discharge plume descended rapidly. This indicates that the discharge pipes are not needed. If confirmed, this would be highly desirable, because a satisfactory method of supporting the pipes had not been devised when testing was terminated.

The proposed designs of grazing OTEC plants do not include water discharge pipes. The cold and warm flows enter the ocean at the bottom of the platform, leaving discharge plumes that sink as the platform moves forward. The small amount of mixing of the discharge plumes with the surface waters will introduce nutrients into relatively sterile tropical ocean water far from land with an expected benefit in increased marine life.

4.3   ENERGY TRANSFER

Practical use of OTEC requires provision of methods and equipment that will allow OTEC power to be delivered to potential users in a form and at a price that will satisfy a commercial demand. As shown in Chapter 3, the U.S. need for nonpolluting fuels that would not have to be imported gives priority to development of OTEC plantships grazing on the tropical oceans that would produce ammonia, methanol, or liquid hydrogen for shipment to world ports. In favorable locations OTEC electric power may be supplied directly to customers from onshore or shelf-mounted plants or from moored OTEC plants via an underwater cable (Schultz et al., 1981).

4.3.1 Electrolysis

In water electrolysis gaseous hydrogen and oxygen are produced, which store the electricity in the form of internal chemical energy. Because electrolysis does not involve a thermal cycle, Carnot's equation does not apply and, ideally, 100% conversion of electrical to chemical energy can occur. In practice, values of 80 to 90% can be achieved; therefore, hydrogen production is a highly efficient way to store electrical energy on grazing OTEC ships. The OTEC hydrogen gas may be liquefied and shipped to world ports or it may be used on board for the manufacture of methanol, of current interest as a replacement for gasoline and diesel fuel, or ammonia, a storable, high-octane-number fuel that would produce only water and nitrogen as combustion products.

Major commercial interest in water electrolysis developed after 1920 in connection with world demands for ammonia fertilizers that could be produced economically using low-cost hydroelectric power to supply hydrogen. Development stopped in the 1940s when natural gas became a lower-cost source of hydrogen for ammonia synthesis. Apprehension about potential depletion of fossil fuel supplies and environmental concerns caused a renewal of research and development in the 1970s that showed that significant improvements in electrolyzer performance were possible compared with the early designs; however, the low cost of hydrogen produced from natural gas has continued to make electrolytic hydrogen noncompetitive for ammonia synthesis and has inhibited growth of a demand for large electrolyzers.

Electrolyzers are designated as unipolar or bipolar. In the unipolar design, a voltage is applied between a positive metal electrode and a negative metal electrode that are suspended in a vessel containing an electrolyte. Hydrogen is evolved at one electrode (the cathode) and oxygen at the other (the anode). A suitable barrier directs the gases into separate ducts. A single cell may hold many electrodes, all of the same polarity being connected in parallel. Thus, a unipolar cell is characterized by low voltage and high current, large size, and simplicity of design.

In bipolar cells the electrodes are formed by depositing or pressing a thin coating of conducting material onto opposite sides of thin sheets of porous material containing the electrolyte. With appropriate packaging each sheet can become an electrolyzer cell, and cells may be assembled in stacks that are held together by tie-bolts, as in a filter press. The positive electrode of one cell is connected to the negative electrode of the next (i.e., in series). Bipolar systems are compact and compatible with voltage and current loadings typical of power generation equipment; however, special electrode–separator constructions and leak-proof gaskets are required that pose problems of reliability, durability, and cost. A concise discussion of the electrochemical and manufacturing status of unipolar and bipolar electrolyzers is presented in LeRoy (1983). A comprehensive review of industrial equipment for water electrolysis is given in Tilak et al., 1981).

Water electrolyzers are commercially available from several manufacturers; however, scale-up and further development are required to provide units of 2- to 5-MWe capacity suitable for OTEC.

4.3.1.1   Electrolyzer, Inc., Unipolar Electrolyzer

The principal developer and supplier of unipolar electrolyzers since the 1940s has been Electrolyzer, Inc., of Canada (Crawford and Benzima, 1986). The electrolyte is concentrated KOH. Development was slow for 3 decades, but in recent years improvements in the cell design have been achieved that boost performance and reduce cost (LeRoy, 1983). Durable surface coatings for the electrodes have been developed that significantly reduce the cell voltage required, as shown in Fig. 4-56. The data provide evidence that electrolyzer efficiency above 85% will be attainable in commercial models. Improvements have also been made in cell resistance and

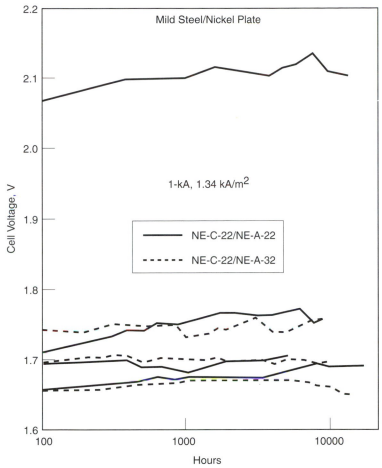

FIG. 4-56. Voltage profiles for improved Electrolyzer, Inc., electrolysis cells (Crawford and Benzima, 1986).

packaging arrangements, as indicated in Crawford and Benzima (1986) The dimensions of the advanced cell, which would use 185 kW of power at 100,000 A and 1.85 V, are listed as 2.1 m long × 1.1 m wide × 1.8 m high. A layout of a 35-MWe electrolyzer plant using 180 of these cells is shown in Fig. 4-57. The plant area is 2750 m^2. Electrolyzer, Inc., has developed plans for electrolyzer installations of 100 MWe or more to use hydroelectric power available in Canada to produce hydrogen for ammonia synthesis or as an energy carrier that could be a cost-effective substitute for electric power transmission to remote users.

4.3.1.2   General Electric Solid Polymer Electrolyte Electrolyzer

The General Electric solid polymer electrolyte (SPE) electrolyzer was developed initially as a fuel cell to furnish electric power for spacecraft. In 1975, the potential of the technology to produce hydrogen with high efficiency and low cost led DOE,

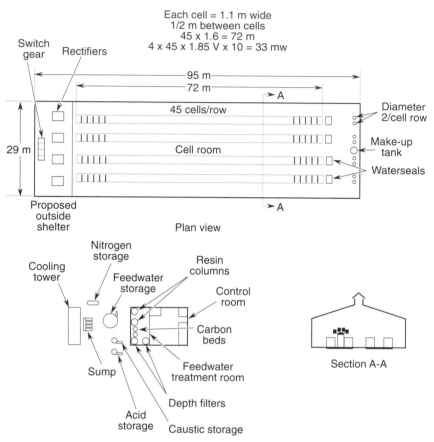

FIG. 4-57. Layout of 35-MWe Electrolyzer, Inc., hydrogen plant (Crawford and Benzima, 1986).

GE, and New York electric power producers to support the development of large units for energy storage and advanced energy systems (Nuttall, 1981). Further impetus for development of large units was provided by the OTEC program, which proposed tests of a 40-MWe baseline plantship as a prelude to larger commercial vessels.

The SPE electrolyzer uses as the electrolyte a sulfonated fluorocarbon plastic film (Nafion™), approximately 1/4 mm thick, which conducts electricity when saturated with water. Electrodes are formed by bonding a thin platinum coating on opposite sides of the film. When a voltage is applied, hydrogen ions pass through the film releasing hydrogen gas at the surface. Electrons are taken from hydroxyl ions at the opposite surface to generate oxygen gas. Since the solid electrolyte is stable and fixed in position, only pure water needs to be supplied for electrolyzer operation. Individual cells are pressed together in a bipolar arrangement with suitable separators to form a complete electrolyzer consisting of 50 to 200 cells. Since the cells are in series, the voltage across a stack of plates is the product of cell voltage times the number of cells. When the electrolyzer operates, water is protonically pumped through the Nafion film at a rate about 8 times the electrolysis rate. The water provides the source of hydrogen and oxygen, as well as cooling to remove heat generated by the ohmic resistance of the Nafion film.

Initial SPE development focused on electrolysis modules comprising 60 cells with Nafion sheets of $1/4$ m^2 area. At $10,000$ A/m^2 the design power input was 275 kW. An example of average cell performance is shown in Fig. 4-58, and the proposed layout of a 4-MWe module is shown in Fig. 4-59. The floor area is 42 m^2, and headroom of 2.7 m is indicated (Nuttall, 1981).

Experiments were conducted with cell area of 1 m^2, which showed the feasibility of scale-up to this size, but problems with gaskets were encountered that had not been eliminated when funding support was withdrawn after 1980 as a result of low fossil fuel prices and the change of DOE administration. It is believed that the emergence of a large demand for electrolytic hydrogen would lead to rapid commercialization of this technology.

4.3.1.3 Teledyne Energy Systems' Alkaline-Electrolyte Electrolyzer

For the past 2 decades the Teledyne Energy Systems has been a commercial supplier of bipolar electrolysis cells that use a concentrated solution of potassium hydroxide as the electrolyte (Kincaide, 1979; Kincaide and Murray, 1981). The commercial units supply a steady industrial demand for packaged electrolyzers that deliver 50–200 standard liters per minute of pure hydrogen. Input power ranges from 10 to 40 kW. Similar systems are also commercially available from several foreign manufacturers: Brown-Boveri (Switzerland), Norsk-Hydro (Norway); and French and Belgian suppliers.

The basic design of these electrolyzers is similar to that of the SPE type of GE. Thin sheets of asbestos of 0.1 m^2 area, saturated with KOH, are used as separators instead of the Nafion film; concentrated KOH solution is circulated instead of pure water for electrolysis and to remove heat. Teledyne was funded in parallel with the

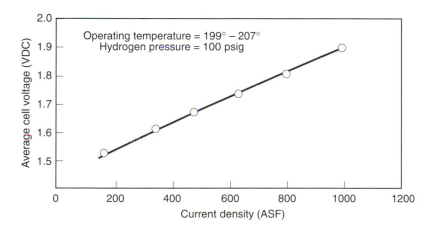

FIG. 4-58. Typical performance of GE SPE electrolyzer (Nuttall, 1981).

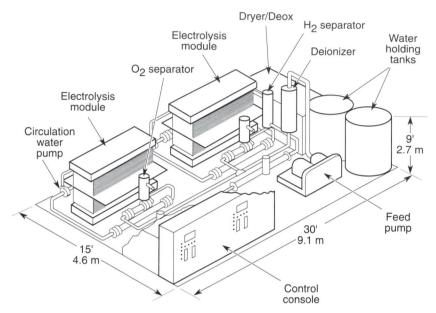

FIG. 4-59. Layout of 4-MWe module of GE SPE electrolyzer system (Nuttall, 1981).

GE program to investigate scale-up of their design to electrolyzer modules in the multi-megawatt range that would be of high efficiency and reasonable cost. Experiments with a complete electrolyzer unit including 4 cells showed the feasibility of scale-up of the cell area to 0.5 m². The cathode in the commercial units is a nickel screen, but much improved performance was demonstrated with new cathodes of proprietary design, as shown in Fig. 4-60. This work provided a basis for estimates of dimensions and costs of multi-megawatt electrolysis systems that would be suitable for OTEC. A cost analysis led to estimated cost in the range of $200–300/kW for electrolyzer units of 4 MWe or larger. As with the GE effort, development support for this program was discontinued after 1980, but it appears that rapid commercialization of large sizes would result if a large demand materialized.

It may be concluded that electrolyzers of multi-megawatt capacity will be produced when definite plans emerge for production of OTEC plantships.

4.3.2 OTEC Hydrogen Production, Storage, and Transfer

The use of hydrogen as an OTEC energy transfer medium involves installation of power conversion equipment on the plantship that will supply DC electricity to the electrolyzer system at optimal values of voltage and current. Also required are hydrogen liquefaction facilities, insulated storage vessels, and special ships equipped to transfer liquid hydrogen from the OTEC plantship to world ports.

The turbogenerator for a 10-MWe (net) (13.5 MWe gross power) OTEC

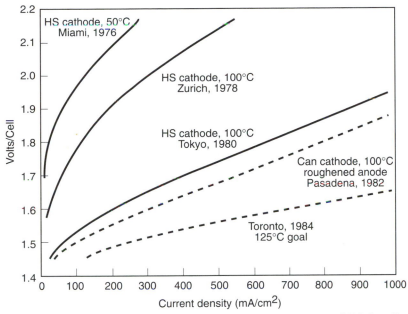

FIG. 4-60. Performance improvements through research and development of Teledyne Energy Systems electrolyzer (Kincaide and Murray, 1981).

module would typically supply alternating current at approximately 15 kV. Transformers and rectifiers must be provided to produce direct current at a voltage in the range of 100 to 600 V, dependent on the type of electrolyzer. Appropriate controls and switch gear must also be supplied. Standard equipment is available for these purposes.

Hydrogen emerges from the electrolyzers as a gas at a temperature near 100°C and pressure ranging from atmospheric (with unipolar electrolyzers) to 25 atm or more. Higher pressures are advantageous to reduce compression requirements for subsequent processes and to minimize piping sizes. The gas then enters the liquefaction plant, if liquid hydrogen is chosen as the energy transfer medium, or goes to the onboard ammonia or methanol plant for conversion to a more tractable energy carrier. The general features of the equipment that would be used for conversion of OTEC energy to a transportable fuel are summarized in the following sections. Further details of systems designs are presented in Section 6.4.

Hydrogen is liquefied by a process that involves compression of the gas, cooling of the compressed gas with liquid nitrogen, and isentropic expansion of the gas against a load so that work is performed. The expanded gas, at a lower temperature, is recycled, causing successive temperature reductions until a temperature is reached at which Joule–Thompson expansion causes a fraction of the gas to liquefy. The remaining cold gas, in countercurrent flow, cools the incoming gas. A schematic of the process is shown in Fig. 4-61 (Perry and Green, 1984, Sec. 12-50). (It is interesting to note that Georges Claude, who built the first OTEC plant,

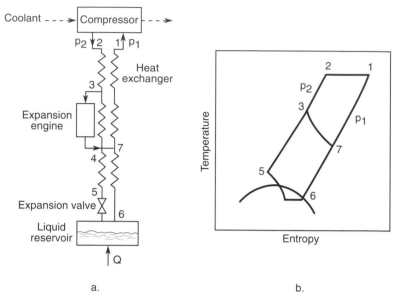

a.   b.

Fig. 4-61.   Schematic of Claude hydrogen liquefaction process (Perry and Green, 1984).

also pioneered the development of this cycle.) The ideal work to liquefy hydrogen is 4.0 kW h/kg. For plants in the kilowatt range the actual requirement is about 25 kW h/kg; at power inputs above 1 MW the requirement is estimated to be 15 kW h/kg. Thus, energy required for hydrogen liquefaction, at a 1-MWe power level, adds 27% to that needed for electrolysis.

After liquefaction the hydrogen flows through insulated lines to storage vessels of Dewar construction that are designed to hold approximately 1 month's output from the liquefaction plant. The design requirements and practical construction of the transfer and storage equipment have been established in operation of the Kennedy Space Center and other sites, and are directly applicable to OTEC design needs (Loyd, 1986). For a 160-MWe OTEC plant a single storage vessel of approximately 40 m diameter, or four vessels of 20 m diameter, would be required. The engineering requirements of transfer lines for liquid hydrogen are discussed by Jones et al. (1983).

The practical problems caused by the low density of liquid hydrogen and the need for cryogenic equipment and procedures make this method of OTEC energy transfer not attractive for near-term applications, unless a large demand should arise for liquid hydrogen per se for space-shuttle or aerospace-plane applications.

4.3.3 *Energy Transfer via OTEC Ammonia Production*

Ammonia (NH_3) is an easily liquefied gas, formed by the combination of three molecules of hydrogen with one molecule of nitrogen, that provides an attractive means for efficiently converting electrolytic hydrogen produced on an OTEC vessel into a form that can be easily stored and transported. It is also an excellent fuel for nonpolluting internal combustion engines (the octane number is 111). With this fuel only water and nitrogen gas are produced as exhaust products. Thus, OTEC ammonia could play a major role in supplying future world needs for a practical fuel that would not contribute to global warming in either its manufacture or its use. The use of OTEC ammonia as an energy transfer medium was proposed in 1975 (Olsen and Pandolfini, 1975). Preliminary design of a 40-MWe OTEC plantship was reported in 1980 (George and Richards, 1980).

The heat of combustion of ammonia based on the hydrogen content is 88% of that of molecular hydrogen, but the density is 0.60 kg/liter compared with 0.071 of liquid hydrogen, and the boiling point is $-33.4°C$ rather than $-258°C$. The heat of combustion per liter is 11.2 MJ versus 8.58 MJ for liquid hydrogen. At 20°C the vapor pressure is 10 atm. These properties combine to make ammonia production an efficient and practical way to store OTEC energy for periodic transport at normal temperatures and reasonable pressure for later use on land.

Ammonia is an important industrial chemical that is produced commercially by steam reforming of methane to supply the hydrogen. In this process nitrogen is prepared either by air liquefaction and distillation or by burning hydrogen in air in a closed vessel to convert the oxygen to water, which is then removed by condensation. The two gases are then combined in a catalytic reactor to form ammonia. For OTEC ammonia manufacture the reactor and distillation plant equipment of the commercial ammonia from natural gas plants are directly applicable. The costly steam-reforming subsystem is eliminated.

The basic reactions in the commercial process are:

$$1.77\ H_2O + 0.88\ CH_4 = 0.88\ CO_2 + 3.53\ H_2$$
$$0.53\ H_2 + air\ (0.266\ O_2 + N_2) = 0.53\ H_2O + N_2$$
$$N_2 + 3H_2 = 2NH_3$$

Thus, 0.44 mole of CO_2 is formed for each mole of ammonia, or 1.3 kg of CO_2 per kg of NH_3.[2]

Ammonia is the primary source material for fertilizers containing nitrogen, which have a major role in international food production. Over 125 million metric tons of ammonia are produced per year worldwide for this and other uses. In consequence, facilities are available worldwide for liquid ammonia storage and transport: in tank cars, trucks, and pipelines, and in ammonia tankers (the liquid is cooled to $-33.4°C$ to allow storage at atmospheric pressure). OTEC ammonia production would be able to use these facilities as soon as commercial operation began, avoiding the slow and costly development of the infrastructure that would be required for liquid hydrogen use as an energy carrier (Avery, 1988).

Preliminary design details of a 40-MWe OTEC baseline ammonia demonstration plantship are presented in Section 6.5.1.

4.3.4 *Energy Transfer via OTEC Methanol Production*

Methanol (CH_3OH) is a liquid fuel that is formed by reaction of two moles of hydrogen with one mole of carbon monoxide. It is produced commercially by catalytic conversion of a feed gas mixture made by steam reforming of natural gas or other hydrocarbons. Oxygen produced by air liquefaction is added to the feed to provide a proper heat balance. The process takes place under equilibrium conditions and involves the basic steps:

$$CH_4 + 2 O_2 = CO_2 + 2 H_2O \quad (\text{process heat})$$
$$CH_4 + H_2O = CO + 3 H_2$$
$$CO + 2 H_2 = CH_3OH$$

A typical composition of the steam-reformed feed gas is 15% CO, 8% CO_2, 74% H_2, 3% CH_4. The process uses approximately 1000 m^3 of natural gas per metric ton of methanol produced and approximately the same volume of pure oxygen.

Use of OTEC for the synthesis of methanol efficiently uses the oxygen as well as the hydrogen generated by water electrolysis in replacing the costly steam-reforming part of the commercial process. On-board coal gasification employing a molten carbonate gasifier can provide a practical method of incorporating carbon in the feed-gas stream. Thus, OTEC methanol synthesis appears ot be the most economically attractive way to produce an energy transport fuel that could be manufactured on OTEC plantships (Avery et al., 1983; Avery, Richards, and Dugger, 1985). A particular attractive feature of this energy transfer option is that it offers a near-term method of supplying an automobile fuel at reasonable cost that can be a direct replacement for gasoline made from imported oil.

Methanol is a liquid at ambient temperatures and boils in the gasoline range; it has an octane number of 108; it burns cleanly and does not contain sulfur. Thus its use reduces pollution compared with gasoline. These features have been successfully demonistrated in California, and programs are now being implemented on a national evel that are designed to achieve commercial production and operation of motor vehicles fueled with methanol instead of gasoline made from imported oil. The preliminary design of a 160-MWe OTEC methanol plantship, including details of the coal gasification process, storage, and shipping, is presented in Section 6.5 (EBASCO, Inc., 1984).

4.3.5 OTEC Aluminum Refining

Primary aluminum production by electrolysis in the United States in 1989 was four million metric tons and consumed approximately 66 billion kilowatt hours of electric power (Aluminum Association of America, 1989). This is equivalent to 2.6% of the U.S. output of the electric power plants in that year. The low bus-bar cost of OTEC electric power suggests examination of the potential economic benefits of using this source of electricity for aluminum refining. A study of the process requirements, siting options, estimated costs, and projected financial returns was made for several alternative scenarios, including on-board aluminum plants and onshore installations receiving electrical power from dedicated OTEC plants moored offshore. The results were compared with data for a baseline plant using the conventional Hall process, built and operated onshore at a Gulf of Mexico site, that would purchase power from a public utility company. Floating OTEC aluminum plantships and shore-based plants using the Hall process were found to be uneconomical. However, the evaluation showed that OTEC plantships equipped with chloride reduction facilities that refined the product from an aluminum chloride production facility onshore, or shore-based plants for aluminum production by the chloride process that received power from dedicated OTEC plants moored offshore, could be operated profitably at sites in the Gulf of Mexico, the Caribbean Islands, and Mindanao. The conclusions depend on favorable financing arrangements and assumptions regarding the cost of purchased power and rates of inflation (Jones et al., 1981).

This method of OTEC energy transfer will not be attractive unless the electric power costs of conventional aluminum production increase significantly.

4.3.6 *Minerals from Deep Ocean Mining*

The discovery of manganese nodules and mineral-rich deposits near sources of volcanic activity in the ocean floor has generated interest in the development of vessels equipped to recover minerals economically from ocean depths of 1000 m or more. OTEC ships could play a significant role by providing low-cost power for these energy-intensive operations. As of 1991 no significant study of OTEC plant requirements and economic potential for this application has been made.

4.3.7 *Electric Power Transmission from Floating OTEC Plants*

OTEC power produced on floating OTEC plants can be transferred to land-based customers via underwater power cables. This requires development of a stationkeeping system and cable arrangement that will place minimum demands on the combined system. A mooring system installed in water depths exceeding 1200 m that will minimize loads on the power cable and that will hold the OTEC plant securely in place in the presence of waves generated by the 100-year hurricane must be developed. As an alternate a dynamic positioning system can be used for stationkeeping. The demands are stringent enough to lead to suggestions of spar or semi-submersible platforms to reduce the loads; however, design studies show that multi-point mooring of barge configurations will be a feasible way to satisfy the mooring requirements and that barge designs are preferable from a total OTEC systems standpoint. Stationkeeping evaluations are discussed in Section 4.4.

Several arrangements have been considered for the power cable, which must be attached to a bobbing, twisting platform at one end, must descend to the seabed, and must be firmly secured as it proceeds toward the shore, where it must be protected to withstand the rigors of the surf zone before finally emerging to be connected to the onshore utility. The OTEC power transmission requirements exceed the capabilities of commercially available cables designed for underwater use. Commercial underwater cables connect one fixed installation to another and are emplaced in a trench to prevent cable flexing and scouring. No high-voltage cables have been laid in depths greater than 560 m (Pieroni, 1980).

The development needs of OTEC power cables were studied in separate programs designed to address the three major problems:

1. The need for durability under repeated flexing of the "riser cable" that would carry power from the OTEC power plant to a firm support on the sea floor or on a submerged buoy supported at a depth low enough to be free from waves and water currents. For plants moored within approximately 15 km of shore, single cables could carry alternating current directly from the OTEC plant to the shore installation. Longer distances would require cables designed for direct current that would involve a bottom-laid cable connected at a junction box to a riser cable of different construction. Methods of predicting loading and bending stresses were developed (Lindman and Constantino, 1981). Two cable designs were identified as candidates for OTEC application: a self-contained, oil-filled design with paper insulation (SCOF) and a type using cross-linked polyethylene as insulation (XLPE). Both types were tested in a special facility designed by Simplex Wire and Cable Co. and operated using facilities provided by Hydro Quebec Institute de Recherche (Kurt, 1981). The construction of the two cable types is shown in Fig. 4-62. In the tests the cables were subjected to repeated cycles of bending and tension under varying loads to 90,000 kg while immersed in seawater in a tank that could simulate water depths to 1800 m (6000 ft). The water temperature was controlled at an average value that would be experienced by a riser cable installation at the Puerto Rico site. Voltage of 200 kV was applied during the test with current flow adjusted to bring the internal temperature to the maximum value predicted for the test regime.

   Sample cables of the two types were subjected to six different testing regimes lasting 1 month representing conditions that could be encountered in moored OTEC operation including 100-year-storm sea states. The tests indicated that cables of the XLPE design would meet requirements for 30-year life under OTEC moored-plant conditions and could be used in water depths over 200 m (Soden et al., 1981).

2. Development of a method for protection of the section of the cable from the anchor position on the seabed at a depth of 1200 m or more to the shore installation. Experience has shown that the cable must be placed in a trench to provide adequate protection. A survey showed that one commercial trenching system (Scarab II of Ocean Search, Inc.) was available that could operate at a maximum depth of 1840 m. A second (Sea Plow IV of AT&T) had a depth capability of 914 m. Both would require operational modifications for OTEC use. Extension of the capabilities to satisfy OTEC requirements appears to be feasible (Nunn, 1981).

3. Identification and development of equipment and procedures for deploying and servicing the system. Alternative riser cable systems that could be used with barge or spar OTEC plants moored at Puerto Rico were evaluated, as shown in Fig. 4-63 (Pieroni et al., 1979). The use of a stand-off submerged buoy arrangement that routes the riser cable from the OTEC plant to the buoy and then to the seabed is favored (Schultz and Dalton, 1981). Design details are presented for a gimballed attachment of the cable to the ship, for disconnect hardware, and for the anchor assembly on the sea bottom. Deployment equipment and costs are also discussed.

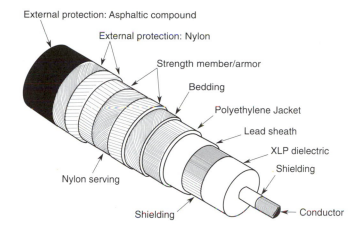

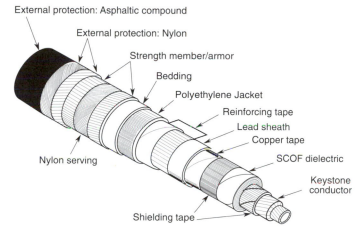

FIG. 4-62. Cable construction for underwater power transmission (Kurt, 1981).

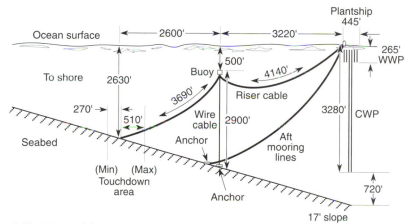

FIG. 4-63. Proposed deployment arrangement for OTEC power cables (Pieroni et al., 1979).

It is concluded that cable feasibility was indicated; however, funding was withdrawn before development and testing of the cables were completed.

4.3.8 *Electrical Energy Transfer via Electrochemical-Bridge Cycles*

This concept is a variant on the method of energy transfer that uses OTEC to produce a fuel that is transported to a land installation where it is used to generate electricity, preferably by fuel cells.

The electrochemical-bridge system stores OTEC energy electrochemically, transfers the product to shore by ship, then generates electric power on shore by reversing the shipboard process. Two processes were investigated: a redox battery system and a lithium–water–air battery system (Biederman, 1978). The author compares the estimated costs of the electrochemical method with other alternatives, including production of hydrogen or ammonia and transfer of molten salts.

The redox battery is a device containing electrodes immersed in an electrolyte containing a salt that can exist in two ionic states, such as ferrous or ferric chloride. Application of a voltage above the ionization potential of the lower state causes migration of the ions to the anode where Fe^{3+} ions are converted to Fe^{2+} ions, with release of an electron to the external circuit. The redox process allows energy to be stored in the electrolyte, which may then be shipped to shore where the reverse process can deliver energy to a user. Quantitative evaluation of the system indicates that the shipping cost of the dilute solution would be too high to make the system of commercial interest.

In the lithium–air–battery process a slurry of lithium hydroxide produced at an onshore plant is transported to an OTEC ship, where OTEC electric power is used electrolytically to convert the slurry to billets of metallic lithium. The lithium is then shipped back to the onshore plant where the lithium becomes the anode of a lithium–air–battery assembly in which concentrated lithium hydroxide is the electrolyte. The battery generates electric power, in the process converting the metallic lithium to lithium hydroxide. Cooling the electrolyte causes lithium hydroxide to precipitate. The precipitate is mixed with water to form a pumpable slurry, which is then returned by ship to the OTEC vessel where the cycle is repeated. The concept has been examined on a small scale and could be economically attractive if the design assumptions were verified in commercial development. The overall efficiency of electric power transfer by this method was estimated to be in the range of 60% (Biederman, 1978).

4.3.9 *OTEC Power Generation by Onshore or Bottom-Mounted Plants*

The need for an underwater cable is avoided by siting the OTEC plant on land or on the continental shelf in shallow water. In this case warm and cold water are conveyed to the plant through bottom-anchored ducts with inlets at appropriate depths. This method allows the cold water to be used for other purposes after flowing through the heat exchangers. Production of fresh water, use of the cold water for air conditioning, and mariculture are all potential benefits. Where the power demand is large and deep cold water is available only far from shore, moored plants stationed at the edge of the continental shelf will be favored. It was estimated that up to 15 GWe of power for southern states in the United States could be produced in this way by OTEC plants sited near the edge of the continental shelf in the Gulf of Mexico. At small tropical-island sites where cold and warm water are available close to shore, power demand is small, but requirements for fresh water and opportunities for profitable mariculture exist. Shore-based OTEC plants will be favored for these installations. The OTEC power from either type of installation could be used for fuel production via water electrolysis, as discussed previously.

4.4 POSITION CONTROL

Floating OTEC systems will require different types of position control, depending on whether the system is designed for grazing operation with onboard production of an energy product or for direct transfer of OTEC power to on-land utilities.

4.4.1 Grazing

OTEC plantships in normal operation will graze at an average speed of 0.4–0.5 knot and a maximum speed of 1.0 knot on the tropical oceans, changing position from day to day to keep the plantship in a region of optimal surface temperature. The propulsion system must also be able to maintain ship heading and stability in waves and winds of the severity predicted for the 100-year storm. Because the forward speed is so low, maneuverability must be achieved by varying the thrust angle of the propulsion system; therefore, the design incorporates rotatable thrusters mounted below the hull at the four corners of the platform. With this arrangement thrust can be oriented as needed to maintain heading in crosscurrents and to counteract the effects of beam seas.

An estimate of the power requirements for OTEC plantships can be made with standard marine engineering procedures based on the geometry of the platform and cold-water pipe. Propulsion equipment (thrusters and controls) of the required performance is commercially available; however, estimates require detailed data on the dependence of drag coefficients (Cd) on grazing speed as affected by winds and waves. The estimated error is in the range of 15 to 25% (SNAME, 1980).

The propulsion requirements for the 40-MWe OTEC ammonia plantship were determined in the baseline design study (George and Richards, 1980). A value of $Cd = 1$ was selected for the platform and $Cd = 0.7$ for the cold-water pipe. The pertinent parameters of the cruising rectangular barge was:

> Bow width (W) = 43 m
> Length (L) = 135 m
> Draft (normal operation) (D_n) = 20 m
> Draft (off-loaded) (D_o) = 14 m
> CWP (OD) = 9.9 m
> CWP (length) = 915 m
> Velocity (calm water) = 0.51 m/s (1 knot)
> Velocity (normal operation) = 0.2 m/s (0.4 knot)

For the maximum cruising speed in calm water the drag is 98,800 kgf.

A typical ship propeller or thruster delivers thrust of 15–22 kg per kilowatt of power (Davidson and Little, 1978). If the motor plus transmission efficiency is 90%, the estimated maximum calm-water cruising power needed for the baseline barge is 5000–7300 kW. For normal grazing at 0.4 knot (0.2 m/s) with average wind and wave conditions, the estimated power requirement is reduced to 2100 kW. The CWP drag coefficient is increased to 1.0 for this estimate.

For plant survival in the 100-year storm, the OTEC power plant is shut down and diesel power is substituted to supply operating needs. The ship is deballasted to reduce the draft to 15 m, with a corresponding reduction in the drag area. In this situation the essential function of the propulsion system is to maintain ship heading to prevent the platform from experiencing beam seas that would cause the ship to roll to angles that would damage the CWP–platform connection. The evaluation of the requirements for the 40-MWe baseline barge was based on the earlier study conduction in connection with the OTEC 20-MWe pilot plant program under NOAA sponsorship. The analysis used a ship control program developed by Honeywell that included data derived from ship experience to estimate drag, current, and wind forces. The combined effects of a 3.2-knot surface current, a 61-knot wind, and seas with significant wave height of 8.8 m, all coincident and acting 45° off the bow, were used to estimate the most severe conditions. The combination yielded a yaw torque of 4.4 million kg m (32×10^6 ft lb). Four thrusters of 18 kg thrust per kW (33 lb/horse power) acting with a lever arm of 52 m were specified. The delivered thrust was assumed to be reduced 4% per knot of current. The required thruster power was then estimated as 1350 kW per thruster.

Reserve power must be provided for thrust modulation in the wave and current environment. Honeywell estimated that this would raise the requirement to 2225 kW per thruster with minimum modulation, but 1850 kW would be acceptable with higher modulation. Funding was not provided for a reevaluation of the requirements for the 40-MWe baseline platform, which would need additional model test data before the final requirements were defined; however, the total power of 8000 kW estimated as necessary for the maximum cruising speed of the 50-MWe baseline barge appeared to be a reasonable choice that would satisfy the thrust requirements for ship survival in the 100-year storm, as well as the extreme needs for normal operation. The thruster requirements for larger platforms can be estimated using Eq. (4.4.1) presented in the next section.

4.4.2 Position Control for Direct Power Transmission from Floating OTEC Plants

Direct transmission of OTEC power to shore requires that floating OTEC plants be maintained at a fixed location by mooring or dynamic positioning. Both procedures have been considered. Engineering design data and equipment can be based on the extensive technology developed in the off-shore oil industry.

4.4.2.1   Stationkeeping

Power transmission from a floating OTEC plant imposes two conditions not encountered in commercial marine engineering practice. A stationkeeping system must be designed to hold the plant within a small circle above the selected geographical position without interfering with the power cable, in water depths exceeding 1000 m. It must also ensure that the platform can be oriented to withstand waves and currents generated in storms of once-in-100-years severity. Mooring and/or dynamic positioning may be used. The requirements and potential designs have been the subject of several investigations.

A general study of OTEC stationkeeping requirements was made in 1976 and 1977 using systems studies from the DOE OTEC program as a source of design data (Davidson and Little, 1978). The study includes assessment of the requirements, equipment options, and costs of stationkeeping systems. Mooring systems and dynamic positioning systems using thrusters, effluent vectoring, and combinations of the two were evaluated. Figures 4-64 to 4-70 are drawn from the report. Characteristics of platform configurations and estimated drag coefficients are given in Fig. 4-64. Marine engineering experience and design information acquired in OTEC plant design studies were used to derive a baseline relationship between OTEC plant displacement and net power output, P(net):

$$\text{Displacement (mt)}/P\text{(net) (MWe)} = 3210P\text{(net)}^{0.18}. \qquad (4.4.1)$$

This equation was used to estimate the stationkeeping loads for OTEC plant configurations ranging in power from 10 to 1000 MWe. Two model environmental situations were selected for evaluation:

1. A moderate environment with surface current of 1.5 m/s.
2. An extreme environment with surface current of 3 m/s.

Drag forces computed for the two environments, which are based on the drag coefficients shown in Fig. 4-64, gave the values shown in Fig. 4-65 for the range of plant powers and configuration types. The shaded areas bounded by the lines labelled "semi-submersible" and "spar" include the ranges for the other platform types. The power requirements for position control are shown in Fig. 4-66.

An estimate of mooring system relative performances and costs showed that moorings could be designed, using state-of-the-art technology, for OTEC power plants as large as 1000 MWe that would have an estimated cost for the extreme environment ranging from about $1000/kW for plants in the 10-MWe range to about $100/kW for the 1000-MWe size (1978 dollars). For a 200-MWe plant, the study predicts a mooring cost of about $200/kW (Davidson and Little, 1978). The data in Chapter 7 show that the estimated mooring system costs would be about 8% of the total plant cost at the 200-MWe power level.

Types of thrusters applicable to OTEC position control are illustrated in Fig. 4-67. As shown in Fig. 4-68, a survey indicated that the delivered thrust of available units ranged from 125 to 225 N/kW (12.5 to 22.5 kgf/kW). For the evaluation study a thruster type designed to produce a thrust of 190 N/kW was chosen as a baseline. With these guidelines the dynamic positioning requirements, without effluent thrust control, were estimated as shown in Fig. 4-68. The corresponding costs are presented in Fig. 4-69.

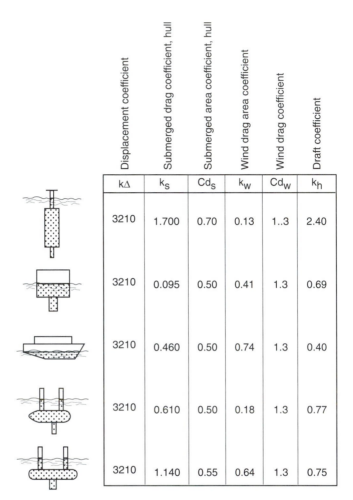

	Displacement coefficient	Submerged drag coefficient, hull	Submerged area coefficient, hull	Wind drag area coefficient	Wind drag coefficient	Draft coefficient
	k_Δ	k_s	Cd_s	k_w	Cd_w	k_h
	3210	1.700	0.70	0.13	1..3	2.40
	3210	0.095	0.50	0.41	1.3	0.69
	3210	0.460	0.50	0.74	1.3	0.40
	3210	0.610	0.50	0.18	1.3	0.77
	3210	1.140	0.55	0.64	1.3	0.75

Cold water pipe drag coefficient, $Cd_p = 0.7$
Cold water pipe depth = 600 m

FIG. 4-64. Dependence of drag coefficients on design for alternative OTEC configurations (Davidson and Little, 1978).

A 100-MWe power plant design combining mooring lines and thrusters is shown in Fig. 4-70. This system offers advantages compared to the use of either thrusters or mooring alone. The thrusters can be used to reduce loads on the mooring lines during stormes but are not required for stationkeeping in calm weather, so are not a constant drain in operating expense. Thrusters would also be valuable during plant deployment and attachment of mooring lines, and would provide redundancy.

A comparison of mooring versus stationkeeping using thrusters shows that the equipment costs for position control in the extreme environment will be similar for mooring and dynamic positioning. If storm conditions required shutdown of the OTEC power plant, diesel power would have to be available as a replacement. The

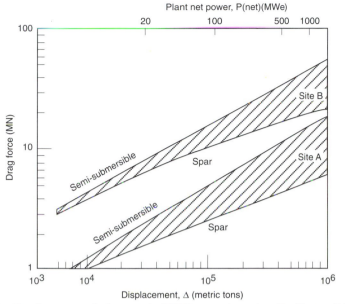

FIG. 4-65. Drag force versus displacement for OTEC system alternatives (Davidson and Little, 1978).

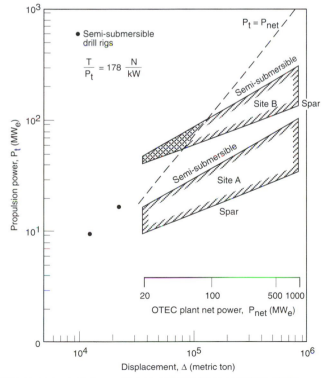

FIG. 4-66. Power requirements for position control (Davidson and Little, 1978).

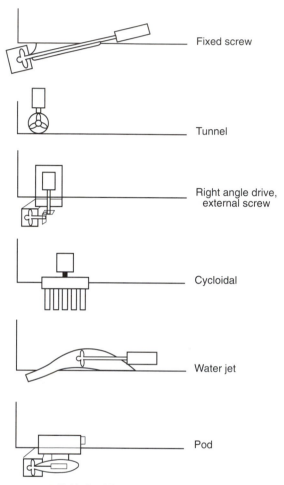

Fixed screw

Tunnel

Right angle drive,
external screw

Cycloidal

Water jet

Pod

FIG. 4-67.   Thruster types available for OTEC position control (Davidson and Little, 1978).

cost of these units was estimated as $0.8/N in the APL baseline barge design. Thus total costs of dynamic positioning equipment that would provide sufficient thrust to ensure survival in extreme storm conditions would be about $280 per kilowatt of OTEC plant power, or roughly 40% higher than mooring costs.

Subsequent to the Davidson and Little study, two detailed evaluations of the stationkeeping requirements of 40-MWe OTEC power plants that would be emplaced either on a barge of the APL baseline design or on a spar were made with DOE support by Lockheed (1979) and M. Rosenblatt & Son (MRS) (1980). A variety of possible systems were studied, and the two most promising designs were identified for each platform type. Designs and costs of 10- and 40-MWe spar plants were also developed by Scott (1979). General reviews were published by Wood and Price (1980) and Ross and Wood (1981). An evaluation of mooring requirements and options was also published by Homma and Kamogawa (1979).

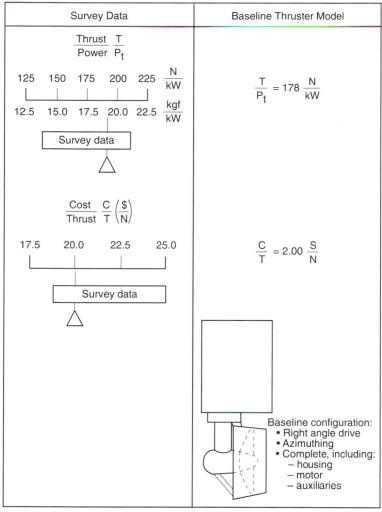

FIG. 4-68. Estimated cost of thrusters for OTEC position control (Davidson and Little, 1978).

Examples of mooring concepts from the Lockheed and MRS studies are shown in Fig. 4-71. In both studies barge and spar platforms were considered. The preferred barge mooring concepts were basically the same, that is, a multileg moor made of chain segments attached to drag embedment anchors at the bottom with wire rope for the intermediate sections. The MRS design brings the wire rope to line handling machinery on the platform. The Lockheed design uses a chain section for the line handling on the platform. The Lockheed design uses a chain section for the line handling on the platform. For the spar configuration both multi-anchor-leg (MAL) mooring and tension-anchor-leg (TAL) concepts were investigated. In the latter scheme the CWP is designed to incorporate tensile elements that can withstand the tension loads.

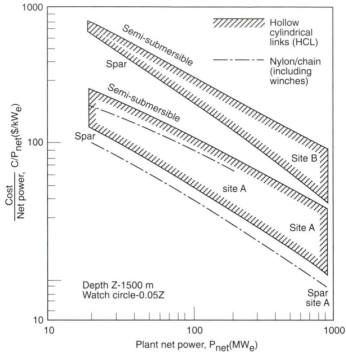

FIG. 4-69. Estimated cost of OTEC position control using thrusters (Davidson and Little, 1978).

- 3 point moor, 14.6 MN HCL line
- 5-2240 kW thrusters
- Platform excursion limited to 5% of depth at T/Tmax = 0.50
- Line cost = 4.0 M$
- Control system cost = 2.0 M$
- Anchor cost = 1.2 M$
- Total cost = 14.1 M$ 141 $/kW$_e$

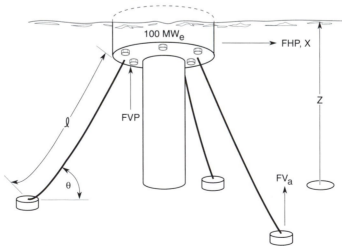

FIG. 4-70. OTEC position control combining mooring and thrusters (Davidson and Little, 1978).

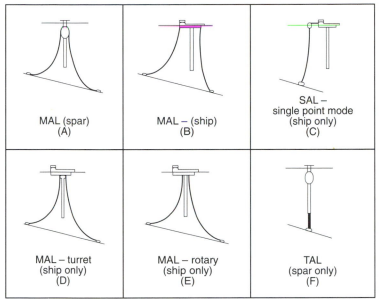

Fig. 4-71. Alternative mooring arrangements for OTEC systems (Wood and Price, 1980).

The mooring procedure for the TAL design involves the following steps (Wood and Price, 1980):

1. A deadweight anchor with buoyant cables attached is floated to the mooring site, lowered to the sea bottom, and filled with concrete or secured by piles driven into the sea floor.
2. The spar platform is positioned over the anchor and ballasted to lower it until the bottom link of the CWP can be attached to the anchor.
3. The spar platform is then deballasted, raising the spar to the operational submergence level.

The TAL concept has several attractive features:

1. The system eliminates the need for onboard anchor-line handling equipment.
2. The simplicity implies minimal operating problems.
3. It can be employed in shallower water than multileg arrangements.
4. It could simplify power cable problems.
5. Life-cycle costs could be lower.

There are two serious disadvantages that were not resolved when the program was terminated:

1. The security of the system is totally dependent on the integrity of the single tension leg, which includes the anchor, the connection to the end of the cold-water pipe, the internal tension member in the CWP, and the universal joint at the top of the CWP. Failure of any one could lead to catastrophic failure of the entire system.

2. The tension loads would exceed commercial design experience by a large factor. Thus the initial installation would involve unacceptable risk unless preceded by a major development and testing effort.

4.5   PLATFORM

The term *OTEC platform* is used here to denote the structure that houses and supports the systems involved in ocean thermal energy conversion, storage, and transfer. It may take many forms depending on size, site, purpose, and subsystems designs. A basic distinction may be made between floating platforms and fixed installations constructed onshore or bottom-mounted in shallow water near the shore. Engineering studies have shown that the design and constuction can use technology already well developed in the shipbuilding and offshore oil industries such as Deep Oil Technology (Brewer, 1979), Gibbs and Cox, Inc. (Scott and Rogalski, 1978), M. Rosenblatt & Son, Inc. (Basar et al., 1979), Gilbert/Commonwealth (Bartone, Jr., 1978), and Mitre Corp. (Roberts, 1977). Some modifications or extensions of U.S. facilities would be necessary. A vigorous OTEC program could revitalize the U.S. shipbuilding industry.

4.5.1 *Floating OTEC Platforms*

The size of the heat exchangers and the demands of the cold-water pipe and water ducting installation require the OTEC platform design to be an integral part of the total system design, if optimum system performance and minimum cost are to be realized. The systems engineering evaluations in Chapter 3 showed that barge configurations are the best choice for OTEC plants that could graze to produce fuels or be moored at suitable offshore sites to provide electric power for onshore utilities. The following discussion concentrates on such designs.

A floating OTEC platform must be large enough to contain the power plant, of which the heat exchangers are a major volume element; it must support the cold-water pipe and warm and mixed water ducting, and must provide space for energy transfer systems, operating personnel, and conventional equipment for seakeeping, safety, and navigation. Floating plants of net power in the range of 100 to 400 MWe are of commercial interest. Standard procedures for design and construction of ships and barges of the required OTEC sizes are available in marine engineering textbooks and handbooks, such as *Principles of Naval Architecture*, J. Comstock, ed. (1980).

A study of construction materials, facilities, and techniques for floating OTEC plants of 20 to 400 MWe was conducted by Brewer (1979) under DOE sponsorship. Forty-two configurations were considered. After preliminary evaluation, eleven plant designs were selected for further study. These included ship and spar configurations with heat exchangers internal or external to the hull, and steel or concrete construction. Dimensions and structural requirements were based on design integration studies by Lockheed (Waid, 1978), Gibbs and Cox (Scott and Rogalski, 1978), M. Rosenblatt & Son (Basar et al., 1978), Gilbert/Commonwealth

(Castellano and van Summern, 1978), and JHU/APL (George and Blevins, 1978). The configurations selected are shown in Fig. 1-11 (Waid, 1978).

The primary conclusion of the study was that all of the configurations studied could be constructed in existing U.S. shipyards—one Atlantic, two Pacific—using either concrete or steel. Facility modifications would be minor; however, for configurations involving draft of more than 28.3 m (80 ft) (spar types), a sheltered deep-water site would be required. Three suitable sites in the U.S. were identified: Puget Sound in Washington, Honolulu Harbor, and Lahaina Roadstead in Hawaii. (Shipyards throughout the world equipped for construction of super tankers and offshore platforms would also be suitable. A listing is presented in Chapter 6.)

For the specific requirements of OTEC, post-tensioned concrete is the favored material for the hull. The cost of construction would be lower than for steel; concrete is durable in seawater without the need for surface protection; concrete is compatible with the use of aluminum heat exchangers. The greater weight and volume of the hull are not a disadvantage because hull drag is not significant at the grazing speed of 0.4 knot, and added weight is necessary, even with the concrete hull, to compensate for the buoyancy of the heat exchangers. The attractiveness of concrete construction for the OTEC barge design was demonstrated in the low-cost construction and deployment of the ARCO barge in 1976 shown in Fig. 6-2. This vessel, 133 m long and 35 m wide with 20 m draft, was constructed in 28 months from the contract date (Mast, 1975; Anderson, 1977). The design studies for this vessel, backed by successful construction experience, formed the basis for the 137 m × 43 m × 27 m deep platform of the 40-MWe OTEC baseline, which is fully described in Chapter 6 (George and Richards, 1980).

To minimize cold-water pipe motions the CWP–platform joint should be located at the center of gravity of the platform, or if more than one pipe is employed, at the longitudinal center of rotation. This requirement and the hull penetrations required for the heat exchangers make it necessary to provide large wing walls at the sides of the platform to meet requirements for longitudinal strength.

Although the platform size and construction requirements are within normal shipbuilding and oil platform technology, the combination of platform and cold-water pipe introduces new features that require evaluation and optimization. Of primary importance is the supporting structure for the cold-water pipe, which must accommodate angular motions up to about 20°, must incorporate a seawater seal, and must be compatible with the proposed method of installing the CWP.

4.5.2 OTEC–CWP Connection

Structures that permit a cold-water pipe to be supported so that it can move freely with respect to the platform, about a fixed center, have been used in three at-sea OTEC tests: DOT X-1 (O'Connor, 1979), OTEC-1 (McHale, 1984a), and NOAA (McHale, 1984b); however, only the OTEC-1 test involved water flow through the gimballed CWP, and a method of sealing the pipe that also permitted pipe angular motion. This design involved a two-axis gimbal that supported the pipe, allowing

30° rotation. In the plane of the gimbal a spherical shell formed the top of the pipe. A cylindrical cap with a flat ring of nylon and Teflon provided a sliding seal between the cold water in the pipe and the surrounding platform structure. A diagram of the assembly, without the sealing equipment, is shown in Fig. 4-72. Operation of the structure was satisfactory during the 5-month at-sea test period, including an occasion when a combination of mooring loads, strong currents, and high seas caused the pipe angle at the platform to reach the design limit of 30°. Impact of the pipe with one of the stops caused minor damage to that structure but did not interfere with the scheduled testing. To prevent further damage the ship was released from the mooring and allowed to drift with the current. On release of the platform from the mooring the pipe angle decreased to 10°. This experience confirms the predicted advantage of the grazing mode in minimizing CWP support problems.

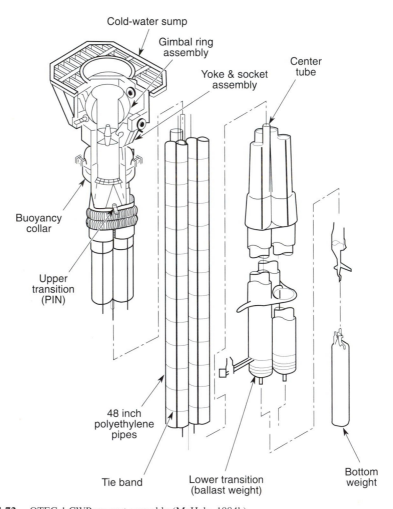

Fig. 4-72. OTEC-1 CWP support assembly (McHale, 1984b).

4.5.3 *Alternative Designs of the Hull–CWP Transition*

4.5.3.1 Support Structure for Flexible CWP

Under NOAA/DOE sponsorship an engineering design study of alternate concepts for CWP–platform joints that would be suitable for diameters of 10 m or more was conducted by TRW as part of a CWP preliminary design project (TRW, 1979). By DOE direction the study was restricted to moored platforms and three candidate flexible pipe designs, all of 9.1 m internal diameter:

1. An FRP sandwich-wall construction,
2. An elastomer construction, and
3. A polyethylene single or multiple pipe.

Static and dynamic analysis showed that, for a safety factor of 1.5 including 100-year-storm loads, the joint should be designed for an axial load of 1.8×10^6 kg (4×10^6 lb). A survey of support concepts led to selection of the three types shown in Fig. 4-73 (TRW, 1979) for detailed studies, all of which were considered feasible to construct with existing technology. The following tradeoffs were considered:

Advantages	Disadvantages
Ball and socket	
No restriction to flow	Possible ball unloading[3]
Freedom of rotation in high seas	Large diameter of mounting
Feasible sealing	
Gimbal	
No restriction to flow	Large box beams to transmit loads
Universal joint	
Minimum structure weight	Some flow restriction
Minimum sealing requirement	High bearing loads
Possible minimum cost	Submerged bearings
Minimum installation diam.	

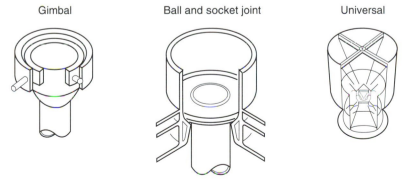

Gimbal   Ball and socket joint   Universal

FIG. 4-73. Types of flexible joints studied for OTEC CWP support (TRW, 1979).

Only the ball-and-socket joint allows freedom of rotation about the vertical axis, which is desirable to avoid torques associated with platform slewing motions; however, analysis showed that the loads would not be excessive for the universal joint. The design concept of the universal joint assembly to support a 9.1-m FRP CWP is shown in Fig. 4-74. The study indicates that optimum design of CWP–platform combinations based on FRP pipes will favor use of cold-water pipes supported at the platform by a universal joint at the center of rotation. A spherical shell segment with the same geometrical center in contact on its outer surface with a ring seal attached to the platform prevents water leakage. The ring seal would be made of replaceable segments. The concept envisions construction of the CWP on shore as a single 9- to 10-m-diameter tube that would be floated to the platform, upended, and inserted from below into the attachment assembly. The concept would be applicable to platforms employing CWP designs of diameters larger than 10–15 m if OTEC development in the future should indicate their desirability.

4.5.3.2   Support Structure for Segmented CWP

An independent preliminary design of a CWP–platform joint was made as part of the baseline design by JHU/APL of a 40-MWe OTEC pilot plant. The concept involves use of a jointed CWP of lightweight concrete that is assembled by successively connecting and lowering 15-m pipe segments through the opening in

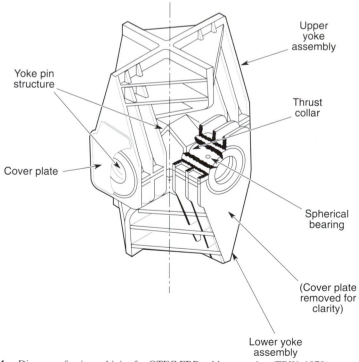

FIG. 4-74.   Diagram of universal joint for OTEC FRP cold-water pipe (TRW, 1979).

the platform. This requires a joint design that allows the pipe to be supported at the joint during the assembly process. After consideration of gimbal and universal joint designs, the ball-and-socket configuration shown in Fig. 4-75 was selected as optimum for this design, because existing technology could be extended with low risk to accommodate both manufacturing and deployment requirements (George and Richards, 1980). An alternate to the design of Fig. 4-75 was subsequently proposed that would employ an inner ball supported by external bearing pads, as shown in Fig. 4-76. This design is preferred because it simplifies problems of fabrication. The large-diameter shell section of Fig. 4-75 would be more difficult to machine and install, without distortion, in the ship structure.

To demonstrate feasibility of the lightweight concrete jointed pipe design, a 3.0-m- (10-ft-) diameter, 1/3-scale pipe, consisting of two pipe sections joined by the planned flexible joint, was constructed and tested (O'Connor, 1980, 1981). Figure 6-4 shows a drawing of the assembly. The joint was tested to failure by applying increasing lateral and longitudinal loads to the two pipe sections. Failure

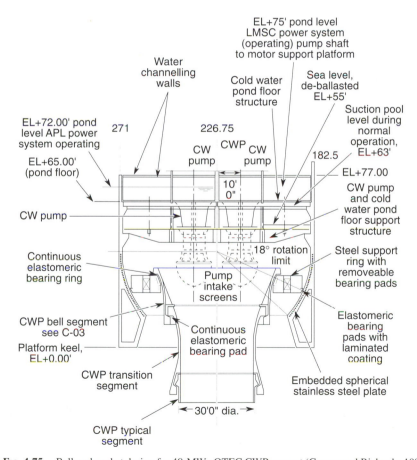

FIG. 4-75. Ball-and-socket design for 40-MWe OTEC CWP support (George and Richards, 1980).

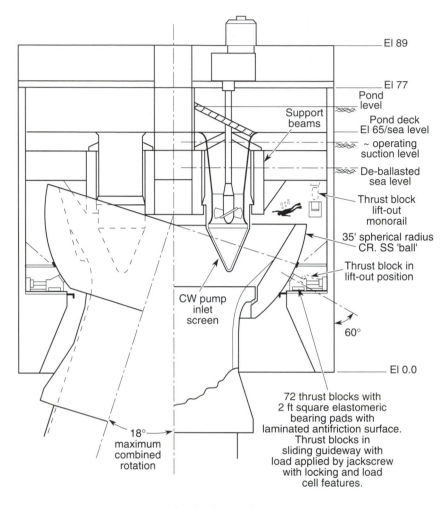

Section A (at 45°)

Fɪɢ. 4-76. Alternative design of ball-and-socket support for 40-MWe OTEC CWP (George and Richards, 1980).

occurred at a load 140% of the design value, based on the loads predicted for the most severe 100-year storm in the equatorial Atlantic Ocean.

Construction of this first-of-a-kind test article at the Concrete Technology commercial facility, using existing equipment, was completed within the planned time and cost and without need for modification of the initial design specifications. The test provided a firm basis for concluding that the jointed CWP made of lightweight concrete would be feasible to construct and would meet operational requirements. Existing handling equipment would be compatible with the planned procedures for deploying the rigid pipe sections.

Both flexible and jointed CWP designs would require platform space along the centerline with an outside diameter of approximately 25 m (80 ft) to support and allow movement of the 9.1-m CWP through an 18° angle.

4.5.3.3 Support Structure for Stockade-Pipe CWP

A third alternative for the CWP–platform joint was proposed by Anderson for support of his novel stockade CWP design (Anderson, 1976). In this concept the pipe is supported by cables connected to an air spring that is designed to have a natural frequency roughly half the minimum wave exciting frequency. This prevents the pipe from resonating with the waves. Thus the pipe remains relatively stationary during the heave motions of the platform. Details of the design and of a later version involving servo control have not been published.

4.5.4 *Platform Layout*

General layouts of OTEC systems on floating platforms were included in the studies discussed previously; however, only the JHU/APL 40-MWe baseline study of moored and grazing OTEC plants included preliminary engineering of the platform (a concrete barge) and details of the structures that would be needed to accommodate all of the subsystems, arranged to form a first approximation to an optimum total system. The design represented a team effort under overall direction of JHU/APL (George and Richards, 1980) including contributions from ABAM Engineers, Inc.; L. R. Glosten and Associates, Inc.; Tokola Offshore, Inc.; Seward Associates; and J. R. Paulling, Inc. This baseline information is suitable for determining weights and dimensions of the components of total OTEC systems, for quantitative estimates of dimensional requirements of larger-scale commercial-size floating OTEC plants and plantships, and for preliminary estimates of system costs.

The dimensions and layout of the 40-MWe baseline barge designed to demonstrate either ammonia OTEC plantship operation or direct power transfer to shore from a moored plant are presented in Section 6.5 (George and Richards, 1980). The baseline design provides data for scale-up of OTEC vessels to sizes above 300 MWe, by considering a 40- to 60-MWe power assembly as a modular unit for larger plants (Richards et al., 1980). Conceptual layouts of ammonia plantships of 40, 60, 120, and 360 MWe (net) capacity, from this article, are shown in Fig. 4-77. It should be noted that area per megawatt decreases from 145 m² (1550 ft²) for the 40-MWe size, to 107 m² (1150 ft²) per megawatt for the 360-MWe size. This implies a significant reduction in cost per megawatt for the larger plant sizes. Further reductions would be expected through optimization of the cold-water flow and incorporation of more compact heat exchangers of the plate type, as discussed in Section 4.2. These considerations suggest selection of a maximum diameter of about 10 m as being optimum, which would imply scale-up in modules of approximately 80 MWe (nom). Detailed designs have not been made.

The dimensions (Z vertical) and weights of the principal subsystems included in the integrated 40-MWe grazing plant were estimated as shown in Table 4-23 (George and Richards, 1980). These components would be typical of power plant

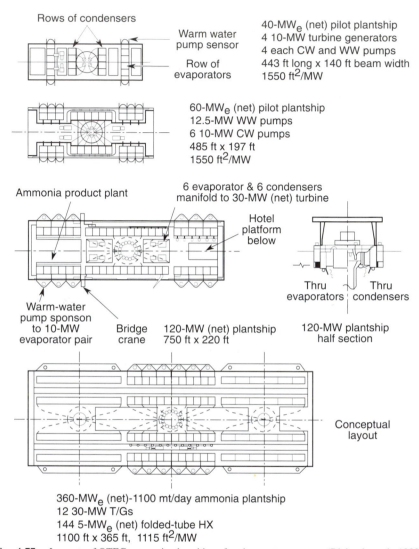

FIG. 4-77. Layouts of OTEC ammonia plantships of various power outputs (Richards et al., 1980).

designs for both floating and stationary systems employing compact 10-MWe modules.

The dimensions of heat exchanger subsystems proposed for OTEC by several contractors are listed in Table 4-22. The advantages of compact HX types compare with the S&T type are evident from the table.

4.5.5 Platforms for Land-Based OTEC Plants

The subsystems of closed-cycle, land-based OTEC plants will be essentially the same as those described in Section 4.5.1; however, the demand for power in the range of 10 MWe, and the opportunity to combine OTEC power generation with mariculture, fresh-water production, and use of the cold water for air conditioning, leads to plant layouts that will be ramified and site specific. Water pumping power will be substantially increased if the plant is placed on the shore at a height above storm-wave limits. Preliminary layouts for plants proposed for Hawaii, Nauru, and Tahiti are summarized in Section 6.4.

NOTES

1. The diagram represents the values of the parameters for the nominal 40-MWe OTEC power cycle discussed in Section 4.1.3.
2. If nitrogen is produced by air liquefaction and distillation, the estimated electrical energy required for the process is 34.3 MJ per kg of nitrogen produced (Perry and Green, 1984). If the energy is supplied by a coal-burning power plant that requires 10,000 kJ per kW h of power produced, approximately 10 kg of CO_2 will be formed per kg of N_2 produced.
3. This would not occur with properly ballasted cold-water pipes.

REFERENCES

Afgan, N. H., and E. U. Schlunder, 1974. *Heat exchangers: Design and theory source book.* Washington, D.C.: Scripta.

Aluminum Association of America, 1988. *Aluminum production and power consumption.*

Anderson, A. R., 1977. "World's largest prestressed LPG vessel." *J. Prestressed Concrete Institute*, **22**, 1.

Anderson, J. H., 1974a. "Ocean thermal energy." *2nd Ocean Thermal Energy Conversion Conf. and Workshop Proc.*, Washington, D.C., Sept.

——, 1974b. "Progress report on a working model of a closed cycle plant." *2nd Ocean Thermal Energy Conversion Conf. and Workshop Proc.*, Washington, D.C., Sept., 55.

——, 1975. "The Andersons' operating model of the ocean thermal energy conversion principle." *Proc. 3rd Workshop on Ocean Thermal Energy Conversion*, Houston, Tex., May, IX.

——, 1976. "Design of cold water pipe for sea thermal power plants." Sea Solar Power, Inc., York, Pa., C00-2961-2.

——, 1978. "A compact heat exchanger concept for ocean thermal power plants." *Proc. 5th Ocean Thermal Energy Conversion Conf.*, Miami Beach, Fla., February, **6**, VI-321.

——, 1986. *Sea solar power business plan.* Sea Solar Power, Inc., York, Pa.

——, and J. H. Anderson, Jr., 1977. "Compact heat exchangers for sea thermal power plants." *Proc. 4th Annual Conf. on Ocean Thermal Energy Conversion*, New Orleans, La., March, **6**, 3.

——, T. D. Dellinger, D. H. Forney, and R. S. Lyman, 1980. "The SSP compact heat exchanger—test results." *Proc. 7th Ocean Energy Conf.*, Washington, D.C., June, **1**, 7.4.

——, and P. B. Pribis, 1979. "Compact heat exchanger design progress." *Proc. 6th OTEC Conf.*, Washington, D.C., **1**, 11.4.

ASHRAE, 1977, 1985. *ASHRAE handbook and product directory fundamentals*, Society of Naval Architects and Marine Engineers, New York.

Ashworth, D. J., and G. Slebodnick, 1980. "Cost and configuration of a closed-cycle OTEC power system using brazed aluminum plate-and-fin heat exchangers." *Conf. Proc. 7th Ocean Energy Conf.*, Washington, D.C., June, **1 & 2**, 7.1.

Avery, W. H., 1988. "A role for ammonia in the hydrogen economy." *Int. J. of Hydrogen Energy* **13**, 761.

——, D. Richards, and G. L. Dugger, 1985. "Hydrogen generation by OTEC electrolysis, and economical energy transfer to world markets via ammonia and methanol." *Int. J. of Hydrogen Energy,* **11**, 727.

——, D. Richards, W. G. Niemeyer, and J. D. Shoemaker, 1983. "OTEC energy via methanol production." *Proc. Intersociety Energy Conversion Conf.*, Orlando, Fla., August, **1**, 346.

Bakstad, P. J., and R. O. Pearson, 1979. "Design of a 10-MWe (net) OTEC power module using vertical, falling-film heat exchangers." *Proc. 6th OTEC Conf.*, Washington, D.C., June, 8.2-1.

Barr, R. A., P. Y. Chang, and C. Thasanatorn, 1978. "Methods for and examples of dynamic load and stress analysis of OTEC cold weather pipe designs." Hydronautics, Inc., Nov.

Bart, J. G., 1983. "OTEC activities by the Netherlands." *Proc. Oceans '83*, Aug. **2**, 716.

Bartone, Jr., L. M., 1978. "Alternative power systems for extracting energy from the ocean: A comparison of three concepts." *Proc. 5th Ocean Thermal Energy Conf.*, Miami Beach, Fla., **3**, VII-68-108.

Basar, N. S., 1978. "OTEC ocean systems evaluation." *Proc. 5th Ocean Thermal Energy Conversion Conf.*, Miami Beach, Fla., **2**, IV-15.

——, J. C. Daidola, and R. C. Sheffield, 1979. "Station keeping subsystem designs for modular experiment OTEC plants." *Proc. 6th OTEC Conf.*, Washington, D.C., **1**, 5.8-1.

Baumeister, T., E. A. Avallone, and T. Baumeister III *Marks' standard handbook for mechanical engineers*, 8th Ed., New York: McGraw-Hill, p. 9.

Berenson, P. J., 1962. "Experiments on pool-boiling heat transfers." *J. Heat Mass Transfer* **5**, 985.

Bergles, A. E., and M. K. Jensen, 1977. "Enhanced single-phase heat transfer for OTEC systems." *Proc. 4th Annual Conf. on Ocean Thermal Energy Conversion*, New Orleans, La., March, **6**, VI-41.

Berndt, T., and J. W. Connell, 1978. "Plate heat exchangers for OTEC." *Proc. 5th Ocean Thermal Energy Conversion*, Miami Beach, Fla., Feb., **6**, VI-288.

——, and T. Hogbom, 1983. "The 1-MWe OTEC pilot plant in Jamaica." *Proc. Oceans '83*, 718.

Biederman, N. P., 1978. "Chemical, electrochemical and thermal bridges for transmission of OTEC power." *Proc. 5th Ocean Thermal Energy Conversion Conf.*, Miami Beach, Fla., **2**, 94.

Blevins, R. W., H. L. Donnelly, and J. T., Stadter, 1980. *Verification test for cold water pipe analysis.* Johns Hopkins Univ. Applied Physics Lab., SR-80-2A,B,C, Oct.

Bretschneider, C. L., 1959. *Wave variability and wave spectra for wind generated and gravity waves.* Beach Erosion Board, TM 118.

——, 1978. "Operational sea state and design wave criteria for potential OTEC sites." *Proc. 5th Ocean Thermal Energy Conversion Conf.*, **2**, IV-178.

Brewer, J. H., 1979. "Land-based OTEC plants—cold water pipe concepts." *Proc. 6th OTEC Conf.*, Washington, D.C., **1**, 5.10-1.

Castellano, C., and J. van Summern, 1978. "OTEC-1 early ocean test project." *Proc. 5th Ocean Thermal Energy Conversion Conf.*, Miami Beach, Fla., **1**, IV-128.

Chang, P. Y., 1977. *Structural analysis of cold water pipes for ocean thermal energy conversion power plants.* Hydronautics, Inc., May.

Chun, K. R., and R. A. Seban, 1971. "Heat transfer to evaporating liquid films." *J. Heat Transfer*, Trans. ASME, Series.

Collier, J. G., 1972. *Convective boiling and condensation*. London: McGraw-Hill.

Combs, S. K., 1979. *Experimental data for ammonia condensation on vertical and inclined tubes*. Oak Ridge National Laboratory.

Comstock, J., ed., 1980. *Principles of naval architecture*. New York: The Society of Naval Architects and Marine Engineers.

Crawford, G. A., and S. Benzima, 1986. "Advances in water electrolyzers and their potential use in ammonia production and other applications." *Int. J. Hydrogen Energy* **11**, 691.

Czikk, A. M., G. W. Fenner, F. Notaro, R. Zawierucha, and D. Mclaughlin, 1975. "Ocean thermal power plant heat exchangers." *Proc. 3rd Workshop on Ocean Thermal Energy Conversion*, Houston, Tex., May, 133.

———, Fricke, H. D., and E. N. Ganic, 1977. "Enhanced performance heat exchangers." *Proc. 4th Annual Conf. on Ocean Thermal Energy Conversion*, March, **6**, VI-71.

———, H. D. Fricke, E. N. Ganic, and B. I. Sharma, 1978. "Fluid dynamic and heat transfer studies of OTEC heat exchangers." *Proc. 5th Ocean Thermal Energy Conversion Conf.*, Miami Beach, Fla., Feb., **6**, 181.

Davidson, Jr., H., and T. E. Little, 1978. "OTEC platform station-keeping analysis." *Proc. 5th Ocean Thermal Energy Conversion Conf.*, Miami Beach, Fla., **11**, IV-237.

Douglass, R. H., 1983. "Ocean thermal energy conversion." in *Handbook of energy technology*, New York: John Wiley & Sons, Chap. 20, p. 877.

Draley, J. E., 1981. *OTEC biofouling, corrosion, and materials program: Progress report for period March 1978 through March 1981*. Argonne National Lab., Sept. ANL-OTEC-BCM-024.

Dugger, G. L., F. E. Naef, and E. J. Snyder III, 1978. "Ocean thermal energy conversion." in *Solar energy handbook*. New York: McGraw-Hill, Chap. 19.

———, D. Richards, E. J. Francis, and W. H. Avery, 1983. "Ocean thermal energy conversion: Historical highlights, status, and forecast." *J. Energy* **7**, 293.

Ebasco Services, Inc., 1984. *Coal-to-methanol OTEC plantship study using the Rockwell molten carbonate gasification system*.

Foust, H., 1981. "Constructability of extended surface heat exchangers for OTEC." *Proc. 8th Ocean Energy Conf.*, June, **2**, 717.

General Electric, 1983. *Closed cycle OTEC power plant final report*. General Electric Co., Schenectady, N.Y.

George J. F., 1979. "System design considerations for a floating OTEC modular experiment platform." *Proc. 6th OTEC Conf.*, June, 5.5-1.

———, and R. W. Blevins, 1978. "Preliminary engineering design of a 5 MWe tropical grazing OTEC pilot plant." *Proc. 5th Ocean Thermal Energy Conversion Conf.*, Miami Beach, Fla., **4**, 148.

———, and D. Richards, 1980. *Baseline designs of moored and grazing 40-MWe OTEC pilot plants*. Johns Hopkins Univ. Applied Physics Lab., SR-80-A & B.

———, D. Richards, and L. L. Perini, 1979. *A baseline design of an OTEC pilot plantship*. Johns Hopkins Univ. Applied Physics Lab., SR-78-3A.

Giannotti, J. G., K. A. Stambaugh, and W. K. Jawish III, 1981. "OTEC cold water pipe hydrodynamic loading and structural response: At-sea measurements, model experiments, and analytical simulations." *Int. Symp. on Hydrodynamics in Ocean Engineering*, The Norwegian Institute of Technology, 963.

Gregorig, V., 1954. *Zeit. fur ang. Math. und Physik*, **5**, 36.

Griffin, A. B., 1979. *CWP test article*. TRW Defense and Space Systems Group. Internal report.

—, 1979. *Cold water pipe preliminary design studies*. TRW Defense and Space Systems Group. Internal report.

Hawaiian Dredging and Construction Company, 1983. *CWP at-sea test article.* TRW Defense and Space Systems Group. Internal report.

Hillis, D. L., J. J. Lorenz, D. T. Yung, and N. F. Sather, 1979. *OTEC performance tests of the Union Carbide sprayed-bundle evaporator.* Argonne National Lab., May, ANL/OTEC-PS-3.

Holman, J. P., 1991. *Heat transfer.* New York: McGraw-Hill.

Homma, T., and H. Kamogawa, 1979. "An overview of Japanese OTEC development." *Proc. 6th OTEC Conf.,* Washington, D.C., **1**, 3.3.

Hoshide, R. K., A. Klein, et al., 1983. *OTEC-1 test operations experience final report.* Energy Technology Engineering Center, Energy Systems Group, Rockwell International, ETEC-82-19.

Jones, M. S., 1979. "Integration issues of OTEC technology to the American aluminum industry." *Proc. 6th OTEC Conf.,* Washington, D.C., **1**, 10.9-1.

———, V. Thiagarahan, and K. Sathjanarayna, 1981. "OTEC aluminum integration studies." *Proc. 8th Ocean Energy Conf.,* Washington, D.C., **1**, 643.

Kajikawa, T., T. Agawa, H. Takazawa, M. Amano, K. Nishiyama, and T. Homma, 1979. "Studies on OTEC power system characteristics and enhanced heat transfer performance." *Proc. 6th OTEC Conf.,* Washington, D.C., June, **II**, 11.5.

———, M. Takazawa, and M. Mizuki, 1983. "Heat transfer performance of metal fiber sintered surfaces." *Heat Transfer Eng.,* **4**, 57.

———, H. Takazawa, H. Nishiyama, T. Agawa, H. Iidaka, and M. Amano, 1980. "Development of enhancement of the working fluid side heat transfer for OTEC." *Conf. Proc. 7th Ocean Energy Conf.* Washington, D.C. Sept., **1 & 2**, 12.2.

Kalvaitis, A. N., F. A. McHale, and L. A. Vega, 1983. "At-sea test of a large-scale OTEC pipe." *Proc. Oceans '83,* Aug., **2**, 734.

Kawano, S., T. Takahata, and M. Miyoshi, 1983. "Performance tests of a condenser for a 100-kW (gross) OTEC plant." *Proc. ASME-JSME Thermal Eng. Joint Conference,* Honolulu, March, **2**, 247.

Keirsey, J. L., J. A. Funk, P. P. Pandolfini, and R. T. Cusick, 1979. "Core unit testing of the JHU/APL shell-less folded tube heat exchanger." *Proc. 6th OTEC Conf.,* Washington, D.C., June, **II**, 11.2

Kincaide, W. C., 1979. "Large alkaline electrolysis systems for OTEC." *Proc. 6th OTEC Conf.,* Washington, D.C., **2**, 10.8-1.

———, and J. N. Murray, 1981. "Advanced alkaline electrolysis systems for OTEC." *Proc. 8th Ocean Energy Conf.,* Washington, D.C., **1**, 685.

Kirkbride, C. G., 1934. "Heat transfer by condensing vapors on vertical tubes." *Trans. AICHE,* **30**, 170.

Kurt, J. R., 1981. "Testing of OTEC riser cable components and models." *Proc. 8th Ocean Energy Conversion Conf.,* Washington, D.C., **2**, 595.

Lachmann, B. A. P., 1978. "European OTEC project." *Proc. 5th Ocean Thermal Energy Conversion Conf.,* Miami Beach, Fla., II-174.

Larsen-Basse, J., 1983. "Effect of biofouling and countermeasures on heat transfer in surface and deep ocean Hawaiian waters—Early results from the seacoast test facility." *Proc. ASME-JSME Thermal Eng. Joint Conf.,* Honolulu, March, **2**, 285.

Lavi, A., 1974. "Heat exchangers and optimum design for solar sea power plants." *2nd Ocean Thermal Energy Conversion Conf. and Workshop Proc.,* Washington, D.C., Sept., 101.

LeRoy, R. L., 1983. "Industrial water electrolysis: Present and future." *Int. J. Hydrogen Energy* **8**, 401.

Lewis, L. G., and N. F. Sather, 1979. *OTEC performance tests of the Carnegie–Mellon University vertical fluted-tube condenser.* Argonne National Lab., May, ANL/OTEC-PS-4.

———, and Sather, N. F., 1978. *OTEC performance tests of the Union Carbide flooded-bundle evaporator.* Argonne National Lab., Dec. ANL/OTEC-PS-1.

Lin, G. W., T. T. Lee, and K. K. Hwang, 1983. "Ocean thermal energy conversion technology development program in Taiwan, Republic of China." *Proc. Oceans '83,* Aug., **2**, 723.

Lindman, R., and R. W. Constantino, 1981. "Methods of predicting loading and bending in an OTEC riser cable." *Proc. 8th Ocean Energy Conf.,* Washington, D.C., **2**, 587.

Little, T. E., J. D. Marks, and K. H. Wellman, 1976. *Deep water pipe, pump, and mooring study: Ocean thermal energy conversion program final report.* Westinghouse Electric Corp., ERDA/C00-2642-3.

Lockheed, 1979. *Preliminary design report for OTEC station-keeping (SKSS).* Lockheed Missiles and Space Co.

———, 1979. *PSD-II final design briefing.* Lockheed Missiles and Space Co., Task 2.

Lorenz, J. J., C. B. Panchal, and G. E. Layton, 1982. *An assessment of heat-transfer correlations for turbulent water flow through a pipe at Prandtl Nos. 6.0 and 11.6.* Argonne National Lab., ANL/OTEC-PS-11.

———, and D. Yung, 1978. "Combined boiling and evaporation of liquid films on horizontal tubes." *Proc. 5th Ocean Thermal Energy Conversion Conf.,* Miami Beach, Fla., Feb., **6**, 46.

———, and D. T. Yung, 1980. "An evaluation of the cost-effectiveness of surface enhancements for OTEC heat exchangers." *Conf. Proc. 7th Ocean Energy Conf.,* Washington, D.C., June, **1 & 2**, 12.3.

———, D. T. Yung, D. L. Hillis, and N. F. Sather, 1979. *OTEC performance tests of the Carnegie–Mellon University vertical fluted-tube evaporator.* Argonne National Lab., July, ANL/OTEC-PS-5.

Loyd, C., 1986, "Hydrogen production processes and facilities." *Hydrogen Technology Conf. Proc.,* NASA Lewis Research Center.

Mailen, G.S., 1980. "Experimental studies of OTEC heat transfer evaporation of ammonia on vertical smooth and fluted tubes." *Conf. Proc. 7th Ocean Energy Conf.,* Washington, D.C., June, **2**, 12.5-1.

Marchand, P., 1980. *Le Energie Thermique des Mer.* IFREMER, 1.

———, 1981. "Travaux francais sur l'energie thermique des mers." *La Houille Blanche,* **4/5**, 315.

———, 1985. "French thermal energy conversion program." *Proc. French–Japanese Symp. on Ocean Development,* Tokyo, Sept., **1**.

Mast, R. F., 1975. "The ARCO LPG vessel." *Conf. on Concrete Floating Ships and Terminal Vessels,* 1.

McGuiness, T., 1982. "Some ocean engineering considerations in the design of OTEC plants." *Proc. IECEC '82 17th Intersociety Engineering Conf.,* **3**, 1423.

———, and A. Griffin, 1979. "Preliminary designs of OTEC cold water pipes for barge and spar-type OTEC plants." *Proc. 6th OTEC Conf.,* **1**, 6.1-1.

McHale, F., and M. Bender, 1979. "Construction and deployment of an operational OTEC plant at Kona, HI." *Proc. 11th Annual Offshore Technology Conf.,* **3**, 1661.

———, 1984a. *OTEC CWP at-sea test program phase II final report.* Hawaiian Dredging and Construction Co.

———, 1984b. *OTEC cold water pipe at-sea test program final report phase II suspended pipe test.* Hawaiian Dredging and Construction Co.

Miyoshi, R., T. Takahata, and H. Mochida, 1977. "Condensation heat transfer of Freon-11."
    Proc. 14th Heat Transfer Symposium (in Japanese).
Moak, K. E., G. M. Hagerman, and J. T. Wu, 1981. "Design and integration of a 40-MWe
    OTEC shelf-mounted pilot plant." *Proc. 8th Ocean Energy Conf.*, Washington, D.C.,
    June, **2**, 133.
Mochida, Y., T. Takahata, and M. Miyoshi, 1983. "Performance tests of an evaporator for
    a 100-kW (gross) OTEC plant." *Proc. ASME-JSME Thermal Eng. Joint Conf.*, Hono-
    lulu, March, **2**, 241.
Mortaloni, L. R., 1983. *CWP at-sea test article.* TRW Internal Report, Hawaiian Dredging
    and Construction Co.
Nunn, G. N., 1981. "OTEC subsea electrical power cable protection systems and technol-
    ogy." *Proc. 8th Ocean Energy Conf.*, Washington, D.C., **2**, 625.
Nuttall, L. J., 1981. "Advanced water electrolysis technology for efficient utilization of ocean
    thermal energy." *Proc. 8th Ocean Energy Conf.*, Washington, D.C., **2**, 679.
Ocean Thermal Corporation, 1982. *Ocean thermal energy (OTEC) pilot plant conceptual
    design study.* Ocean Thermal Corp.
——, 1984. *OTEC preliminary design engineering.* Ocean Thermal Corp., **2.2.1**.
O'Connor, J. S., 1979. *Lightweight concrete development program, phase I.* Johns Hopkins
    Univ. Applied Physics Lab., SR-79-1.
——, 1981. *Lightweight concrete cold water pipe tests phase II.* Johns Hopkins Univ.
    Applied Physics Lab. , SR 80-5A.
Olsen, H. L., and P. P. Pandolfini, 1975. *Analytical study of two-phase-flow heat exchangers
    for OTEC systems.* Johns Hopkins Univ. Applied Physics Lab., AEO-75-37.
Otteman, L. G., R. C. Crooke, and R. Shoemaker, 1985. "Offshore oil and gas technology
    assessment." *Proc. Oceans '85*, 4.
Owens, W. L., 1978. "Correlation of thin film evaporation heat transfer coefficient for
    horizontal tubes." *Proc. 5th Ocean Thermal Energy Conversion Conf.*, Miami Beach,
    Fla., Feb., **6**, 71.
——, 1979. "OTEC system response and control analysis." *Proc. 6th OTEC Conf.*, June, 8.8-
    1.
——, 1983. "Optimization of closed-cycle OTEC plants." *Proc. ASME-JSME Thermal Eng.
    Joint Conf.*, Honolulu, March, **2**, 227.
——, and L. C. Trimble, 1980. "Mini-OTEC operational results." *Conf. Proc. 7th Ocean
    Energy Conf.*, Washington, D.C., June, **2**, 14.1-1.
Panchal, C. B., 1984. "Heat transfer with phase change in plate-fin heat exchangers." *Proc.
    22nd ASME-AICHE National Heat Transfer Conf.*, Niagara Falls, N.Y., August.
——, D. Hillis, L. Seren, D. Yung, J. Lorenz, A. Thomas, and N. Sather, 1981. "Heat
    exchanger tests at Argonne National Laboratory." *Proc. 8th Ocean Energy Conf.*,
    Washington, D.C., June.
——, D. L. Hillis, and A. Thomas, 1983. "Convective boiling of ammonia and Freon 22 in
    plate heat exchangers." *Proc. ASME-JSME Thermal Eng. Joint Conf.*, Honolulu,
    March, **2**, 261.
Pandolfini, P. P., W. H. Avery, and J. Jones, 1977. "Effect of biofouling and cleaning on the
    external heat transfer of large diameter tubes." *OTEC Biofouling and Corrosion Conf.*,
    Battelle Seattle Research Center, Seattle, Wash., Oct.
——, G. L. Dugger, F. K. Hill, and W. H. Avery, 1980. "ALCLAD-aluminum, folded-tube
    heat exchangers for OTEC." *Proc. 3rd Miami International Conf. on Alternative
    Energy Sources*, Miami Beach, Fla., Dec.

——, J. L. Keirsey, and J. A. Funk, 1981. *Testing of a shell-less folded-aluminum-tube heat exchanger core unit as an evaporator and a condenser.* Johns Hopkins Univ. Applied Physics Lab., AEO-81-122.

——, J. L. Keirsey, and J. L. Rice, 1978. "Tests of the JHU/APL heat exchanger concept." *Proc. 5th Ocean Thermal Energy Conversion Conf.*, Miami Beach, Fla., Feb., **6**, 366.

Paulling, J. R., 1970. "Wave induced forces and motions of tubular structures." *Proc. 8th Symp. on Naval Hydrodynamics*, 1083.

——, 1979. "Analytical modelling of the coupled dynamics of the OTEC cold water pipe and platform." *DOE-NOAA Technical Workshop on OTEC Cold Water Pipe Technology Development Program*, Jan., 1.

——, 1980. *Theory and users manual for OTEC cold water pipe programs: ROTECF and SEGPIP.* J. R. Paulling Associates, Berkeley, Cal.

Perry, R. H., and D. W. Green, 1984. *Perry's chemical engineers' handbook*, 6th Ed., New York: McGraw-Hill.

——, 1980. "Material evaluation and testing program for OTEC riser cable." *Proc. 15th Intersociety Energy Conversion Engineering Conf.*, Seattle, **1**, 360.

Pieroni, C. A., R. T. Traut, D. O. Libby, P. Riley, and T. F. Garrity, 1979. "The development of riser cable systems for OTEC power plants." *Proc. 6th OTEC Conf.*, Washington, D.C., **1**, 7.21.

Price, D., and W. Wood, 1980. "The design and station keeping subsystem for OTEC pilot plants." *Proc. 7th Ocean Energy Conf.*, Washington, D.C., June.

Richards, D., E. J. Francis, and G. L. Dugger, 1980. "Conceptual designs for commercial OTEC-ammonia product plantships." 3rd Miami International Conf. on Alternative Energy Systems.

Roberts, R., 1977. "Selected OTEC power plant designs." *Proc. 4th Annual Conf. on Ocean Thermal Energy Conversion*, New Orleans.

Rohsenow, W. M., 1952. "A method of correlating heat transfer data for surface boiling liquids." *Amer. Soc. Mech. Eng.*, **74**, 769.

Rosenblatt and Son, 1980. *OTEC SKSS preliminary design.* M. Rosenblatt & Son, Inc.

Ross, J. M., and W. A. Wood, 1981. "OTEC mooring system development: Recent accomplishments." *Proc. 8th Ocean Energy Conf.*, Washington, D.C., **2**, 85.

Rothfus, R. R., and G. H. Lavi, 1978. "Vertical falling film heat transfer: A literature summary." *Proc. 5th Ocean Thermal Energy Conversion Conf.*, Miami Beach, Fla., Feb., **6**, 90.

——, and C. P. Neuman, 1977. "The OTEC program at Carnegie–Mellon University—Heat transfer research and power cycle transient modeling." *Proc. 4th Annual Conf. on Ocean Thermal Energy Conversion*, New Orleans, March, **6**, 55.

Sabin, C. M., and H. F. Poppendiek, 1977. "Heat transfer enhancment for evaporators." *Proc. 4th Annual Conf. on Ocean Thermal Energy Conversion*, New Orleans, March, **6**, 93.

——, and H. F. Poppendiek, 1978. "Film evaporation of ammonia over horizontal round tubes." *Proc. 5th Ocean Thermal Energy Conversion Conf.*, Miami Beach, Fla., Feb., **6**, 237.

——, and H. F. Poppendiek, 1979. "Ammonia vaporization and condensation investigations related to OTEC heat exchangers." *Proc. 6th OTEC Conf.*, Washington, D.C., June, **1**, 11.9.

Sather, N. F., L. G. Lewis, J. J. Lorenz, and D. Yung, 1978. "Performance tests of 1-MWt OTEC heat exchangers." *Proc. 5th Ocean Thermal Energy Conversion Conf.*, Miami Beach, Fla., Feb., **6**, 1.

Schultz, J. A., T. C. Dalton, and R. Eaton, "Developmental design status of riser cable systems for OTEC pilot plants." *Proc. 8th Ocean Energy Conf.*, Washington, D.C., **2**, 607.

Scott, R. J., 1979. "Conceptual designs and costs of OTEC 10-40-MWe spar platforms." *Proc. 6th OTEC Conf.*, Washington, D.C., **1**, 5.6-1.

——, and W. W. Rogalski, 1978. "Considerations in selection of OTEC platform size and configuration." *Proc. 5th Ocean Thermal Energy Conversion Conf.*, Miami Beach, Fla. **2**, IV-77.

SNAME, 1980. Society of Naval Architects and Marine Engineers, New York.

Snyder, III, J. E., 1976. *Energy utilization aboard OTEC platforms.* Ocean and Energy Systems.

Soden, J. E., D. O. Libby, and J. R. Spiller, 1981. "A study of ocean cableships applied to submarine power cable installation." *Proc. 8th Ocean Energy Conf.*, Washington, D.C., **2**, 617.

——, R. Eaton, and J. P. Walsh, 1982. "Progress in the development and testing of OTEC riser cables." *Oceans '82 Conf. Records*, Marine Tech. Soc., Washington, D.C., **E**, 587.

Stevens, H. C., 1982. "ANL research and development programs at the seacoast test facility." *Proc. Marine Technology Workshop*, Washington, D.C., Sept., 56.

——, 1986. "Updated conceptual design of a 10-MW shore-based OTEC plant." *Proc. Energy Sources Technology Conf. and Exhibition*, New Orleans, Feb.

Stevens Institute, 1975. *Hydrogen storage and transfer.* Stevens Institute of Technology, AD-A016 25.

Suematsu, H., 1974. "Condensation heat transfer on fluted tubes." *Proc. Kakagu Kokaku 8th Shuki* (in Japanese), 402.

Takazawa, H., K. Nishiyama, and T. Kajikawa, 1983. "Condensing heat transfer enhancement on vertical spiral double fin tubes with drainage gutters." *Proc. ASME-JSME Thermal Engineering Joint Conf.*, 269.

Thomas A., D. Hillis, C. B. Panchal, J. J. Lorenz, D. Yung, and N. D. Sather, 1980. "Experimental tests of 1-MWt OTEC heat exchangers at Argonne National Laboratory." *Conf. Proc. 7th Ocean Energy Conf.*, Washington, D.C., June, **2**, 12.1-1.

——, J. J. Lorenz, and D. L. Hillis, 1979. "Performance tests of the 1-MWt shell-and-tube heat exchangers for OTEC." *Proc. 6th OTEC Conf.*, Washington, D.C., June, **2**, 11.1-1.

Tilak, P. W., T. W. Lu, J. E. Colman, and S. Srinivasan, 1981. "Electrolytic production of hydrogen." in *Comprehensive treatise of electrochemistry*, Vol. 2, New York: Plenum Press.

TRW, 1979. *Ocean thermal energy conversion cold water pipe preliminary design project final report.* TRW Energy Systems Group.

——, 1982. *Ocean thermal energy conversion (OTEC) three-dimensional design methodology.* TRW Defense and Space Systems Group. Vol. 1, Technical manual, Vol. 2, User's manual.

Uehara, H., H. Kusuda, M. Monde, T. Nakaoka, and H. Sumitomo, 1983. "Shell-and-plate type heat exchangers for an OTEC plant." *Proc. ASME-JSME Thermal Eng. Joint Conf.*, Honolulu, March, **2**, 253.

——, K. Masutani, and M. Miyoshi, 1978. "Heat transfer coefficients of condensation on vertical fluted tubes." *Proc. 5th Ocean Thermal Energy Conversion Conf.*, Miami Beach, Feb., **6**, 146.

Vachon, R. I., G. H. Nix, and G. E. Tanger, 1968. "Evaluation of constants for the Rohsenow pool-boiling correlation.' *J. Heat Transfer*, May, 239.

Vega, L. A., and G. C. Nihaus, 1985. *OTEC cold water pipe at-sea test program data analysis project, comparisons between measured and predicted barge and pipe response: Evaluation of the NOAA/ROTECF and NOAA/TRW computer models.* Vega and Associates, Honolulu, HI.

Vilain, R. H., M. A. Spielrein, and J. P. Elcano, 1985. "Structure and material for the CWP for OTEC Tahiti experimental plant." *Proc. Ocean Engineering and the Environment (Oceans '85),* San Diego, Nov., **2,** 1217.

Waid, R. L., 1978. "OTEC platform design optimization." *Proc. 5th Ocean Thermal Energy Conversion Conf.,* Miami Beach, Fla., **2,** IV 1-15.

Webb, R. L., 1978. "A generalized procedure for the design and optimization of fluted Gregorig condensing surfaces." *Proc. 5th Ocean Thermal Energy Conversion Conf.,* Miami Beach, Fla., Feb, **6,** 123.

Westinghouse, 1978. *Ocean thermal energy power system development phase I PSD-I final report.* Westinghouse Electric Corp. Power Gen. Div.

Wood, W. A., and D. Price, 1980. "The design of stationkeeping subsystems for OTEC pilot plants." *Conf. Proc. 7th Ocean Energy Conf.,* **1,** 3.4-1.

Wright, D., W. Wagner, J. Shoji, and J. Campbell, 1978. "Summary of analysis and testing on advanced Variflux OTEC scaled evaporators and condenser components." *Proc. 5th Ocean Thermal Energy Conversion Conf.,* Miami Beach, Fla., **6,** 345.

Wu, C., 1987. "A performance bound for real OTEC heat engines." *Ocean Engineering,* **14,** 349.

Yung, D. T., D. L. Hillis, J. J. Lorenz, and N. F. Sather, 1979. *OTEC performance tests of the Union Carbide enhanced-tube condenser.* Argonne National Lab., ANL/OTEC-PS-2.

—, J. J. Lorenz, D. L. Hillis, and C. B. Panchal, 1981. *OTEC performance tests of the Johns Hopkins University Applied Physics Laboratory folded-tube heat exchanger.* Argonne National Lab., ANL/OTEC-PS-9.

Zener, C., and A. Lavi, 1973. "Drainage systems for condensation." *Proc. Solar Sea Power Plant Conf. and Workshop,* Carnegie–Mellon University, Pittsburgh, July, 60.

5

OPEN-CYCLE OTEC

The historical development leading to the proposal by Claude to generate power by producing steam in flash evaporation of warm seawater has been discussed in Chapter 2. In this chapter, the thermodynamic fundamentals of the open-cycle concepts are discussed, leading to a detailed review of state of the art and commercial prospects of the process.

There are several variations on the standard OTEC open-cycle (OC) system. The three major variations are "hybrid cycle" (Bartone, 1978), "mist lift cycle" (Ridgway, 1977), and "foam lift cycle" (Beck, 1975; Zener et al., 1975). These are advanced concepts that offer certain attractive features and are being investigated. The three cycles will be discussed in Sections 5.3, 5.4, and 5.5, respectively. The standard OTEC open cycle is discussed in the following.

5.1   OPEN-CYCLE SYSTEM PERFORMANCE

The modest but nearly steady temperature difference that exists between the warm surface water and the much colder water at great depth in some tropical regions of the world has attracted the attention of many thermodynamicists from the time that these temperature differences were first observed. From the thermodynamicist's view, any significant temperature difference can be used to produce power.

The open or Claude cycle is the forerunner of various OTEC cycles. The open cycle refers to the use of seawater as the working fluid. A schematic diagram of the system, which comprises a flash evaporator, vapor expansion turbine and generator, steam condenser, noncondensables-removing equipment, and deaerator, is shown in Fig. 5-1 (Chen, 1979).

The cycle is a basic Rankine cycle for converting thermal energy of the warm surface water into electrical energy. In the cycle, the warm seawater is deaerated and then passed into a flash evaporation chamber, where a fraction of the seawater is converted into low-pressure steam. The steam is passed through a turbine, which extracts energy from it, and then exits into a condenser. This cycle derives the name "open" from the fact that the condensate is not returned to the evaporator as in the "closed" cycle. Instead, the condensate can be used as desalinated water if a surface condenser is used, or the condensate is mixed with the cooling water and the mixture is discharged back into the ocean.

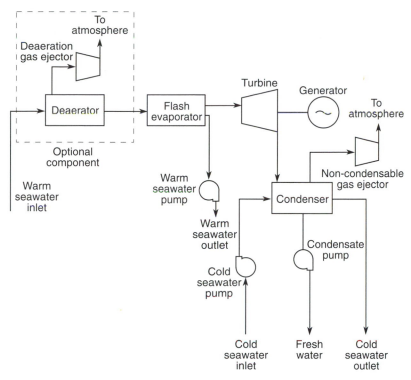

Fig. 5-1. Open-cycle OTEC schematic diagram (Chen, 1979).

A thermodynamic temperature—entropy diagram of the open cycle is illustrated in Fig. 5-2, where the numerals are the thermodynamic states that correspond to those shown in Fig. 5-1. State 1 is the warm surface seawater at about 27°C (80°F) and atmospheric pressure. The seawater flows to the inlet of a deaerator at state 2, where the pressure is suddenly reduced to a value slightly above the saturated vapor pressure at the corresponding temperature. The majority of the dissolved gases, which have a high partial pressure in the water, will be released due to the sudden pressure drop. The vapor release in the deaerator will be minimized because the pressure in the deaerator is kept above saturated vapor pressure. The deaerated seawater then flows into the flash evaporator at state 2, where ambient pressure is dropped to the saturated vapor pressure equivalent to the outlet steam temperature; this pressure drop is the driving force for evaporation. A small amount of working fluid is flashed into steam with flashdown temperature drop corresponding to the vapor latent heat requirement. The bulk thermodynamic state in the flash evaporator is represented by state 3 in Fig. 5-2. The vapor phase of the working fluid at state 3_g with a mass flow rate of $x\dot{m}$ (where $\dot{m}$ is the total warm seawater mass flow rate and x is the vapor mass fraction) expands in a turbine, where the thermal energy is converted into mechanical work. The slightly concentrated seawater with a mass

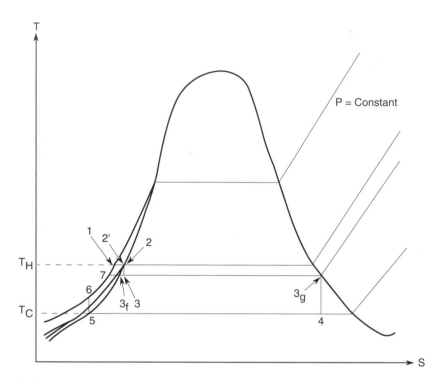

FIG. 5-2. Typical open-cycle OTEC *T-S* diagram.

flow rate of $(1 - x)\,\dot{m}$ at state 3_f is pumped back to the ocean at state 7, where its temperature is lower than the inlet warm seawater temperature. This temperature difference is the "flashdown" of a flash evaporator. The turbine exhaust at state 4 is condensed by giving up heat to the cold seawater and ends up at state 5 as saturated condensate before it is pumped back to the surroundings at state 6. The condensate in a surface condenser is fresh water, which can be utilized as a by-product of an open-cycle OTEC power plant. If the condensate is put back into the ocean, it will eventually be mixed with the seawater and travel along the process path from states 6 and 7 to state 1 through a natural convection mechanism originated from the absorption of solar thermal energy by the earth's atmosphere and the oceans.

   The system temperature and mass flow rate distribution in the open cycle consisting of the flash evaporator, turbine/generator, and surface condenser is shown in Fig. 5-3.

   Neglecting the small sensible heat effect in the steam production and combining the energy and mass balances in the flash evaporator, turbine, and surface condenser gives the rates of heat transfer and power production in the basic open-cycle OTEC system in the following equations:

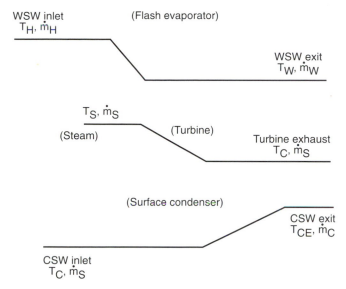

FIG. 5-3.   System diagram and mass flow rate distributions.

$$\dot{Q} = \dot{m}_H C_p (T_H - T_W) \tag{5.1.1}$$

$$P_g = \eta_T \dot{Q}_{in}(1 - T_c / T_H) \tag{5.1.2}$$

$$\dot{Q}_{out} = \dot{Q}_{in} - P_g \tag{5.1.3}$$

$$P_n = P_g - P_{CSW} - P_{WSW} - P_{misc} \tag{5.1.4}$$

$$\eta = P_n / \dot{Q}_{in} \tag{5.1.5}$$

where $\dot{m}_H$ is the warm water mass flow rate, T_H is the warm-water inlet temperature, T_W is the warm-water exit temperature; T_c is the steam condensing temperature; C_p is the specific heat; $\dot{Q}_{in}$ is the rate of heat transfer to the system; $\dot{Q}_{out}$ is the rate of heat removal from the system; η is the efficiency of the system; η_T is the turbine efficiency; P_n is the net power; P_g is the gross power; and P_{CWS}, P_{WSW}, and P_{misc} are the parasitic power loss in cold- and warm-water loops and miscellaneous equipment, respectively.

The principal disadvantages of the open-cycle OTEC system are the low Δp available to operate the turbine, 2.8 kPa (0.15 psi) compared with 270 kPa (49 psi) for the closed-cycle system using ammonia working fluid, and the large specific volumes that must be used by the steam turbine. A low cycle efficiency due to the small temperature difference combined with the large specific volumes requires a very large turbine. So far, no full-scale turbines for plants 1 MW and greater have been designed and operated. A further disadvantage of open-cycle operation is the need to provide vacuum pumps to remove fixed gases that are removed from the seawater along with the steam that forms the working fluid to drive the turbine. If allowed to accumulate in the condenser, these fixed gases would rapidly degrade the operation of the condenser. The power needed to operate the vacuum pumps can significantly reduce the net power output.

On the other hand, for a given temperature difference, the open-cycle OTEC system is thermodynamically superior to the closed-cycle system because a larger portion of the temperature difference between the hot and the cold water is available to produce net power. The reason is that in the direct-contact heat transfer mode of the evaporator and condenser used in the open-cycle system, the steam in the vacuum chamber can approach the temperature of the warm-water discharge from the evaporator (Kreith and Bharathan, 1987). At the same time, the exhaust steam from the turbine can be condensed at a temperature approaching the condenser exit temperature of the cold water from the deep ocean. In the closed-cycle system, the working fluid passing through the turbine is separated from the warm and cold waters of the ocean by a solid wall. Hence, in both the evaporator and the condenser, a first temperature drop is necessary to transfer the heat from the ocean water through the wall to produce the vapor in the boiler, and then another temperature drop is required to condense the vapor of the working fluid at the outlet from the turbine by cold seawater. Consequently, the heat exchangers for closed-cycle OTEC plants require very large surface areas to minimize the thermodynamically required temperature losses. These exchange surfaces are continuously exposed to corrosion and biofouling in the ocean environment. A large fraction of the heat exchanger cost is saved if the Claude process is used. Moreover, in the open cycle one need not consider the possible loss of volatile or hazardous working fluids. A supporting reason for building OTEC power plants with open-cycle power generators (Hagen, 1975) is the additional use of cold water for fish farming. However, all OTEC designs make use of massive amounts of cold deep ocean water; therefore, no one design possesses any significant advantage over any other with respect to mariculture considerations.

An analytical model of the power cycle and seawater supply systems for an OC OTEC power plant that allows ready examination of the effects of system and component operating points on plant size and parasitic power requirements has been developed by Parsons et al. (1984, 1985). Johnson (1978) derived a formula to compute the maximum amount of work that can be extracted from a given combined mass of warm and cold ocean water in OC OTEC operation.

To summarize, the Claude-cycle OTEC offers the following advantages:

1. Less loss in temperature due to the use of heat exchangers (increased power production);
2. Reduced heat exchanger expenses, depending on whether or not fresh water is produced (requiring surface condensers);
3. Relative immunity to biological fouling of surfaces.

While the past work of researchers in this area shows the overall feasibility of such a system, a need for a more detailed technical updating of the system and its components exists. Even though almost 50 years of engineering progress have elapsed since the time of Claude, many of the problems that he faced will not be overcome by a mere updating of his work.

5.2   OPEN-CYCLE RESEARCH AND DEVELOPMENT STATUS

The following presentation is a summary of the status of open-cycle research and development that identifies specific technical problem areas on which future work must concentrate. This future research may circumvent the problems with present designs.

The major components of OC OTEC systems are:

1. Evaporator;
2. Condenser;
3. Turbine;
4. Fixed gas exhaust systems;
5. Water pumps;
6. Control systems.

Each of these components is discussed in the following sections.

5.2.1 Evaporator

The evaporator supplies the OC OTEC plant with low-pressure steam produced by vaporizing warm seawater. The evaporator is a vacuum chamber in which the pressure is maintained below the saturation pressure (1/20 atmosphere pressure) of the warm seawater. As the superheated seawater enters the low-pressure chamber, some of it flashes to steam.

Assuming an adiabatic evaporator and that the system comes to thermodynamic equilibrium, energy balance of the flash process gives

$$\dot{m}_s H_s + (\dot{m}_H - \dot{m}_s) C_p T_W = \dot{m}_H C_p T_H \qquad (5.2.1)$$

$$H_s = C_p T_E + H_{fg} , \qquad (5.2.2)$$

where $\dot{m}_H$ is the warm seawater mass flow rate, $\dot{m}_s$ the steam mass flow rate, C_p the specific heat capacity of seawater, T_H the warm seawater inlet temperature, T_W the warm seawater exit temperature, T_E the saturated temperature at the evaporator pressure, H_{fg} the steam enthalpy of evaporation at T_E or at p_E, and H_S the enthalpy of the steam leaving the evaporator, as shown in Fig. 5-4.

The mass rate of steam flow generated by the evaporator is

$$\dot{m}_s = \frac{\dot{m}_H C_p (T_H - T_W)}{H_{fg} - C_p (T_W - T_E)} . \qquad (5.2.3)$$

Because the available temperature drop of the seawater is limited by the overall cycle temperature difference, the ratio of $\dot{m}$ is very small. For $(T_H - T_W)$ equal to 3, approximately 0.005 kg of steam is produced for each kilogram of warm seawater. The required volumetric flow of warm seawater is therefore quite large,

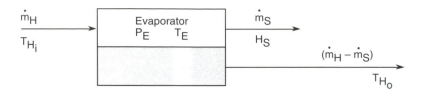

FIG. 5-4.   Flash evaporator.

and it is important that the water be used effectively and that the liquid head losses or pressure drops be minimized. A well-designed evaporator delivers steam with an acceptably low liquid content to the turbine and/or condenser(s), while minimizing the overall cost. The vacuum boundary of the evaporator is usually incorporated into the structure of the plant. A large proportion of the cost of the evaporator system is the cost of the structural pressure boundary required to enclose it.

The water distribution system enhances evaporation effectiveness while limiting head losses and platform area. Generally, evaporation is enhanced by a large interfacial area at which vapor can be formed. Adequate superheating of the inlet water (relative to saturation conditions in the evaporator) must also be allowed in order to yield a high effectiveness. The effectiveness is defined as:

$$\eta_E = \frac{(T_H - T_W)}{(T_H - T_E)}, \qquad (5.2.4)$$

where T_E is the saturation temperature at the evaporator pressure. An effectiveness as high as 0.95 can be achieved (Sam and Patel, 1984; Bharathan et al., 1984).

Head losses in the feed water flow path must be limited so that parasitic losses do not overly degrade the system performance. High-pressure nozzles, for example, are not appropriate in the OC OTEC evaporator. Head losses of 1 m of water or less are most desirable.

Head losses in the steam flow path are a function of steam velocity and pressure loss, and entrainment of seawater droplets. SERI has conducted a preliminary study of the drop size distribution in OC OTEC evaporators (Bharathan et al., 1983). Droplet diameters range from fractions of a millimeter to several millimeters.

Several types of flash evaporators including horizontal open channels, vertical falling films, vertical falling jets, and vertical spouts have been considered by the recent open-cycle OTEC studies.

The Colorado School of Mines study (Watt et al., 1978) suggested bubble caps or tubes as a possible flash evaporation method. This type was never employed in saline water conversion plants, and very few prediction techniques exist in the open literature.

The Westinghouse study (1978) selected a flash evaporator for its OTEC power module. The flashing brine enters the flash chamber under a sluice gate. It then flows in a nearly horizontal channel to the exit. For the single-stage flash evaporator, no sealing is required. As a result, the brine free falls from the channel to a brine collector and is discharged back into the ocean at sufficient depth to preclude mixing with inlet flows. This concept has been successfully used in multistage flash evaporators and is known to be a low-pressure-drop option. However, horizontal channels are unlikely to perform at nearly theoretical effectiveness due to rapid suppression of evaporation caused by hydrostatic pressure in the channel. The effectiveness of the Westinghouse evaporator was estimated to be about 80%.

The Westinghouse flash evaporator is a toroidal open-channel flow type. The turbine is located on the center of the vertical axis and above the plane of the toroid. Steam released in the flashing process flows vertically upward from the open channel and then inward and down to the turbine, as shown in Fig. 5-5.

Equations of the sluice gate volumetric flow, empirical discharge and contraction coefficients, mass balance, energy balance, pressure summation, temperature summation, generated steam vapor-brine ratio, and generated steam vapor—air ration of the flash evaporator of the Westinghouse 100-MW OTEC design have been derived. The pressure drop, the steam vapor production flow, the air production flow, and the size of the multistage flash evaporator have been calculated and

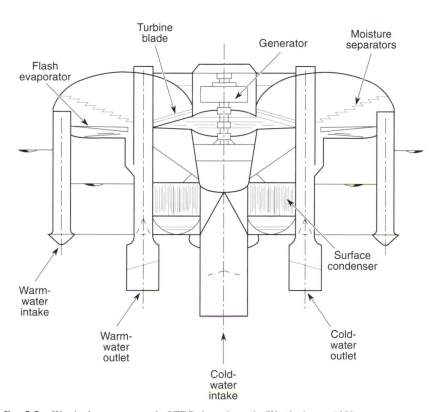

FIG. 5-5. Westinghouse open-cycle OTEC plan schematic (Westinghouse, 1978).

designed based on those equations, and several control factors including specific mass velocity, interfacing constraints, a moisture carryover removal device, and the degree of thermal nonequilibrium in the flash evaporation process. At the low pressures and temperatures experienced in an OC OTEC plant, flashing can be conservatively conceived as a surface evaporation process. Assuming a surface evaporation process, Westinghouse obtained results for a rectangular, open-channel flash evaporator with a baseline design tray length of 20 m (65.6 ft). An overall platform cost and volume as functions of flash evaporator specific mass flow rate was developed by Westinghouse, who estimated the platform cost as $495/kW (1977 value) and the approximate platform outside diameter as 106.7 m (350 ft).

The University of Massachusetts study (Boot and McGowan, 1974) selected a vertical falling film flash evaporator for the open-cycle power module. The disadvantages of this design are the large free-fall height and the high turbine cost. The free fall is lost head and must be compensated by pumping power. The effectiveness of the falling film evaporators was estimated by Bugby et al. (1983) and ranged from 0.75 to 0.95.

The vertical spout evaporator originally suggested by Claude combines the unobstructed vapor path of the open-channel configuration with the large interfacial surface area generated by the falling-jets type. Effectiveness ranging from 0.85 to 0.97 was measured by Bharathan et al. (1983) and Ghiaasiaan and Wassel (1983).

The channel-flow flash evaporator (Lewandowski et al., 1982) and the vertical spout evaporator performance analysis have been performed by Lewandowski et al. (1982) and Block (1984), respectively.

Evaporation from falling superheated water jets for application to OC OTEC power plants was considered by Wassel and Ghiaasiaan (1985), Green et al. (1981), Kogan et al. (1980), and Fournier (1985). Analyses and experimental results were performed to show that the interfacial resistance was of no importance to evaporator design and that evaporation was liquid-side controlled. The heat exchanger performance was presented in terms of its effectiveness and change of bulk temperature.

Mass transfer (Leninger, 1983) and experimental (Ofer, 1983) evaporation from round turbine water jets of varying velocity, diameter, and length have been studied. Experimental data for liquid-side mass and heat transfer coefficients have been presented.

A heat exchanger evaluation of old and new concepts was carried out by Bell (1983) to update information on the OTEC heat exchanger.

A sketch (Kreith and Bharathan, 1987) of the typical spout evaporator is shown in Fig. 5-6. The sequence of spout appearance (Larsen-Basse et al., 1986) is shown in Fig. 5-7. In the initial stage, before boiling has begun, the water exits the spout to form an umbrella-shaped sheet (Fig. 5-7a). At lower vacuum levels bubbles begin to develop in the spout upcomer, and at a certain point bubbles also form in the umbrella-shaped sheet to the point where the sheet has a honeycomb structure (Fig. 5-7b). As the pressure drops slightly farther, vigorous boiling sets in, starting in the upper part of the upcomer, and droplets are ejected into the chamber at substantial velocity (Fig. 5-7c).

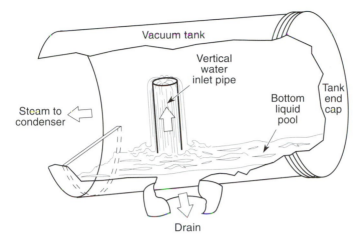

FIG. 5-6.   Sketch of a single spout evaporator (Kreith and Bharathan, 1987).

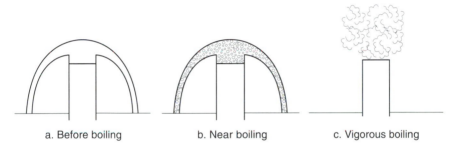

a. Before boiling   b. Near boiling   c. Vigorous boiling

FIG. 5-7.   Sequence of spout flow patterns (Larsen-Basse et al., 1986).

5.2.2 *Condenser*

After expanding through the steam turbine of the open-cycle OTEC system, the steam is condensed in either a surface condenser or a direct-contact condenser by cold water from the ocean depth. Typically, the surface condensers are shell-and-tube configurations with cold water flowing inside the tubes and steam condensing on the outside of the tubes. Data on the performance of the surface condenser are available. For the expense of adding a surface condenser, the open-cycle OTEC plant can be configured to provide fresh water in addition to power. The surface condenser consists of a heat transfer surface across which the heat released by condensing steam is transferred to the cold seawater. A potential advantage of the open-cycle OTEC systems is that it allows the use of direct-contact condensers instead of the conventional surface heat exchangers required in closed-cycle OTEC systems. Direct contact heat transfer results in a lower temperature difference between the water vapor working fluid and the seawater, which implies a higher conversion efficiency, lower seawater flow requirement, and lower capital cost. However, this system cannot provide fresh water as a product. These two types of condensers are discussed in the following sections.

5.2.2.1 Direct Contact Condenser

Direct contact condensers were used with steam engines and turbines to obtain low exhaust pressures and high thermal efficiencies; however, with the adaption of closed-cycle boiler water systems and the need for boiler feedwater treatment, they were replaced with surface condensers. The use of surface condensers became standard with the introduction of high-temperature steam turbines around 1900. From that time, the importance of direct-contact condensers in power plants steadily decreased.

There are little general design data for direct-contact condensers. They are mostly designed using empirical data for particular applications, sizes, and fluids. Design work has generally been carried out by "rules of thumb." The open-cycle OTEC application of a direct-contact condenser is unique because of the large amount of noncondensable gases, low-temperature driving force, and low pressure level.

The direct-contact condenser for the open-cycle system may be a "jet"-type or a "spray"-type condensing unit. There is no surface separating the steam from the coolant, and inefficiencies associated with the thermal resistance of a heat exchanger surface are eliminated. In addition, the expense of a heat exchanger surface is eliminated in the direct-contact condenser. Because the condensed steam is mixed with the cold seawater, the direct-contact condenser does not produce fresh water.

As the cold seawater absorbs the latent heat of the condensing steam, its temperature rises, and an approximate heat balance yields

$$\dot{m}_L = \frac{\dot{m}_c C_p (T_L - T_c)}{H_{fg}},$$
(5.2.5)

where $\dot{m}_L$ is the mass rate of steam condensed into liquid, $\dot{m}_c$ is the mass flow rate of cold seawater, C_p is the heat capacity of cold seawater, T_L is the exit temperature of cold seawater, T_c is the inlet temperature of cold seawater, and H_{fg} is the latent heat of vaporization of steam.

If the coolant temperature rise is 5°C (9°F), roughly 100 kg (220 lb) of cold water is required for each kilogram of steam condensed. In a typical application, cold seawater is available at 5°C (41°F), and the steam enters the condenser with a saturation temperature between 10°C (50°F) and 15°C (59°F) [corresponding to a vapor pressure between 1.2 kPa (0.1741 psi) and 1.7 kPa (0.2467 psi)].

The condenser performance (Block, 1984) can be characterized by an effectiveness defined as

$$\eta_c = \frac{T_L - T_c}{T_s - T_c},$$
(5.2.6)

where T_s is the saturation temperature of the vapor in the condenser. As η_c approaches unity, the cold seawater has absorbed as much energy from the condensing steam as possible since, as T_L approaches T_s, condensation will cease.

Condenser effectiveness is a function of the condenser pressure, subcooling, water flow rate, and steam load, as well as of the condenser geometry. The concentration of noncondensables in the condenser will also have an impact because of the tendency for a film of noncondensable gas to build up on the condensing interface. This layer of noncondensable gases poses an additional resistance to heat transfer between the steam and cold seawater. It is important that noncondensable gases be swept to a region of the condenser from which they can be removed from the system. Noncondensable gases will be present in the steam and will also be evolved from the cold seawater as it enters the condenser.

In addition to high effectiveness, a well-designed direct-contact condenser should have a minimal head loss in the cold seawater flow in order to minimize the required pumping power.

The Westinghouse 100-MW OC OTEC study selected a spray-droplet-type direct-contact condenser, as shown in Fig. 5-8. The flash evaporator and direct-contact condenser are at essentially the same elevation, which is at the barometric height above sea level. The volumetric heat transfer coefficient, U, and the maximum vapor velocity, U_g, correlation are given by Coons (1957)

$$U = 1500\Delta T(p/p_0)^{0.2} , \tag{5.2.7}$$

where U is the volumetric heat transfer coefficient (kWe/m^3 °C); ΔT the temperature driving force (°F); and p/p_0 the condenser pressure/atmospheric pressure ratio,

$$U_g = K_e \left(\frac{\rho_l - \rho_g}{\rho_g} \right)^{1/2} ; \tag{5.2.8}$$

where K_e is the entrainment coefficient (usually 0.05–0.15 m/s), ρ_l the liquid density (kg/m^3), and ρ_g the vapor density (kg/m^3).

The initial design height of the Westinghouse direct contact condenser is 0.884 m (2.9 ft) for a steam flow rate 1.36×10^6 kg/h (3×10^6 lb/h), and a 0.278°C (0.5°F) terminal temperature difference with a cold seawater flow rate of 270×10^6 kg/h (594×10^6 lb/h). The inlet and saturation temperatures of the cold seawater are 5°C (40°F) and 7.78°C (46°F), respectively. Note that the cold seawater flow rate required is reduced (compared with that of a surface condenser) because of the much smaller terminal temperature difference.

Gases from the flashed hot seawater, from air leakage into the OTEC OC vessel, and from air liberated in the direct-contact condenser must be considered in determining the size of the unit. A large portion of the cost of the unit is the cost of the vacuum boundary required to contain it. Thus the unit should be small to be cost effective.

Steam condensation on falling water jets has been studied by many investigators in recent years. Experimental programs have produced correlations relating the

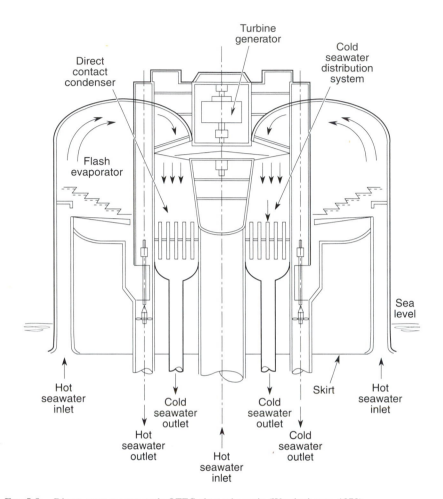

Fig. 5-8.   Direct-contact open-cycle OTEC plant schematic (Westinghouse, 1978).

Stanton number to such parameters as the jet Reynolds number, Prandtl number, Weber number, jet length and thickness, coolant subcooling, and air content of the steam. The Stanton number is defined as:

$$St = \frac{h}{\rho_c v_j C_p},$$
(5.2.9)

where, h is the heat transfer coefficient based on jet surface area (W/m^2 °C), ρ_c the coolant density (kg/m^3), V_j the jet velocity (m/s), and C_p the coolant heat capacity (J/kg °C).

With a constant saturation temperature in the vapor phase, the condenser effectiveness can be related to the Stanton number by:

$$\eta_c = 1 - \exp(-2 \, \mathrm{St} \, L/t), \tag{5.2.10}$$

where L is the jet length (m) and t the jet thickness (m).

Bharathan (1981) and Sam et al. (1982) have studied falling jet condensers under OC OTEC conditions. Bharathan et al. measured the performance of multiple planar jets 3.2 mm thick and 76 cm high with concurrent vapor flow. The Stanton number was a weak function of the jet Reynolds number, coolant temperature, or steam loading. With no noncondensables present, the data showed:

$$\mathrm{St}_0 = 0.0017 \tag{5.2.11}$$

or a condenser effectiveness equal to 0.55.

Sam et al. (1982) studied planar jets with vapor in cross flow. Performance of the jets was insensitive to coolant inlet temperature, vapor pressure, subcooling, or salinity. Performance was unaltered in a test of multiple jets. Heat transfer increased with the jet velocity. The data were correlated, for zero noncondensable content, as:

$$\begin{aligned} \mathrm{St}_0 &= 0.075 \, \mathrm{Re}^{-0.35} \, \mathrm{We}^{0.3}, \quad \mathrm{We} < 1 \\ &= 0.075 \, \mathrm{Re}^{-0.35}, \quad 1 < \mathrm{We} < 130, \end{aligned} \tag{5.2.12}$$

where $\mathrm{Re} = \rho_c V_j 2t/\mu_c$, $\mathrm{We} = \rho_{stm} t V_{stm} 2/\sigma$, V_j is the jet velocity, t is the jet thickness, ρ_c and μ_c are the cold seawater density and viscosity, respectively, ρ_{stm} and V_{stm} are the steam density and velocity, respectively, and σ is the surface tension of the seawater in steam. At typical test conditions this correlation yields

$$\mathrm{St}_0 = 0.002, \tag{5.2.13}$$

or a condenser effectiveness equal to 0.32.

Heat transfer of the condensation of steam on turbulent water jet was investigated by Benedek (1980). Condensation of steam on subcooled water jets was reviewed by Saha (1979).

An experimental investigation of OC OTEC direct-contact condensation and evaporation processes was performed by Sam and Patel (1984). Falling turbulent jets and films were tested at typical operating conditions. Condenser heat transfer coefficients of the order of 27 kW/m² °C were achieved with jets, which were higher than those obtained with films. Empirical correlations were developed for the condenser data.

These various correlations in the form of dependence on heat transfer, hydraulic, and geometric parameters were developed to predict the performance of the direct-contact condenser. These nondimensional numbers imply a compact package, low fluid head loss, and high heat transfer effectiveness for a well-designed direct-contact condenser.

5.2.2.2 Surface Condenser

The surface condenser consists of a heat transfer surface across which the heat released by condensing steam is transferred to the cold seawater. Typically, power plant surface condensers are shell-and-tube configurations, with cold water flowing inside the tubes and steam condensing on the outside of the tubes (shell side). Thus the surface condenser configuration (Block, 1984) would be a shell and tube with condensation occurring on the shell side. The steam pressure in a typical OC OTEC surface condenser is on the order of 1.500 kPa (0.22 psi), corresponding to a saturation temperature of approximately 12°C (54°F). The surface condenser heat transfer can be characterized by the log mean temperature difference, or LMTD, defined as:

$$\text{LMTD} = (T_L - T_c) / \ln[(T_s - T_L)/(T_s - T_c)], \qquad (5.2.14)$$

where T_L is the cold seawater exit temperature, T_c the cold seawater inlet temperature, and T_s the steam saturation temperature.

With cold seawater available at 5°C (41°F), and assuming a coolant temperature rise of 5°C, the LMTD in the OC OTEC surface condenser is roughly 4°C (7.2°F), compared to 5 or 6°C in conventional power plant condensers. Thus the driving force for heat transfer is reduced by 20 to 33% in the OC OTEC surface condenser. Because of the small available temperature difference, a large surface area is required for heat transfer and the ratio of cooling water to steam condensed is quite high.

Noncondensable gases degrade condensation heat transfer by collecting around condenser tubes as the steam condenses. This implies an added resistance to heat and mass transfer through the layer of noncondensable gases. The impact of noncondensables must be added to the thermal resistance of the condensate film, the tube wall, and the coolant inside the tubes.

Fouling of condenser tubes has been anticipated as an added resistance to heat transfer. Typical fouling resistances when seawater at normal temperature is used as a coolant are on the order of 0.0001 W/m^2 K (Fraas and Ozisik, 1965). Cold seawater, typical of OTEC operation, has been shown to produce negligible fouling compared to typical shore-based power plants because of its purity and lack of biological activity. Larsen-Basse (1983) observed no fouling with cold deep-ocean water in a test that extended for 1000 days.

The choice of tube material in the surface condenser must be made on the basis of cost and anticipated durability. Enhanced tube surfaces, employing internal fins and/or corrugations on the condensate film side, may be cost effective for OC OTEC systems. The tradeoff is one of enhanced heat transfer performance versus increased capital costs and increased parasitic losses caused by additional friction in the tubes.

Large-scale steam surface condensers have been employed in the power industry for many years. There exists a large base of condenser performance data and design experience. In part, standard design considerations can be applied to the OC OTEC surface condenser. Manufacturers typically include steam lanes to direct the vapor flow into the tube bank and to ensure that noncondensable gases are swept to a point of removal. Baffles are commonly employed to direct the vapor through the coldest region of the condenser just prior to the point of noncondensable removal. By rule of thumb, steam velocities of at least 30 m/s (98.4 ft/s) are maintained in order to effectively sweep the noncondensables to the exit.

In other respects, the OC OTEC surface condenser lies outside the ordinary design limits for power plant condensers. The vapor pressure is no greater than 1.500 kPa (0.22 psi), compared to 3.4–6.8 kPa (0.50–1.0 psi) in more standard low-pressure steam applications. The total temperature difference between steam and cooling water is 5 to 7°C (9 to 12.6°F) at OC OTEC conditions, while 12°C (21.6°F) or more is typical in the power industry. Thus the driving force for heat transfer is over 50% lower than in standard steam condensers. Similarly, the available coolant temperature rise is roughly halved, implying a higher mass flow rate of coolant. The OC OTEC condenser is also faced with a higher-than-normal noncondensable mass fraction in the inlet steam (0.2 to 0.4% anticipated).

No full-scale steam surface condenser performance data are available at OC OTEC conditions. Most experimental programs have been conducted outside the extremely low pressure and low temperature difference range of OC OTEC systems. A brief review of analytical approaches and experimental results for the OC OTEC surface condenser design is available (Block, 1984).

An analytical approach to condenser analysis or design includes the thermal resistance of the condensate film and the tube wall, forced convection to the cooling water, and the effect of noncondensable gases in the vapor phase. The standard analytical technique is to use Nusselt's theory and the Colburn-Hougen method to calculate forced convection inside the tubes using an equation such as the Dittus–Boelter correlation.

Experimental results by Sklover and Grigorev (1975), Fuks and Bernova (1970), Bobe and Malyshev (1971), Buglaev et al. (1971), Mills et al. (1974), Henderson and Marchello (1969), etc. have correlated the mass transfer Nusselt number and friction factor for steam in air as a function of Reynolds number and Prandtl number at various conditions.

The water-side heat transfer and pressure drop can be calculated using standard thermohydraulic equations discussed in detail in Chapter 4.

Eight surface enhancement arrangements were considered as promising for the surface condenser in the Westinghouse 100-MW OC OTEC system design. Tube materials selected were 90-10 Cu-Ni, aluminum, and titanium. Enhanced surfaces considered were applied coatings, fins, corrugations, and flutes. Only those surfaces that had comparable test data available and had been manufactured in sufficient quantity were selected by the Westinghouse project. The eight steam condensers are listed in the report from Westinghouse Electric Co. (1978). The inside heat transfer coefficients and friction factors, the tube metal wall heat resistance, the outside heat transfer and pressure drop, surface area calculation, weight, and cost of the OC OTEC surface condenser are also discussed (Westinghouse Electric Co., 1978).

The University of Massachusetts study (1974) chose a small 1-in.-diameter tube with a low inside velocity of 3.05 m/s (10 ft/s). The shell diameter is 14.8 m (48.57 ft); total number of tubes is 154,276; the core length is 12.4 m (40.76 ft); the overall heat transfer coefficient is 3.98 kW/m² °C (686.3 Btu/h ft² °F); the total heat area is 201,200 m^2 (2,165,00 ft^2); the core pressure drop is 45.9 kPa (6.80 psi); and the cooling water pumping work required is 13.6 MW. The turbine exhaust mass flow rate is 1309 kg/s (2.88 × 10^3 lb/s); and the mass flow rate of the cold ocean water is 2.70 × 10^5 kg/s (5.95 × 10^5 lb/s) for the 100-MW OC OTEC preliminary design project.

Results of experiments and analyses that determine jet flow distribution from slotted pipes of dimensions typical for OC OTEC condensers are described by Olson (1981). Apparatus for OC OTEC heat and mass transfer experiments, including single- and three-vertical spout, deaeration, and scoping test of a surface condenser at the sea coast test as applied to OC OTEC condensers, is also given by Panchal (Panchal and Bell, 1984; Panchal and Stevens, 1986). The state of the art of heat and mass transfer related to condensation of steam at low pressure in OC OTEC is summarized by Bharathan et al. (1984). These experimental data on steam condensation can be extended to the conditions encountered in OC OTEC systems. Design of the condenser can be based on established analytical techniques such as the Nusselt analysis.

5.2.3 Turbine

The large volumetric rate of steam flow necessary in the open-cycle OTEC system results in a unique large and expensive turbine. The turbine is a significant element of the open-cycle OTEC system (Westinghouse Electric Co., 1978). Several investigators have stated that the cost of the turbine alone made the open-cycle OTEC power system uncompetitive. However, through the recent research, introduction of unique design concepts, development of materials, fabrication, and other advances in technologies, turbines for the open-cycle OTEC system could be both economically and technically feasible.

The turbine system consists of the turbine rotor, stator, bearing, generator, and power conditioning equipment. The steam flows from the evaporator through a fixed blade row (stator), which imparts tangential velocity to the flow. The angular momentum is extracted by the rotor, and the nominally axial discharge flow passes through a diffusing section to the condenser.

A thermodynamic analysis of turbine performance and geometry is described in detail by Lewandowski et al. (1980).

Selection of a particular type of turbine for a specific application depends on the maximum efficiency potential for the turbine type at the operating conditions.

The specific speed, N_s, and specific diameter, D_s, in terms of rotational speed in rpm (N), exit volumetric flow in m^3/s (Q_e), tip diameter in m (D), and isentropic enthalpy drop across the machine ΔH in m, are given by Coleman et al. (1981)

$$N_s = 2.44 N Q_e^{1/2} (\Delta H)^{3/4} \qquad (5.2.15)$$

$$D_s = 0.743 D (\Delta H)^{1/4} (Q_e)^{-1/2}. \qquad (5.2.16)$$

For an OC OTEC turbine operating with nearly saturated steam at 21°C inlet temperature and an 8 to 10°C temperature drop, axial flow turbines are chosen

(Parsons et al., 1985) because the most efficient turbine at the OC OTEC operating condition is an axial flow turbine with an efficiency of around 80%. There is also a large specific speed range over which high efficiencies are possible with axial flow turbines.

The aerodynamic design of axial turbines is a mature technology. Sophisticated design methods, based on fundamental physics and buttressed with experimental factors, have been developed by the major turbine manufacturers. The boundaries of these design methods are set by the major nondimensional groups: specific speed, specific diameter, Mach number, and Reynolds number. If the operating conditions, overall diameter, and speed fall inside these boundaries, a detailed blade design effort will succeed in producing an aerodynamic efficiency exceeding 80% (Block, 1984).

The mechanical design of turbines large enough for OC OTEC systems remains speculative. The largest turbine today is the last low-pressure stage of a nuclear power plant. These steel alloy (stainless steel) blades are up to 1.1 m long and are attached to a disk of approximately 2.0 m diameter. Such a stage run at typical OC OTEC conditions would generate about 1.6 MWe. (In the typical dual flow configuration it would generate about 3 MWe.)

For larger OTEC plants, Westinghouse (1978) has suggested fiberglass-reinforced plastic (FRP) rotor blades. Their study covered OC OTEC turbines to 140-MWe gross power output. Extensive review of aerodynamics, aeroelasticity, stress, fabrication techniques, and cost has uncovered no barrier to the production of such blades. However, construction of an FRP turbine must be regarded as a major development effort.

The Westinghouse (1978) turbine design uses three basic building blocks that form its OC OTEC turbine model. The three blocks are turbine performance model, turbine mechanical model, and turbine cost model. The turbine mechanical model is used to determine the basic geometric turbine parameters, including the blade diameter, blade height, thickness, etc. To meet the thermodynamic and fluid dynamic requirements, the turbine mechanical model determines stresses such as the blade bending stress, centrifugal stress, and tangential stress. The model makes sure that these stresses are within acceptable levels at industry accepted factors of safety. The turbine cost model predicts the overall cost of potential design alternatives. Exercising the three models over a wide range of pertinent design parameters, Westinghouse provided the information to develop an excellent conceptual turbine design (Coleman et al., 1981).

The University of Massachusetts (Boot and McGowan, 1974) turbine design is based on a 1.36-kPa (0.2-psi) pressure head across the turbine and a steam mass flow of gross output power. An axial flow turbine with a specific speed of 120 and specific diameter of 1.2 is designed to rotate at 139 rpm and has a turbine blade diameter of 58 m (191 ft) for a net power of 100 MW per OC OTEC plant. A radial inflow turbine with a specific speed of 80, specific diameter of 1.4, turbine blade of 68 m (223 ft), and 92.8 rpm rotating speed was also investigated. Since the size of one turbine for the OC OTEC cycle is extremely large, it might be necessary to have several smaller turbines instead of one large turbine (Lewandowski et al.,

1980; Watt et al., 1978). Analysis has shown that the relative size of five individual axial and radial turbine sizes are 26.2 m (86 ft) and 30.5 m (100 ft), respectively.

Rosard (1978) examined turbine sizes, speeds, and efficiencies for various candidate working fluids for OC OTEC power systems. The turbine performance and design limits were found to be strongly influenced by blade stress criteria, which had been ignored previously by investigators.

The design of a low-pressure OC OTEC plant is described by Coleman et al. (1981). Small-scale turbine rotors made from composites (Penney, 1986) offer several technical advantages for an OC OTEC plant. Penney (1986) designed a composite turbine rotor using state-of-the-art analysis methods for large-scale (100-MW) OC OTEC applications. Successful near-term tests using conventional low-pressure turbine blade shapes with composite materials would demonstrate feasibility and modern credibility of the OC OTEC power plant. Application of composite blades for low-pressure turbo-machinery potentially improves the reliability of conventional metal blades affected by stress corrosion. A partial rotor assembly of the Penney design is shown in Fig. 5-9.

Preliminary studies were being conducted by University of Pennsylvania, Texas A & M, and Massachusetts Institute of Technology in 1989 to identify and evaluate advanced turbine concepts for the open-cycle power system. These steam turbines have the potential to reduce the cost of OTEC electrical power plants

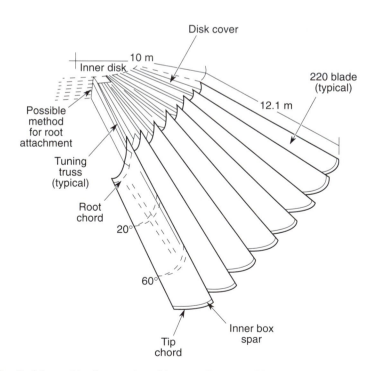

Fig. 5-9. Partial assembly of composite turbine rotor (Penney, 1986).

significantly. The three concepts are (1) a cross-flow turbine, (2) a vertical-axis, axial-flow turbine, and (3) a double-flow, radial-inflow turbine with mixed-flow blading. In all cases, the concept involves the use of light-weight, composite plastic blading and a physical geometry that facilitates efficient fluid flow to and from the other major system components and reduces the structural requirements for the turbine or the system vacuum enclosure, or both. These concepts were all found to be workable. Key issues with respect to cost reduction remain on overall system integration and use of low-cost composite material for blading.

5.2.4   *Fixed Gas Exhaust Systems*

Dissolved oxygen and nitrogen gases are present in seawater. Some of these gases evolve when the seawater enters a containment power module under vacuum. Air leakage also exists because the containment is under vacuum. These gases significantly lower the heat transfer coefficients on the vapor side in the condenser and decrease the pressure drop across the turbine. This problem must be overcome by continuous operation of the vacuum pump in the condenser chamber of the system containment structure. These gases must also be removed to maintain the vacuum of the containment power module. The power requirement of the vacuum pump is also significant.

Oxygen and nitrogen are present in the seawater at normal OTEC operating temperature at 6.4 and 10.8 ppm, respectively (Dukler, 1971; Watt et al., 1977). The air leakage rate is assumed to be 1820 kg/h (4000 lb/h) regardless of the power module size (Boot et al., 1974).

For the open cycle with a surface condenser, conventional shell-and-tube condensers are used. The gases in the cold seawater will remain in solution since the cold seawater always remains in closed pressurized conduits. Only the gases in the hot seawater and the air leakage must be removed. For the open cycle with a direct-contact condenser, the gases liberated by the cold seawater also must be removed.

The gases in the warm seawater can be removed in one or more deaerators located upstream of the flash evaporator. The remaining gases plus the air leakage must pass through the system along with the steam and must be removed continuously from the condensers. Sophisticated design is necessary to avoid loss of steam along with the noncondensable gases being expelled. For the open cycle with the direct-contact condenser, deaerators can also be used before the condenser in the cold-water loop.

The exhaust system consists of a series of stages. Each stage has a deaerator, a vent condenser, and a compressor, as shown in Fig. 5-10 (Lewandowski et al., 1980). Aerated seawater, consisting of dissolved air and water, enters at stage N. The vent condenser maintains a pressure above the stage larger than the vapor saturation pressure at the water temperature, drawing air out of the seawater. Each successive deaerator stage operates at a lower pressure. The pressures in the stages are fixed by the vent condensers. Air and water vapor enter the first vent condenser from the last stage in the staged system. Much of the liberated water vapor is

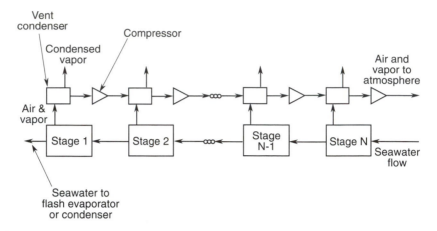

FIG. 5-10.   Gas exhaust system (Lewandowski et al., 1980).

condensed by the intercoolers to reduce the amount of gas that must be compressed. A compressor pumps the remaining mixture to the next vent condenser, fixing its operating pressure and the stage 2 deaerator operating pressure (higher than stage 1). Air and water vapor are added from the stage 2 deaerator, compressed to the next vent condenser, and so on down the line. The main purpose of a staged system is to reduce the amount of air removed at low pressure. Compression efficiency decreases with decreasing pressure; therefore a greater penalty is applied to the low-pressure stages.

A four-stage system consists of four compressors with three vent condensers located between the compressors. There is no vent condenser before the first stage, since it operates at the same temperature as the incoming water vapor.

The total mass flow rate through the first compressor, $\dot{m}_{t,1}$, is the sum of air flow $\dot{m}_a$ and uncondensed vapor $\dot{m}_{r,o}$

$$\dot{m}_{t,1} = \dot{m}_a + \dot{m}_{r,o} .\tag{5.2.17}$$

The compression ratio is given by Lewandowski et al. (1980) by

$$\gamma^n = \left[p_o + \Delta p \left(\sum_{i=1}^{n} r^i \right) \right] / p_{s,1} ,\tag{5.2.18}$$

where γ is the compression ratio per stage, $p_{s,1}$ is the pressure in stage 1, p_o is the outlet pressure of the multistage system, Δp is the vent condenser pressure drop, and n is the number of stages.

The flow of vapor through subsequent compressors is reduced by the vent condensers. The vapor flow for ith-stage compressor, $\dot{m}_{r,1}$, is

$$\dot{m}_{v,i} = \dot{m}_a (P_{vc,in}) / [(M_a / M_w)(P_{a,i})],\tag{5.2.19}$$

where M_a and M_w are the molecular weights of air and water vapor, respectively, $P_{vc,in}$ is the saturated water vapor pressure at inlet water temperature, and $P_{a,l}$ is the air pressure at the inlet of the ith-stage compressor.

The total flow through the ith-stage compressor, $\dot{m}_{t,i}$, is

$$\dot{m}_{t,i} = \dot{m}_a + \dot{m}_{v,i} .$$ (5.2.20)

The power required to compress this flow for the ith stage can be computed. The total power required for the system is the sum of the power for each compressor.

The University of Massachusetts study (1974) assumed that all the liquid sprayed into the evaporator gave up its noncondensable gases. This flow of gas was pumped from the condenser, where it was at 10°C (50°F) and 1360 Pa (0.2 psi), to the ocean's surface, where it was expelled at atmosphere pressure. The deaeration pump work was found to be 8.2 MW (Hagan, 1975), or a fraction of 5.8% of the gross power.

The Westinghouse study (1978) selected compressors with intermediate vent condensers to exhaust the gases to the environment. The optimum number of compressors determined by the study is between three and five. Two types of compressors (roots and axial flow) were considered. The study concluded that axial compressors were superior to the roots type because of the large rate of gas flow that must be handled. The basis for this conclusion was the substantial reduction in the power consumption with only a marginal cost increase. Vent condensers located between the compressors were used to reduce the gas volumetric flow rate to the following compressor. The impact of the vent condensers on the power and cost of the air removal system was dramatic. The power required was reduced from 10.4 MW without vent condenser to 4.25 MW with three vent condensers, and the cost was reduced from 3.50×10^6 to 2.4×10^6 (1978 value) for the Westinghouse design. A thin film mechanism for promoting desorption to achieve deaeration of seawater was selected by the Westinghouse design. The packing deaerator design for a warm seawater flow rate of 4.54×10^8 kg/h (10^9 lbm/h) was estimated at a cost of 2.84×10^6 (1978 value), or $28.4 per net output kW.

A test loop for OTEC gas desorption was constructed by Oak Ridge National Laboratory (Chen and Golshani, 1982), and experiments were carried out in three areas: (1) vacuum deaeration in a packed column using plastic pall rings, (2) deaeration in a barometric water intake system, and (3) disposal of noncondensables through hydraulic air compression.

Subsystem analyses based on the experimental data have shown that a 10% reduction both in deaeration cost and pumping power can be realized with a combination of barometric intake and packed column deaeration.

Bugby et al. (1983) examined bubble nucleation and growth in the evaporator, condenser, upcomers, and feedwater distribution systems of OC OTEC systems. The phenomenon that has the most impact on OTEC system design is cavitation in the warm-water feed near the entrance of the evaporator. The critical bubble size for cavitation is about 105 μm. Sources of bubbles in the warm-water feed are those entering from the ocean, those nucleating on suspended particles, and those nucleating on the upcomer wall. A debubbler upstream of the evaporator entrance is recommended.

Preliminary gas desorption data on a vertical spout, direct contact evaporator, and multiple condenser geometries were measured by Penney and Althof (1985). Results indicate that dissolved gas can be substantially removed before the seawater enters the heat exchange process, reducing the uncertainty and effect of inert gas on heat exchanger performance.

Performance characteristics of vacuum predeaeration hardware under low pressure and high gas/liquid loading rates typical for OC OTEC conditions are still not well established (Parsons et al., 1985).

5.2.5 *Water Pump*

The general configuration of a high-efficiency pump is a function of operating point flow and head. The tremendous large water flow rate and low head OTEC seawater pump operating characteristics lead directly to the choice of an axial flow propeller type unit. The number of units used determines the pump size and unit costs.

The appropriate pump size, speed, and power can be calculated (Bartone, 1978; Parsons et al., 1985) using the same method employed in Chapter 4. The cost per unit flow capacity in 1977 dollars was presented (Parsons et al., 1985) as:

$$\text{Cost } \$/Q = (253,000)Q^{-0.616}. \tag{5.2.21}$$

An early study indicated that the size of the water pump could permit the circulation pumps to be housed within the water inlet or outlet pipe.

5.2.6 *Control Systems*

Control systems are essential to OTEC plant operation. The major control systems are the overspeed protection control, the load control, and the speed control.

Turbine overspeed protection is necessary for a full load drop. Because of the large volumetric steam flow, conventional inlet control and stop valves are too large for practical application. Westinghouse (1978) chose a flow diversion with a hinged door system and injection of warm water to the condenser system to be the primary and secondary protection systems for reliability. The secondary protection system acts if the primary protection system fails. The hinged doors stop the turbine steam flow at the inlet and bypass the flow around the turbine at the exit. The injection of warm-water spray into the flow path between the turbine exit and the condenser inlet reduces overspeed.

A large change in steam flow is required to raise the turbine load from no load to full load or vice versa. Westinghouse (1978) used variable-speed warm-seawater pumps to control the steam flow by varying the rate of warm water fed to the flash evaporator. It is a much slower process than opening and closing huge valves but is a feasible approach because changing load needs to be done relatively slowly.

The control of speed, frequency, and synchronization of the generator to the line requires both a rapid and a precise control system. Because of the large steam and water flow rate, it is difficult to make valves large enough and fast enough to ensure that the turbine steam flow rate, power, and torque stay within safe limits. Westinghouse (1978) considered an electrical turbine-generator control system involving feed-forward and feed-back modulation of a voltage regulator for speed control.

5.2.7 *Seawater Ducting*

The function of the seawater ducting is to provide warm surface seawater to the evaporators and cold deep seawater to the condensers, and return the effluent streams to the sea. Although four seawater piping systems are required for operation of an OC OTEC plant, it may be desirable to combine the warm and cold discharge streams in one pipe.

5.2.7.1 Cold-Water Pipe

The majority of the closed-cycle OTEC power plant designs are sited on various types of floating structures. However, most of the open-cycle OTEC power plant designs are land based, which offers an attractive alternative to floating plants for certain applications in specific locations (particularly islands). The general requirements of the cold-water pipe (CWP), including structure, design, and static and dynamic forces acting on the CWP, at sea test, scale-model test, and material selection, have been discussed in Chapter 4 on closed-cycle OTEC plants. Most of these discussions also apply to open-cycle, land-based OTEC plants. The installation of the cold-water pipe of a land-based OTEC plant involves unique construction aspects. Both the large-diameter requirement (nominally 10 m for a 40-MW plant) and the depth of water (1000 m) for installation exceed the current demonstrated capabilities (Brewer, 1979) of the offshore pipe laying industry. Therefore the design and deployment of the cold-water pipe is one of the most difficult and challenging engineering tasks of the OC OTEC plant.

A variety of materials, construction techniques, and deployment methods have been suggested (Brewer, 1979) for the cold-seawater pipe design based on the plant type (i.e., land-based, shore-mounted, and shelf-mounted). A short list of candidate pipe materials and the estimated cost (1979 value) for manufacturing an 1800-m length of pipe for a 40-MW land-based OTEC plant is presented in Table 5-1. The comparison is for the deep water, steeply sloped, portion of the cold-water intake pipe only. There is no advantage of buoyant pipe in the shallow-water zone because the pipe has to be trenched and buried.

The best method for installing the cold-water pipe in the deep water and steep slope zone depends on many factors (Creare, Inc., 1984), which include:

1. Whether the pipe is installed in a single length or multiple sections;
2. Whether the pipe is transported to its final position along the sea floor or on the surface;
3. Whether the pipe deployment proceeds toward the deeper water or toward the shallower water;
4. The sea floor conditions along the pipe line route.

Table 5-1. Comparative cost for cold-water pipe materials

Type	Material	10-m Diameter unit length cost ($/m)
Heavy	Concrete	8.47
Heavy	Steel	19.27
Buoyant	Nylon	29.33
Buoyant	Fiberglass	13.13

The installation method may be quite different for different situations (Brewer, 1979).

A schematic of the cold-water system of the Westinghouse floating (1978) OC OTEC plant is shown in Fig. 5-11. The pump head, W_p, required for the system can be analyzed by the energy equation:

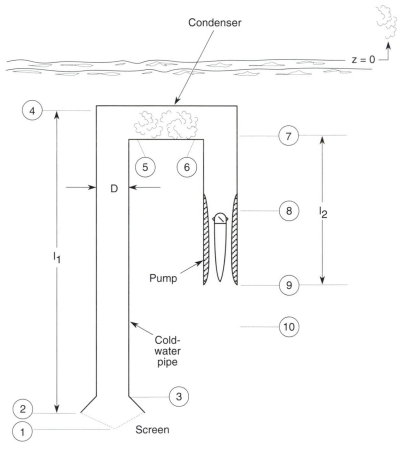

Fig. 5-11. Schematic of cold-water system (Westinghouse, 1978).

$$W_p = H_Z + H_V + H_F + H_M + H_C, \qquad (5.2.22)$$

where H_Z, H_V, H_F, H_C, and H_M are the potential head, velocity head, piping frictional head, minor head, and condenser head losses, respectively. These head losses are expressed by the following equations.

$$H_Z = \int_{z_1}^{z_9} (1 - \rho/\rho_i) g \, dz \qquad (5.2.23)$$

$$H_V = (V_9^2 - V_1^2)/2g \qquad (5.2.24)$$

$$H_F = (fl/D)V^2/2g \qquad (5.2.25)$$

$$H_M = (\Sigma K_M)V^2/2g, \qquad (5.2.26)$$

where ρ_i and ρ_o are the seawater inside and outside densities, z_1 and z_9 are the inlet and discharge depth, V_9 and V_1 are the seawater velocity at states 9 and 1 shown in Fig. 5-11, f is the frictional factor of the pipe, l and D are the length and diameter of the pipe, and K_M is the minor loss factor. The Westinghouse 100-MW OC OTEC floating plant (1978) assumes a water velocity of 2 m/s in the pipe, $V_1 = 1$ m/s at the screen inlet, $V_9 = 2$ m/s at the pump discharge, a condenser head loss of 2.3 m, a cold water inlet depth of 940 m, and a discharge depth of 100 m. The minor cold-water-path viscous dissipation loss factors are: $K_{screen} = 3$, $K_{area\ contraction} = 0.05$, and $K_{90°\ turn} = 1$. The head loss calculation shows that $H_z = 0.41$ m, $H_V = 0.2$ m, $H_F = 0.11$ m, $H_M = 0.56$ m, $H_C = 2.3$ m, and $W_p = 3.37$ m. Each of the four pumps requires a power of 3490 kW and delivers a flow rate of 102 m³/s.

The cold-seawater pipe length of a land-based OC OTEC power plant will be much longer than that of a floating plant. For example, the pipe proposed for an OC OTEC plant at Hawaii (Brewer, 1979) is 1768 m in length and 6.7 m in diameter. The pump power required to deliver a flow rate of 35 m³/s is a large portion (0.29) of the total gross output power of the plant.

The cold seawater pipe cost (C) is assumed (Creare Inc., 1984) to be proportional to the product of the length (L), diameter (D), and a cost factor K without consideration of set-up and other related costs. The factor K is determined from the seawater pipe data given by Creare, Inc. (1984):

$$C = KLD. \qquad (5.2.27)$$

5.2.7.2 Warm-Water and Mixed Effluent Pipes

Warm water enters the OC OTEC plant directly from the ocean surface through the warm-water intake structure as shown in Fig. 5-12. The intake is below the mean sea level, and its opening is covered by many vertically positioned stainless steel screens that minimize transverse flows. The flow velocity at the inlet face should be low to maintain hydraulic efficiency and minimize ingestion of fish and floating debris. The water flows inward at low velocity (1 m/s or 3.3 ft/s) in the pipe to the

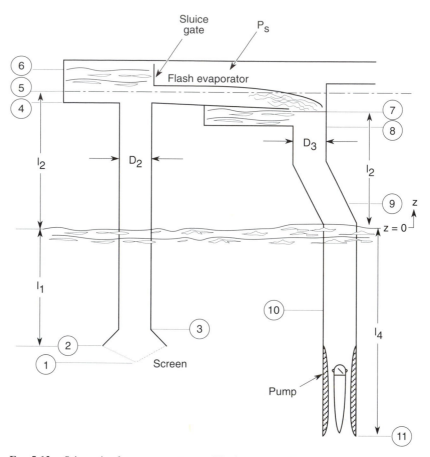

FIG. 5-12. Schematic of warm-water system (Westinghouse, 1978).

warm-water distribution channel and manifold (if multimodules are used), then to the evaporator heat exchangers. The warm-water pipe of a land-based 40-MW closed-cycle OTEC pilot plant designed by Ocean Thermal Corporation (1983) is a concrete pipe of 5.2 m (17 ft) inside diameter to provide about 1.5 million gal/min of warm water to the plant.

Analysis of the warm-water system can be based upon the energy equation from the inlet state 1 to evaporator outlet 7. The evaporator elevation l_2 provides the pressure, velocity, potential, and frictional head losses as

$$l_2 = H_p - H_V - H_Z - H_L - H_H , \qquad (5.2.28)$$

where H_p, H_V, H_Z, H_L, and H_H are the pressure head, velocity head, potential head, frictional head, and evaporator losses, respectively. The head losses are expressed by the following equations:

$$H_p = (p_s - p_a)/\rho g \tag{5.2.29}$$

$$H_V = (v_7^2 - v_1^2)/2g \tag{5.2.30}$$

$$H_Z = l_1 \tag{5.2.31}$$

$$H_L = (fl/D + \Sigma K_M)(v^2/2g) = H_f + H_M , \tag{5.2.32}$$

where p_s and p_a are absolute pressure at the free surface inside (state 6) and outside the warm-water system, ρ is the seawater density, v_7 and v_1 are the velocity at the outlet of the evaporator and inlet of the pipe, l_1 is the inlet depth, f is the frictional factor of the pipe, l and D are the length and diameter of the pipe, v is the water velocity in the pipe, and K_M are the minor loss factors of the system including inlet, screen, turn, etc.

The Westinghouse 100-MW OC OTEC floating plant design (1978) has p_s = 2.76 kN/m^2 abs. (0.40 psia), $v_1 = 1$ m/s, $v = 2$ m/s, $D = 3.6$ m, $f = 0.0116$, K_{screen} = 1.5, $v_7 = 0$, $K_{area\ contraction} = 0.05$, $K_{expansion} = 1$, $l = l_1 + l_2 = 40$ m, $H_Z = 0.005$ m, H_V = 0 m, $H_M = 0.39$ m, $H_H = 0.2$ m, $H_p - l_2 = 0.60$ m, and $l_2 = 9.45$ m.

The warm-water discharge from the evaporators of the OC OTEC plant shown in Fig. 5-12 can also be analyzed by the energy equation

$$W_p = H_E + H_Z + H_V + H_L + H_P , \tag{5.2.33}$$

where W_p is the pump head, H_E is the head required to extract water from the low-pressure evaporator chamber, H_Z is the potential head loss as water flows from state 7 to sea level, H_V is the velocity head loss from state 7 to 11, H_L is the frictional head loss, and H_P is the pressure head.

The Westinghouse study (1978) gives $V_{pump\ discharge} = 2$ m/s, $H_E = 2.1$ m, $H_P = 10.05$ m, $H_Z = 7.35$ m, $H_L = 0.14$ m, $H_V = 0.2$ m, and $W_p = 3.1$ m.

The warm-water discharge and cold-water discharge from the evaporators and condensers of the OC OTEC plant may be mixed after passage through the heat exchangers. The mix effluents are then discharged to a certain depth, where they sink to the equivalent density depth.

5.2.8 Structure Requirements

The primary functions of the structure in an OC OTEC plant are to provide a vacuum boundary of low permeability; to support and locate the plant subcomponents; to interface with seawater and product delivery conduits; to provide intercomponent flow paths and internal distribution and collection of seawater, fresh water, and noncondensable streams; to sustain the rigors of site-specific environment with minimal outage for 20 years or more; and, for a floating installation, to provide barometric elevation and pitch/roll/heave stability as required by flash evaporator and direct-contact condenser design details.

Most of the literature concerning the plant structure derives from closed-cycle work and focuses on dynamic, survivability, and station-keeping requirements of floating plants (see Chapter 4).

The Westinghouse (1978) 100-MW OC OTEC power plant structure consists of a central core connected to a 900-m-long cold-water pipe around which are grouped one to five identical modules. Each module consists of a vacuum chamber, condenser, apron, turbine shaft, generator room, cold-water intake, cold-water outlet with pump shaft, warm-water intake, warm-water outlet with pump shaft, etc. The modules are interconnected by a core. The floating plant platform is made of prestressed concrete. The position of the plant is normally fixed. The basic design and problems of the floating structure have been discussed in Chapter 4.

The smaller OC OTEC power plants are usually designed to be shelf mounted, shore mounted, or land based.

Shelf-mounted plants at depths of approximately 100 m have been proposed as a means of avoiding some high-cost or high-uncertainty aspects of floating plant designs. In this approach the supporting structure incorporates warm-seawater intake and effluent pipes and the upstream end of the cold-water pipe at the seabed, from which point it follows the bottom to the cold-water resource. Such designs might be built horizontally at existing facilities, towed to the site, upended, and secured to the bottom. Platform dynamic response is eliminated by virtue of shelf-mounting, along with anchor and mooring requirements, although the supporting structure must be designed to withstand storm extremes. Some CWP uncertainties and risk of loss are reduced at the expense of increased pipe length and substitute uncertainties in the laying and securing of bottom-mounted pipe. Cabling and piping distances to shore for the power and fresh-water products may be substantially reduced relative to floating plants.

Shore-mounted plants move the shelf-mounted concept to water depths 10–30 m closer to shore, preferably in sheltered harbors or lagoons. The support structure is reduced to pilings that may be inserted through the platform into preformed holes and trussed, allowing shipyard construction and tow out of shallow-draft plant designs without upending operations. Wave action in the surf zone may be more severe, but the shallow water facilitates construction of breakwaters, and islands offer excellent shelter in their lee for appropriate sites. Cold-water and effluent pipes must be trenched and buried out to water depths of approximately 30 m to protect them from surf damage throughout the length of the bottom-mounted effluent pipe. Easier plant access to the shore reduces the maintenance and operating expenses of crew transport.

A true land-based plant avoids structural problems of direct wave loads and wave response considerations, but is subject to potential seismic loads and storm damage. Warm, cold, and effluent pipes must be trenched through the entire surf zone to their respective terminal points. On-site construction of the plant is likely. Barometric constraints and flow circuit head loss dictate the placement of flash evaporators at approximately 10 m above sea level, and surface condensers must be entirely below this elevation to avoid flashing in the tubes and excessive pressure drop. Nominal grades above sea level therefore reduce the head room available for plant components and require a larger plant plan area or deeper foundation excavation. Tidal variations impose a requirement for variable head seawater pumps, as is also the case for shelf-mount and shore-mount options.

In summary, floating plants offer potentially the lowest structure and pipe costs, offset by high mooring costs for plants designed for electric power transmission to shore. Tradeoffs exist between uncertainty and risk vis-a-vis CWP installation and storm survival. Shelf mounting exchanges substantial foundation costs for high mooring costs, reduces survival uncertainty, increases CWP length and cost, and raises new uncertainties relative to deployment of bottom-mounted CWPs. Shore mounted or surf-zone sites reduce foundation costs, further reduce survival uncertainty, and increase pipe length, number, and installation cost. Land basing increases construction and foundation costs and pumping costs, along with pipe length and installation costs, but virtually eliminates platform risk and uncertainty and eases operation.

Both concrete and steel have been extensively studied for OC OTEC plant construction materials. Reinforced concrete provides lower installation cost and substantially lower maintenance cost, due to corrosion of steel in the marine environment. Concrete of high quality and minimum water content shows excellent marine durability. Care must be exercised to provide adequate coverage of reinforcing steel, to minimize chloride penetration and depassivation of the steel (Browne and Bury, 1981). Air permeability is low and may be kept low by periodic application of external sealant coats on exposed surfaces while plant vacuum is maintained. Use of plasticizers may be beneficial to enhance the workability of low-water concrete mix. Installed costs may be accurately estimated on the basis of concrete volume, assuming conventional shapes and reinforcement (Barr and Chang, 1977). The structure cost model used a value of $1,540/m^3$ (1983$) for installed reinforce concrete.

The unit area cost estimation of the structure (Z/A) in 1983 dollars per unit plant area in square meters given by Creare R and D, Inc. (1984) is roughly a constant due to the need for ballast weights:

$$Z/A = 10,520. \qquad (5.2.34)$$

For a small 1-MW OC OTEC power plant with a plan area of 117 m^2 designed by Penney (1984), the cost estimation is $9831 per unit area.

5.2.9 *Energy Delivery Systems*

Electrical energy produced by offshore OC OTEC plants could be transported to shore by submarine cables. Alternatively, storage processes such as ammonia or methanol production could be carried out on a floating platform. These systems have been discussed in Chapter 4.

Electrical energy generated by shelf- or land-based OC OTEC plants could be sent via a high-voltage step-up transformer and cable directly to utilities on shore. Energy transfer equipment of the system may include the generator switch gears, motor control centers, transformers, substations, cables, bus, switchyards, and necessary instrumentation and controls. All this equipment is conventional and commercially available. The closed-cycle OTEC (see Chapter 4) land-based plant proposed by Ocean Thermal Corporation at Hawaii (1983) generates electric power at 13.8 kV and is sent via a 56-MVA, 13.8- to 138-kV step-up transformer to the electric utility. The transmission voltage of 138 kV is selected to match the transmission voltage of the local electric power company. Transformers and switch gears should be housed in enclosures and supplied with force ventilated, filtered air to reduce corrosion. Conventional oil-filled cables are selected for transmitting power to the onshore grid.

5.3   HYBRID CYCLE

5.3.1 *System Analysis*

The hybrid-cycle OTEC system is one of the variations on the standard open-cycle OTEC system. It is a blend of both the open and closed cycles designed to produce both electric power and fresh water as OTEC products. The hybrid cycle is an attempt to combine the best features and avoid the worst features of the open and closed cycles.

The focus of the U.S. OTEC program was shifted in 1980 from large (100- to 400-MW) floating plants to small (10- to 40-MW) shore-based plants for early commercialization. Small plants have potential applications for the island market. Moreover, some islands can use desalinated water, which can be produced by a hybrid OTEC plant. Small hybrid OTEC plants for simultaneous production of desalinated water and power could have greater potential for islands and deserts than the closed and open cycles for power production alone.

In the hybrid cycle (Panchal and Bell, 1987), the warm seawater is flash evaporated under vacuum to steam (as in OC OTEC systems), as shown by process 1-2 of the *T-S* diagram in Fig. 5-13. The heat in the resulting low-pressure steam is then transferred via a heat exchanger to ammonia in a conventional closed Rankine cycle system and condensed in the ammonia evaporator as shown by process 3-4. The drop in the condensation temperature from state 3 to state 4 is due to the combined effects of noncondensable gas evolved from the seawater and steam-side pressure drop. Ammonia, evaporating in process 5-6, is passed through the turbine in process 6-7 and is condensed in a surface condenser in process 7-8. Cold seawater is sensibly warmed in passing through the condenser, process 9-10.

The schematic diagram of the hybrid OTEC power system is illustrated in Fig. 5-14. Warm seawater is pumped from a depth of about 10 to 15 m. It is flash evaporated, and about 0.5% of the water flow is converted into low-pressure steam. During the flash evaporation, dissolved gases evolve. The low-pressure steam flows to the ammonia evaporator, where about 95% of the steam is condensed. All of the steam cannot be condensed due to the presence of noncondensable gases that reduce the condensation temperature. Therefore, an optimum fraction of steam that should be condensed in the ammonia evaporator needs to be determined for a given set of operating conditions.

The major part of the remaining steam is condensed in the condenser, where ammonia from the condenser is used as a coolant. Alternatively, cold seawater can be used as a coolant in the vent condenser. However, ammonia is the better choice for the following reasons:

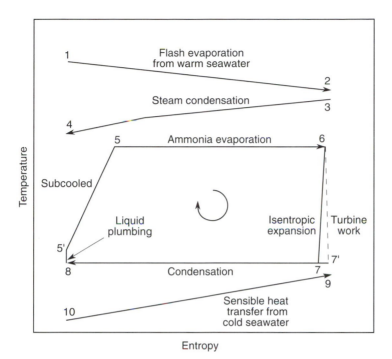

FIG. 5-13.   *T–S* diagram for the hybrid-cycle OTEC.

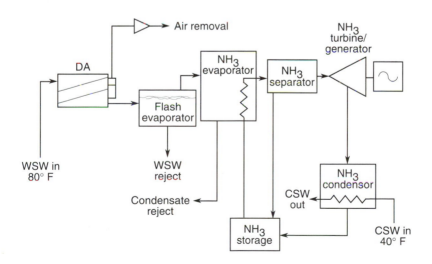

FIG. 5-14.   Hybrid-cycle OTEC power system.

1. Boiling ammonia gives a higher heat transfer coefficient and isothermal coolant condition in the heat exchanger.
2. By not using seawater, fouling and corrosion problems are eliminated. As a result, a relatively inexpensive and efficient brazed aluminum plate-fin heat exchanger can be used.

The steam/gas mixture is pumped out from the vent condenser using a multistage compressor. The condensed water vapor from the ammonia evaporator and vent condenser is collected and stored for further processing.

The ammonia cycle is identical to that for the closed-cycle OTEC system. Ammonia liquid is separated from the evaporator before it is allowed to expand in the turbine and generate power. Exhaust vapor from the turbine is then condensed in the condenser, which is cooled by the deep cold seawater.

The difference between the hybrid cycle and the open cycle is the addition of a shell-and-tube heat exchanger, where condensation of the water vapor generates the ammonia vapor, which circulates in a closed ammonia loop as in the closed cycle.

The hybrid cycle can be compared to the open-cycle system with a surface condenser, because both systems produce desalinated water and power simultaneously. The key comparisons are:

1. The hybrid cycle eliminates the larger low-pressure turbine. No turbine has yet been designed and built to operate in this pressure range, and it therefore should be regarded as a high-risk component.
2. By eliminating the large low-pressure turbine, the vacuum enclosure is substantially reduced in size and structural complexity.
3. Since the ammonia turbine is a small component in the total system, it can be easily modularized without sacrificing the efficiency or incurring cost penalties. Therefore, it is relatively easy to design a hybrid-cycle OTEC plant for a given ratio of desalinated water to power generation that suits local needs.
4. Condensation of steam in the hybrid cycle takes place at a higher pressure (20 mm Hg) than in the open cycle (10 mm Hg). It is, therefore, possible to condense a larger fraction of the steam and reduce the steam-to-gas ratio before the gas/steam mixture is pumped out. Moreover, the compression ratio is reduced from about 76 to 38. It should be noted that exhaust system pumping power for the open cycle could be in the range of 8 to 20% of gross power, depending upon the condenser efficiency and degree of desorption of gases from the seawater. On a general basis, it should be easier to handle noncondensable gases in the hybrid cycle than in the open cycle.

The ammonia evaporator and condenser are efficient heat exchanger designs selected from the heat exchanger programs described in Chapter 4.

The overall heat transfer coefficients of the heat exchangers are usually estimated. The total heat transfer surface areas required for each component are estimated based on the estimated heat transfer coefficients, the log mean temperature differences resulting from the baseline temperature conditions, and heat rates resulting from the hybrid power cycle requirements.

$$A_E = Q_E/U_E(\text{LMTD})_E \qquad\qquad (5.3.1)$$

$$A_C = Q_C/U_C(\text{LMTD})_C , \qquad\qquad (5.3.2)$$

where A_E and A_C are evaporator and condenser surface areas, Q_E and Q_C are heat transfers in the evaporator and condenser, U_E and U_C are overall heat transfer coefficients of the evaporator and condenser, and $(\text{LMTD})_E$ and $(\text{LMTD})_C$ are the log mean temperature differences between the ammonia and seawater in the evaporator and condenser, respectively.

$$Q_E = \dot{m}_{wf} C_p (T_3 - T_4) = \dot{m}_A (H_6 - H_5) \qquad\qquad (5.3.3)$$

$$Q_c = \dot{m}_c C_p (T_9 - T_{10}) = \dot{m}_A (H_7 - H_8) , \qquad\qquad (5.3.4)$$

where $\dot{m}_{wf}$, $\dot{m}_A$, and $\dot{m}_c$ are the mass rates of flow of the warm flashed seawater, ammonia, and cold seawater, respectively; C_p is the heat capacity of seawater; T is the temperature of the seawater; and H is the enthalpy of the ammonia at various thermodynamic states shown in Fig. 5-13.

The basic heat exchange analysis and design factors of the hybrid ammonia closed cycle are very similar to those of the closed cycle, which have been discussed in detail in Chapter 4.

Vacuum deaerators are required in the hybrid cycle. The presence of noncondensable gases, particularly oxygen, in the hybrid-system ammonia evaporator creates the potential for excessive corrosion. In addition, performance degradation of the system results from the blanketing of heat transfer surface by the noncondensable gases in the hybrid system ammonia evaporator. The basic flash evaporator and deaerator analysis and design factors of the hybrid system are similar to those of the open system discussed in Section 5.2.1.

The ammonia turbine and pump of the hybrid system are related to the thermodynamic states (Fig. 5-13) by the following equations:

$$\dot{W}_T = \eta_T (\dot{m}_a)(H_6 - H_7) \qquad\qquad (5.3.5)$$

$$\dot{W}_p = \dot{m}_a (H_5 - H_8)/\eta_p , \qquad\qquad (5.3.6)$$

where $\dot{W}_T$ is the actual turbine power delivered, $\dot{W}_p$ is the actual pump power needed, and η_T and η_p are the turbine and pump efficiencies of the ammonia closed cycle, respectively.

The analysis and design factors of the ammonia turbine and pump have been discussed in Chapter 4. The flash evaporator has been discussed in Section 5.1.1. The mass rate of steam flow generated by the evaporator is

$$\dot{m}_{WF} = \frac{\dot{m}_W C_p (T_1 - T_2)}{H_{fg} - C_p (T_2 - T_E)} , \qquad\qquad (5.3.7)$$

where $\dot{m}_W$ is the warm seawater mass flow rate, C_p is the specific heat capacity of seawater, T_1 is the warm seawater inlet temperature, T_2 is the warm seawater outlet temperature, T_E is the saturated temperature at the evaporator pressure, and H_{fg} is the steam enthalpy of evaporation at T_E.

A hybrid OTEC system performance model has been developed by Panchal and Bell (1987). For a given set of input parameters, the model takes input from the power system and property routines and calculates operating temperatures and pressures for the working fluid that gives maximum gross power.

The rate of desalinated water production is about 2.2 million liters per day per MW of net power generation. About 2.8% of the gross power is used for pumping out noncondensable gases, which is significantly less than a value of about 10% that might be required (Penney et al., 1984) for an OC OTEC plant. Table 5-2 shows typical component specifications of the hybrid system design (Panchal and Bell, 1987).

Except for the seawater evaporation system, all components of the hybrid-cycle plant can be designed and built with state-of-the-art technology.

Table 5-2. Power system parameters for 10-MW hybrid-cycle OTEC plant

Vertical spouts flash evaporator	
Rate of heat transfer	639 (MW)
Water flow rate	42,560 (kg/s)
Rate of steam generation	261 (kg/s)
Water velocity in spouts	1.83 (m/s)
Number of spouts	1799
Rate of desorption of gases	0.68 (kg/s)
Brazed aluminum plate-fin ammonia evaporator and steam condenser	
Rate of heat transfer	607 (MW)
Fraction of steam condensed	0.95
Heat exchanger area	84,160 (m²)
Steam-side channel size	44.5 (mm)
Ammonia-side channel size	7.1 (mm)
Brazed aluminum plate-fin vent condenser	
Rate of heat transfer	30.8 (MW)
Fraction of steam condensed	0.9
Heat exchanger area	8400 (m²)
Steam-side channel size	31.7 (mm)
Ammonia-side channel size	6.35 (mm)
Titanium shell-and-tube ammonia condenser	
Rate of heat transfer	620 (MW)
Tube diameter	25.4 (mm)
Tube length	15.2 (m)
Heat transfer area	80,532 (m²)
Gas compressor system	
Number of stages	5
Inlet steam-to-gas ratio	0.7
Inlet pressure	2.58 (kPa)
Compressor efficiency	0.8

5.3.2 *Water System and Fresh Water Production*

The seawater system for a typical 10-MW hybrid cycle plant site representing an island OTEC application is shown in Fig. 5-15 (Panchal and Bell, 1987). Cold water is drawn from a depth of about 650 m using a buoyant fiberglass-reinforced plastic (FRP) pipe. In the shallow-water region, cold, warm, and effluent water pipes are bottom mounted and held by weight anchors. For water depths less than about 21 m, pipes are buried for protection against waves. All pipes are channeled through the land region to the plant. Table 5-1 shows that 29% of the gross power is used for pumping seawater. The rate of desalinated water production is about 2.2 million liters per day per MW of net power generation.

For shore-based hybrid-cycle OTEC plants, it is essential to minimize water requirements, especially the deep cold water, because the seawater piping represents a major cost in the overall system. As a result, larger water temperature changes and corresponding losses of temperature difference available for energy conversion reduce the output of the power system.

The surface warm water is pumped to the vacuum chamber, and it rises vertically to the flash evaporator, which consists of vertical spouts. Although other flash-evaporator configurations can be used, the vertical spout has desirable features for OTEC applications. The flash evaporator is a free-floating platform guided by a sliding system on the wall and vertical guide bars. The free-floating

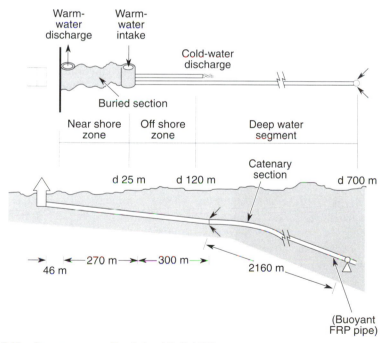

FIG. 5-15. Seawater system (Panchal and Bell, 1987).

device with associated control systems should dampen the effects of ocean dynamics on the flash-evaporation process and the resulting steam pressure.

The low-pressure steam, generated in the flash evaporator, flows to the ammonia evaporator. About 95% of the steam is condensed in the ammonia evaporator. It is an aluminum plate-fin heat exchanger, assembled using a brazing technique. Aluminum has proved to be suitable for closed-cycle heat exchangers, which do have contact with seawater. Steam condenses in horizontal channels formed by the fins while ammonia evaporates on vertical channels formed by serrated channels. The condensate is pumped to a fresh-water storage tank.

The hybrid OTEC plant discussed (Panchal and Bell, 1987) produces 2.25 million liters per day of desalinated water per MWe of electric power generation. If the ratio of water to power production needs to be other than the above value, the hybrid-cycle plant can be designed accordingly. On the other hand, for a certain range of operating conditions, this plant can be operated off design conditions for production of additional desalinated water at reduced net power generation. This can be done either by lowering the flash-down temperature (i.e., increasing temperature change for warm water) or bypassing the ammonia turbine (i.e., allowing the ammonia vapor to flow directly to the condenser). It is, however, necessary to install a control system for such off-design operations. For a modular plant system, one of the turbine modules can be shut down and the ammonia vapor allowed to flow directly to the condenser, resulting in a 60 to 80% increase in water production for that module. It should be noted that the net power generation decreases by shutting down a module.

Since the available temperature difference for power production is small, it is not possible to increase production of desalinated water significantly from a plant that was optimized for power generation. However, certain design features can be incorporated into the original design for required water production at a given power generation level. Among other design options, provision for varying the warm seawater flow rate seems to be an efficient and least expensive approach. In this case, the warm-water pipe, flash evaporator, and ammonia evaporator are designed for varying water flow and the corresponding rate of steam generation. The rest of the system, including the cold-water flow system, remains essentially the same as that for optimum power production.

Three different hybrid plant configurations have been studied by Creare, Inc., as shown in Fig. 5-16. One configuration is a single-stage (Plant 1.0) and two are double-stage (Plants 2.0 and 2.1) plants. In the single-stage plant, all of the steam is produced in one flash evaporator and condensed in a single surface condenser, allowing for the production of both electricity and fresh water. Double-stage plants have potentially higher efficiency than the single-stage plant and also can provide greater flexibility of operation. Steam produced in the first-stage flash evaporator is routed directly to a surface condenser to obtain fresh water, and steam from the second flash evaporator is expanded through a turbine to produce power. Analyses have been performed to determine the optimal plant configurations for the produc-

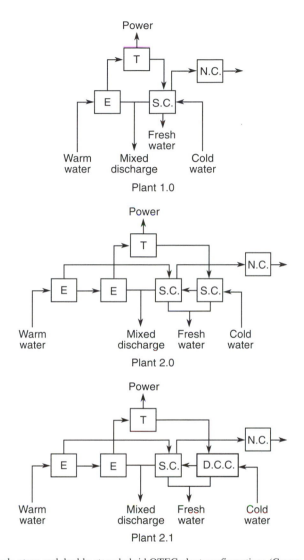

Fig. 5-16. Single-stage and double-stage hybrid OTEC plant configurations (Creare, 1984).

tion of fresh water alone (a zero net power output), for combined power and fresh water production, or for power production only. Results (Creare, Inc., 1984) show that there was little cost difference between single- and double-stage plants in generating power, but a double-stage plant was consistently more economical for the production of water. The capital costs in 1983$/(m³/s) for the plants optimized for fresh-water production were found to be $120 and $170 for double stage and single stage, respectively, at a fresh-water production rate of 4 m³/s.

5.4 MIST LIFT CYCLE

An OTEC mist lift cycle may be considered (Chen and Michael, 1980) as an open cycle that uses a hydraulic turbine for power generation instead of the very low-pressure steam turbine used in the Claude open cycle. It offers an alternate approach to the utilization of ocean thermal energy. The lift-cycle power plants have many different system options depending on the lift process employed. Foam (Beck, 1975) and mist (Ridgway, 1977) lift processes are currently in the technical feasibility development stage.

The development of the lift-cycle concept was inspired by Beck's (1975) (Zener and Fetkovich, 1975) lift water concept, in which warm seawater is introduced through a lift generating device at one end of a lift tube and a steam—water two-phase mixture is created in the lift tube, providing the necessary hydraulic head to elevate the water to a higher potential energy state. Thermodynamically, it can be modeled by an internal transfer of the work produced during the isentropic expansion of the vapor to isentropically compressed liquid. The energy is then removed by a hydraulic turbine, as shown by the Beck cycle in Fig. 5-17.

Two methods of implementing the Beck cycle have been proposed. Both proposals use the gravitational potential as a means of internally storing the energy being transferred. The idea is to use the thermal energy extracted from the warm ocean seawater vapor to lift the working fluid against gravity. The difference in enthalpies per unit weight generated by the expansion is equal to the height to which unit weight is lifted in the ideal process.

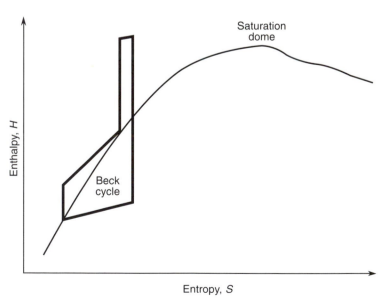

FIG. 5-17. Ideal Beck cycle.

Ridgway (1977) (Beck, 1975) proposed to do this by a two-phase vertical flow process of a very fine mist of droplets suspended in their vapor. The process is called *mist lift*, which is somewhat analogous to pneumatic transport commonly used in the chemical industry to convey fine solid particles in suspension in a gas-phase process. The cycle applying to OTEC is the mist lift cycle.

Zener and Fetkovich (1975) (Ridgway, 1977) suggested lifting the water in the form of a foam. The foam lift OTEC cycle will be discussed in Section 5.5.

The difference between the mist lift and foam lift comes from whether the lift process incorporated in the OTEC power system is pre-lift or post-lift in configuration, in respect to the hydraulic turbine, as shown in Fig. 5-18.

5.4.1 *Concept*

The Beck cycle is a modified Rankine cycle. Over a small range of temperatures involved in OTEC (typically 22°C) applications, constant property curves for steam are nearly straight lines, as shown in the following *T–S* and *p–v* diagrams (Fig. 5-19). Therefore, the *T–S* diagram of the Beck cycle is essentially a triangle.

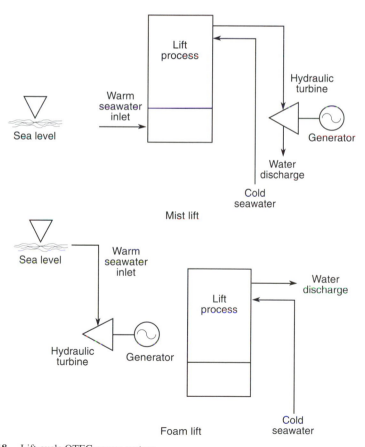

FIG. 5-18. Lift-cycle OTEC power system.

Warm seawater at state 4 enters a two-phase lift generator to create a two-phase mixture that proceeds along an isentropic line to state 5 in a lift tube. The steam at state 5 is condensed by cold seawater along an isothermal line to state 2 in a condenser.

Let the efficiency of the Carnot cycle over the same temperature range of the Beck cycle be η_c. The theoretical heat added and work output of the Carnot cycle are shown by areas 1234561 and 23452 in Fig. 5-19. Similarly the heat added and work output of the Beck cycle are area 124561 and 2452, respectively. Let the efficiency of the Beck cycle efficiency be η_B. Then

$$\eta_B = \frac{Area\ 2452}{Area\ 124561} = \frac{(1/2)\,(Area\ 23452)}{Area\ 1234561 - Area\ 2342}$$

$$= \frac{(1/2)\,(Area\ 23452)}{Area\ 1234561 - (1/2)\ Area\ 23452} = \frac{(1/2)}{1/\eta_c - (1/2)} \qquad (5.4.1)$$

$$= \frac{\eta_c}{2 - \eta_c} = \frac{\eta_c}{2} \quad (\text{if } \eta_c < 2).$$

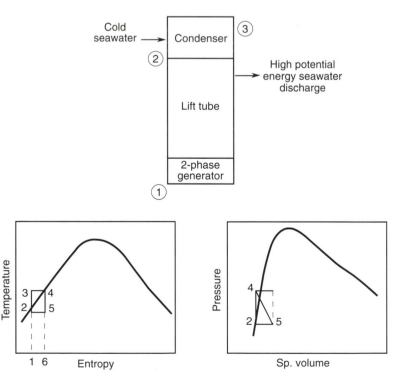

FIG. 5-19. Beck cycle over a small temperature range.

Since η_c of the OTEC cycles is of the order 0.08, we find that the ideal Beck cycle is very nearly one-half of the maximum possible Carnot cycle efficiency. The p–v diagram of the Beck cycle is also a triangle. Since a rectangular cycle area on the p–v diagram represents the maximum possible extraction of work over a certain pressure range, the Beck cycle is found to produce one-half the maximum work, too. It was claimed that the Beck cycle was effective compared to other power cycles when applied to OTEC.

A schematic diagram of the mist lift OTEC plant and its corresponding T–S diagram are illustrated in Figs. 5-20 and 5-21. Warm seawater at state 1 is taken from near the ocean surface, filtered, and allowed to descend through a penstock to a hydraulic turbine inlet at state 2. There is a working pressure drop across the hydraulic turbine, which is represented by the difference in level of states 1 and 2. The exit water at state 3 from the turbine is sprayed upward in tiny jets into the bottom mist generator (lifting duct), where the ambient pressure is slightly less than the saturation vapor pressure of the injected water. Some of the water evaporates

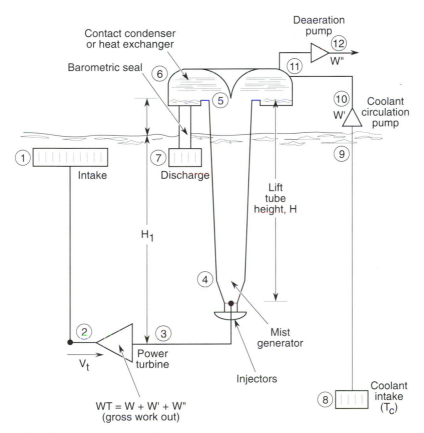

FIG. 5-20. Schematic diagram of a lift cycle.

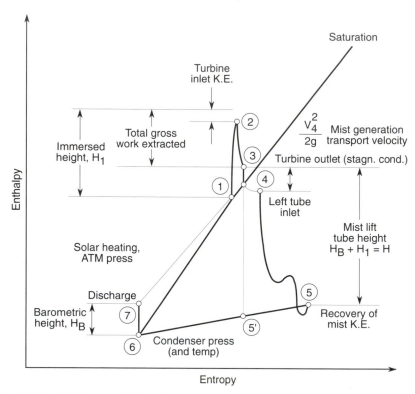

FIG. 5-21. *H-S* diagram of a lift cycle.

to form vapor at state 4, and the expansion of this vapor raises the droplets to the top of the mist generator at state 5 against the force of gravity. When the flow reaches the top, it is turned radially outward to the condenser where cold water at state 11 is sprayed into the flow to condense the vapor. Then the mixed flows at state 6 are discharged at state 7 to the ocean at a depth where their temperature matches the local ocean temperature in order to minimize the environmental impact and to prevent the used water from being recirculated through the plant. Cold deep seawater at state 8 is drawn into the cold-water pump at state 9. Noncondensable gases are drawn from the direct-contact condenser by the deaeration pump at state 12 to the atmosphere.

Analysis (Charwat, 1978) of the mist lift process and the entire cycle is rigorously described by the conservation of mass and energy, and momentum equations. They are

$$\dot{m} = A_r V_r = A\, V/v \tag{5.4.2}$$

$$H_0 = H + h + \frac{V^2}{2g} = H_r + h_r + \frac{V^2}{2g} \tag{5.4.3}$$

$$\dot{m}V_r + p_r A_r + \frac{p_r + p}{2}(A - A_r)$$

$$= \dot{m}V + pA + \left(\frac{A_r + A}{2}\right)\left(\frac{2}{V_r + V}\right)h, \tag{5.4.4}$$

where the subscript r is the reference state chosen for convenience; H_0 is the stagnation enthalpy; H is the enthalpy; h is the vertical height; V is the average velocity; A is the cross-sectional area; v is the specific volume; $\dot{m}$ is the rate of mass flow; and p is the pressure of a state.

The two-phase mixture is assumed to be in thermodynamic and dynamic equilibrium. It is completely described by three properties: either (v, H, p) or H_0, v_r, x), where x is the quality of the mixture.

Mechanical shaft work, which can be extracted from the water of high potential energy by a hydraulic turbine, between a heat source temperature T_H and a condensing temperature T_C is expressed as

$$W = C_p(T_H - T_C)^2 /(T_H + T_C). \tag{5.4.5}$$

The theoretical specific work in the unit of water lift height for warm seawater at 25°C as a function of condensing temperature is shown in Figure 5-22. For a condensing temperature of 10°C, the theoretical lift height is 165 m.

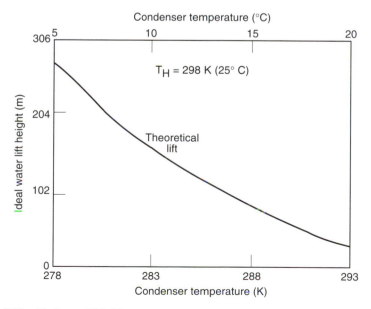

FIG. 5-22.   Ideal water lift height.

5.4.2 *Design Analysis*

Since the theoretical thermodynamic analyses done by several investigators (Ridgway, 1977; Ridgway et al., 1980; Davenport, 1982; Charwat, 1978) have shown the energy conversion potential of the lift cycle concept, development efforts (Lee and Ridgway, 1983; Chen and Michael, 1980) have been directed towards establishing the technical feasibility and understanding of the physical process of the lift concepts. Mist lift power system designs and cost estimates have been carried out by R & D Associates (1978) supported by UCLA (Charwat et al., 1979) in the performing of experiments and Dartmouth College in simulation (Wallis et al., 1980).

A conceptual 10-MW mist flow plant and platform design is shown in Fig. 5-23. The central structure is the large evacuated duct in which the water is lifted

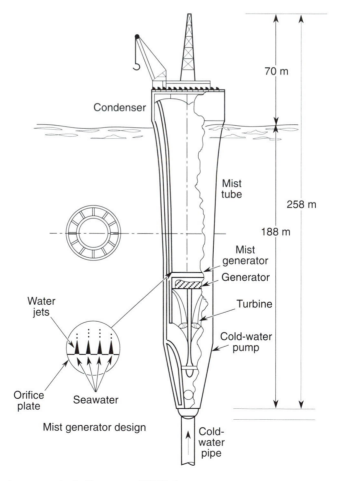

FIG. 5-23. A conceptual mist flow power OTEC plant.

a distance of 200 m by the conversion of its thermal energy into work. The warm ocean surface water enters through the filter and then descends to the water turbine to produce power. Low-pressure warm water is sprayed upward into the bottom of the duct. The bottom 20 m of the duct comprises the mist acceleration zone. Cold water is introduced into the accelerated mist in the form of a circumferential converging sheet in the coast zone, which is also the condenser. A vacuum pump removes noncondensables at the top. The pressure difference between the start and the end of the acceleration zone drives the mist to a velocity of nearly 50 m/s. The converging cold-water sheet merges with the accelerated mist to form a single jet, which coasts upward against gravity and overflows into a basin from which it then drains into the ocean. The reference characteristics of the plant and losses expected in the mist flow process are listed in Table 5-3 and Table 5-4, respectively.

The details of the characteristics and losses are discussed by Ridgway and Hammond (1978), Charwat et al., (1979), and Wallis et al. (1980). The power plant design is based on the results of an experimental mist lift cycle at UCLA (Charwat et al., 1979).

The major findings in the experiments of Lee, Ridgway, Charwat, and Hammond have verified the mist lift theory and demonstrated that water droplets (200 μm diameter) could be lifted to substantial heights (50 m) by their own vapor produced in flashing over temperature differences typical of the tropical seas. The coupling between the vapor and droplet was found to be excellent. These results imply that a larger extrapolation mist flow OTEC power plant could be designed (Lee and Ridgway, 1983) and built. Because the mist lift cycle is a relatively new concept, further research and development efforts are needed to verify its potential for OTEC power.

The net cost comes in at a little under $1000/kW (1978 value), which is very low compared to that of the closed and open cycles previously discussed. The costs quoted by Ridgway (1977) and Charwat (1978), however, seem to be lower-end estimates.

5.5 FOAM LIFT CYCLE

Foam lift is another open-cycle concept that uses a hydraulic turbine for power generation. It is a modification of the Beck concept (Zener and Fetkovich, 1975) using intentional foaming of seawater. They believed that the foam lift cycle will have better efficiency than an open cycle and be less costly than a closed-cycle plant. Research on the foam process has been sponsored by the U.S. DOE.

5.5.1 Concept

A foam is defined as a mixture of liquid and vapor in which the overwhelming volume percentage is in the vapor phase. The overwhelming mass percentage is in the liquid phase, and the vapor is contained in cells bounded by liquid films.

The Beck, mist, and foam lift systems operate upon the same thermodynamic principles, which can be shown on the T-S diagram (Fig. 5-24).

Table 5-3. Reference characteristics of 10-MWe mist flow ocean thermal plant

Warm-water temperature (°C)	25
Cold-water temperature (°C)	5
Injection nozzle diameter (cm)	0.01
Injection nozzle volocity (m/s)	20
Ratio of cold- to warm-water flow	3:1
Mixed-water temperature (°C)	15
Droplet size (cm)	0.02
Hull diameter at surface (m)	60
Depth at juncture of hull to cold-water conduit (m)	200
Displacement (metric tons)	189,000
Freeboard to upper deck level (m)	25
Structure weight (metric tons)	50,000
Equipment weight (metric tons)	10,000
Ballast (metric tons)	129,000
Warm-water intake flow (metric tons/s)	13
Filter area (m^2)	130
Cold-water intake flow (metric tons/s)	39
Cold-water conduit diameter (m)	7
Cold conduit depth (neutral buoyancy pipe) (m)	1,500
Pumping load (kWe)	2,500
Housekeeping electric load (kWe)	1,000
Turbine-generator gross rating (kWe)	14,000
Net output (kWe)	10,000

Table 5-4. Summary of losses expected in mist flow process

Theoretical lift height (m)	164
Lift-tube losses (m)	
Flashdown loss	11.5
Slip loss	23.0
Exit loss	10.2
	−44.7
Mechanical losses (m)	
Filter and deaerator losses	10.0
Cold-water pumping losses	20.0
Turbine-generator losses	10.0
	−40.0
Net output (m)	79.3

The data are based on a mist flow plant using 25°C warm water, 5°C cold water, and 10°C mixed outlet temperature, (cold/warm flow 3:1).

Warm surface water at A suffers an isentropic drop in pressure to C. This pressure drop is inevitably associated with a temperature drop. The transition from C to B is accomplished by a cold deep seawater spray. The cold water is the only single exhaust of the system. The exhaust consists of both the warm-water intake, which has isentropically cooled to T_C, and the input cold water, which absorbs the heat of condensation of the vapor at C.

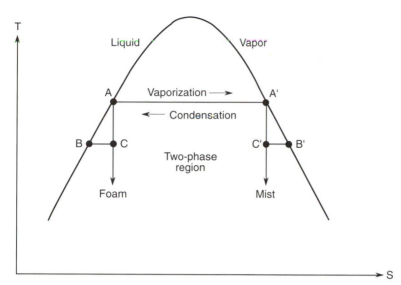

FIG. 5-24. Lift cycle *T-S* diagram.

Mist and foam lift are two types of single-exhaust OTEC systems. The liquid and vapor phases in the two-phase region may be distributed either having droplets of water in a continuous vapor phase (mist) or having bubbles of vapor in a continuous liquid phase (foam). The enthalpy released by the warm water in process AC may also be changed either directly as potential energy (foam lift) or as kinetic energy (mist lift). In the potential energy case, the water automatically rises against gravity. The foam lift cycle proposed by Zener and Fetkovich (1975) raises the liquid water first and then allows it to fall and releases its stored potential energy to a hydraulic turbine. In contrast, the mist lift cycle discussed in the last section let the warm water first fall into a deep hole to convert the released energy directly into kinetic energy; the kinetic energy is then transformed into useful work by an impulse turbine.

Several advantages of the foam lift (and mist lift) system over the conventional OC OTEC and closed-cycle OTEC systems are (Zener and Fetkovich, 1975):

1. Closed-cycle double-exhaust OTEC plants require heat exchangers. These constitute a major cost of such plants. Single-exhaust systems have no heat exchangers.
2. Open-cycle double-exhaust systems require power production by a very low vapor pressure turbine. In contrast, single-exhaust OTEC systems require only high head hydraulic turbines. Hydraulic turbines are less expensive per unit power output than very low pressure vapor turbines.
3. All double-exhaust OTEC systems eject slightly chilled warm water. Special care must be taken to avoid this warm exhaust water being sucked back into the warm-water intake.

4. Whereas double-exhaust systems only skim the available enthalpy of the warm-water intake, single-exhaust systems use the major part of this enthalpy. The power obtained from a given warm-water intake is therefore much greater from a single-exhaust plant than from a double-exhaust plant.
5. In addition to the power output of a single-exhaust OTEC plant being 5 times greater for the same warm-water input than for a double-exhaust OTEC plant (Zener and Fetkovich, 1975), the single-exhaust OTEC plants may be spaced much closer together without interference than can double-exhaust OTEC plants. The power density per unit area can therefore be greater than that of double-exhaust plants. This greater power density is especially important when the power is to be transmitted to shore from offshore plants. The factor of 5 increase quoted here, however, may be a very high-end estimation.

A schematic diagram of the foam lift OTEC plant designed to generate a hydrostatic water head is shown in Fig. 5-25. Points A, B, and C correspond to those similarly labeled in Fig. 5-24. Consider that the air is removed from the device, so that only water (liquid and vapor) exists within. The warm surface seawater intake at 25°C with a saturation vapor pressure of 3.17 kPa (0.458 psi) takes place at point A, which is above sea level. The foam breaker at C separates the liquid from the vapor in the foam. At B the vapor is condensed by thermal contact with deep ocean

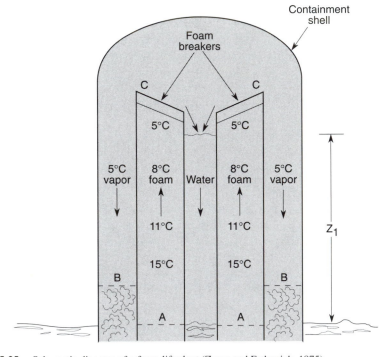

Fig. 5-25. Schematic diagram of a foam lift plant (Zener and Fetkovich, 1975).

water at 5°C [saturation pressure 0.87 kPa (0.126 psi)]. There is a flow of vapor from the warm water (25°C) to the condenser (5°C). The water between 25 and 5°C exists as a foam. The flow of vapor from the warm region to the cool region is accompanied by a flow of the whole foam structure. The foam structure consists of the individual foam walls, the cell edges where two walls meet, and the cell corners. The liquid and vapor can be separated by the foam breaker at the top of the tower. The liquid is channeled into the central water column, where it is used to drive the turbine and then discharged back into the sea.

A maximum lift height of 296 m (972 ft) is attained only when the available enthalpy is completely converted into potential energy. Under this condition, the foam would rise with a zero velocity and, hence, does not produce any power. The maximum power per unit horizontal area is generated (Zener and Fetkovich, 1975) at 17.4 kW/m^2 (1.62 kW/ft^2) when the foam breaker is at two-thirds of its maximum height, 197 m (648 ft). The time required for the foam to reach the foam breaker at 197 m is 21 s. However, surfactant (e.g., Neodol) at a concentration of the order of 200 ppm is required to create a stiffer foam to prevent premature rupture during the lift process, and antifoaming chemicals may be required to break the foams in the condenser. Effects of the chemical additives upon the operating cost and the environment are unique to the foam lift power system and will have an important bearing on the success of this concept.

5.5.2 Design Analysis

The foam lift systems development has been carried out at Carnegie–Mellon University (Zener et al., 1975) complemented by work at the University of Puerto Rico (Kay, 1979), where seawater experiments have been conducted. A conceptual design (Zener and Fetkovich, 1975) of a foam lift OTEC plant is presented in Fig. 5-26. A dome encloses the power plant above the sea level. Inside the dome, the pressure lies between the saturation vapor pressures of warm surface seawater and cold deep seawater. The dome wall will be made of reinforced concrete strong enough to withstand the hydrostatic pressure of the outside atmosphere. The wall thickness will be proportional to the dome diameter. A wall 0.61 m (2 ft) thick will suffice for a dome radius of 244 m (800 ft). Such a radius is appropriate for a 1000-MW plant.

The dome is supported by a barge, which has the shape of a doughnut sliced horizontally. Such a shape allows the dome's weight to be supported on the outer edge of the barge with a minimum of bracing structures. The machinery and pipes are supported on the inner edge of the barge.

The low pressure inside the dome sucks warm water through the warm-water input duct from near the surface onto a foam-generating tray. A suitable surfactant is mixed with the warm water. The low pressure inside the dome also sucks cold water from the ocean depth through the vertical pipe into the spray condenser. The low vapor pressure of this cold water (compared to the vapor pressure of the warm water in the foam generator) sucks the foam upward. In the rising foam, the motion of the vapor inside each foam cell is tied to the motion of the liquid of the enclosing

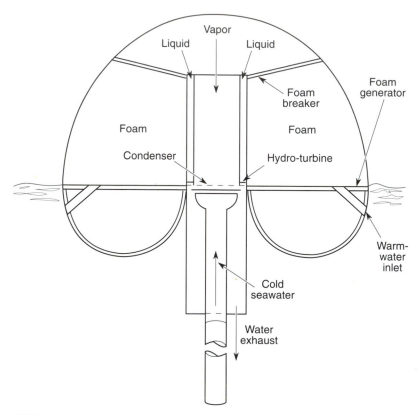

FIG. 5-26. A conceptual 1000-MW foam lift OTEC plant design (Zener and Fetkovich, 1975).

cell walls and cell corners. This close tie-in is destroyed by the foam breaker. The liquid runs down into standing pipes and thence through the hydro-turbine. The freed vapor descends the pathway into the condenser. The condensation of this vapor somewhat warms the cold water entering the spray. Together they leave the spray chamber to join the exit water from the turbines. This exit water is directed through a pipe that ends above the level where the ocean temperature matches the temperature of the exiting water.

Based on a temperature differential of 20°C (from 25 to 5°C) and at maximum power generated per unit of foam generator area condition, the design dome radius is 295 m (970 ft). Allowing one-fourth of the area enclosed by the dome for the condenser and three-fourths for the foam generator, the area is 206,000 m² (2,217,000 ft²). The net system efficiency is estimated at 33% of its ideal efficiency. This is lower than that of the closed cycle.

The basic costs of the concrete hull structure and power machinery (including hydraulic turbine, generator, water pump, and vacuum pump) are estimated (Chen and Michael, 1980) to be $131 and $135/kWe (1979 value), respectively. The cost estimate is comparable to those of the closed- and the open-cycle systems.

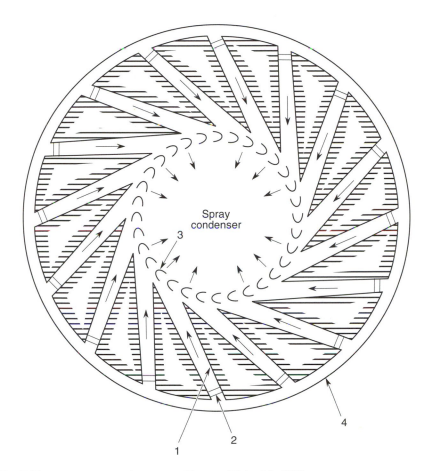

FIG. 5-27.   A vapor supersonic generator (Zener and Fetkovich, 1975).

A conceptual design (Zener and Fetkovich, 1975) of a supersonic foam generator is given in Fig. 5-27. Warm water enters one of a multitude of diverging channels (*1*). At the entrance to each channel is a foam generator (*2*). The foam leaving each channel is supersonic; hence, the diverging channel. In the channel the pressure drops from 3.17 to 0.87 kPa, corresponding to a temperature drop from 25 to 5°C. The foam leaving the channel is traveling at 76 m/s (250 ft/s), corresponding to an enthalpy drop of 2.8 J/g. This high-velocity foam strikes the impulse blades (*3*) of a DeLaval turbine moving at a velocity of about 38 m/s (125 ft/s). Water, liquid, and vapor leaving the blades have an essentially zero tangential velocity. The vapor is converted back to water by a spray condenser formed from cold water brought up from the ocean depths. The condensed water has a temperature somewhat higher than that of the cold-water intake. This condensed water, together with the liquid water originally within the foam, is ejected downward back into the ocean.

The whole assembly is surrounded by a shell (*4*), which allows the desired low pressure to be maintained within it. The conceptual design shown in Fig. 5-27 is independent of absolute dimensions.

The foam lift OTEC power plant is a relatively new concept. Further investigation on technical issues includes (1) demonstration of stable foam lift flow concentration in the order of 10 ppm or less, (2) development of size scale-up relations, (3) handling of noncondensables, and (4) system and platform design to verify its potential for OTEC power and industrial applications.

5.6   ESTIMATED COSTS AND MARKET OPPORTUNITIES

Some early calculations of both the initial capital cost and the cost per kilowatt-hour imply a favorable economic future for OC OTEC technology. Although the first unit built will cost considerably more per kilowatt than fossil-fueled or nuclear power plants, expected savings due to technological improvements, no fuel expense, and long-life operation may allow OC OTEC plants to become economically competitive with more conventional energy sources in the near future. Estimates of the capital cost of an OC OTEC power plant vary widely since the different plant designs assumed different temperature differentials, different design parameters, different sites, different materials for heat exchangers, different based (floating or fixed) plants, and different deployment methods. Some of the cost estimates are given in the following.

A power module cost comparison between a 100-MW open-cycle and a closed-cycle OTEC plant has been presented (Westinghouse Electric Co., 1978) for major cost items. The open cycle uses a surface condenser with 90–10 Cu–Ni tube bundles and no warm-water deaeration. The major cost items in the closed-cycle power module are the two heat exchangers, but the ammonia turbines are relatively low cost contributors. The major cost items in the open-cycle module are the steam turbines, condenser, and staged air-removal system.

The introduction of the OC OTEC technology into the marketplace will not be simply a consequence of its successful demonstration and its achieving competitive cost. It also depends on resource availability, potential markets, market penetration, net energy considerations, etc.

Factors other than technology and economics (for example, legal, institutional) also confront the commercialization of open-cycle OTEC power plants. These factors will be discussed in a later chapter.

5.7   CURRENT OC OTEC STATUS

OTEC research and development efforts led to many important accomplishments in the late 1970s and early 1980s. At-sea testing with Mini-OTEC in 1979 and OTEC-1 in 1981 demonstrated the technical feasibility of closed-cycle OTEC systems and provided data on heat exchanger thermohydraulics, environmental impacts, cold-water pipe deployment, and systems operations among others.

OTEC research prior to 1981 was directed toward large floating plant systems that would provide power to islands, certain parts of the mainland, and large plantships that produce energy-intensive products. Closed-cycle OTEC plants were believed to be closer to commercial readiness. Open-cycle plants were thought to have significant long-term potential but were less developed and thus offered less potential for near-term production. Funding for OTEC development was curtailed in 1980, and the limited funds available were shifted to longer-term research and redirected from closed-cycle to open-cycle OTEC research. Current OTEC research focuses primarily on assessing and solving the technological problems leading to the development of land-based or near-shore open-cycle systems ranging in size from 2 to 15 MW. Researchers are looking at ways to make OC OTEC plants more efficiently by designing smaller plants that produce more power, using smaller volumes of water, and incorporating less expensive materials and components.

A number of recent discoveries (1983–1988) have enhanced the economic prospects for the OC OTEC systems:

1. Plentiful and relatively inexpensive aluminum alloy can be used in lieu of the more expensive titanium for making OTEC heat exchangers.
2. Cost-effective turbines for OC OTEC systems are being developed.
3. A flash evaporator with 97% efficiency has been developed to convert warm seawater into low-pressure steam for use in an OC OTEC steam turbine.
4. Desalinated water has been produced by using the open-cycle process at prototypical conditions.
5. A seawater supply system has been deployed to provide the warm and cold seawater for OC OTEC electrical energy production, with a cold-water mariculture option.
6. A direct-contact condenser used in OC OTEC systems has made significant advances, with thermal effectiveness over 95% at OTEC operating conditions being tested.

REFERENCES

Barr, R., W. Deucher, J. Giannotti, R. Scotti, J. Stadter, J. P. Walsh, and R. Weiss, 1980. "Report of the Ad Hoc OTEC CWP Committee: An assessment of existing analytical tools for predicting CWP stresses." *Proc. 7th Ocean Energy Conf.*, Washington, D.C., June, 2, 14.5.

Barr, R. A., and P. Y. Chang, 1977. "Some factors affecting the selection of OTEC plant platform/cold water pipe designs." *Proc. 4th Annual Conf. on Ocean Thermal Energy*, New Orleans, La., March, V-11.

Bartone, L. M., Jr., 1978. "Alternative power systems for extracting energy from the ocean: a comparison of three concepts." *Proc. 5th Ocean Thermal Energy Conversion Conf.*, Miami Beach, Fla., Feb., V-11.

Beck, E. J., 1975. "Ocean thermal gradient hydraulic power plant." *Science,* 189, 293.

Bell, K. J., 1983. "Heat exchanger evaluation for OTEC program." Department of Energy Report, DOE/ET/20312-5, Feb.

Benedek, S., 1980. "Heat transfer at the condensation of steam on turbulent water jet." *International J. of Heat and Mass Transfer,* **19**, 448.

Bharathan, D. F., 1981. "Measured performance of direct-contact jet condensers." *AIAA Second Terrestrial Energy Systems Conf.,* Colorado Springs, Col., AIAA-81-2594-252-1437, Dec.

——, F. Kreith, and D. Schlepp, 1984. "Heat and mass transfer in OC-OTEC condenser." *ASME Trans. J. Heat Transfer Engineering,* **5**, 287.

——, F. Kreith, and W. L. Owens, 1983. "An overview of heat and mass transfer in open-cycle OTEC systems." *Thermal Engineering Joint Conf. Proc.,* Honolulu, Hawaii, March, **2**, 301.

Block, D. L., 1984. *Thermoeconomic optimization of OC-OTEC electricity and water production plants.* Florida Solar Energy Center Report, FSCE-CR-108-84, Sept.

Bobe, L. S., and D. D. Malyshev, 1971. "Calculating the condensation of steam with a cross-flow of steam gas mixture over tubes." *Teploenergetika,* **18**, 84.

Boot, J. L., and J. G. McGowan, 1974. *Feasibility study of a 100-MW open cycle ocean thermal difference power plant.* National Science Foundation report, NSF/RANN/SE/GI-34879/TR/74/3, Aug.

Brewer, J. H., 1979. "Land-based OTEC plants—cold water pipe." *Proc. Ocean Thermal Energy for the 80's Conf.,* Washington, D.C., June, **1**, 4B-5.

Brewer, J. H., J. Minor, and R. Jacobs, 1979. *Feasibility design study, land-based OTEC plants.* DOE report, 600-4431-02, Jan.

Brown, C. E., and L. Wechsler, 1975. "Engineering analysis of an open cycle power plant for extracting solar energy from the sea." *Proc. 7th Ocean Energy Conf.,* Washington, D.C., June, **1**, 1.

Browne, R. D., and M. R. C. Bury, 1981. "Practicalities of concrete for OTEC applications." *Proc. 8th Ocean Energy Conf.,* Washington, D.C., June, **1**, 117.

Bugby, D. C., A. T. Wassel, and A. T. Mills, 1983. "Bubble nucleation and growth in OC-OTEC subsystems." *ASME Transactions J. of Solar Energy Engineering,* **105**, 119.

Buglaev, V. T., M. M. Andreev, and V. P. Kleschevnikov, 1971. "Investigation of steam condensation in the presence of air on a horizontal tube bank under vacuum conditions." *J. Thermal Engineering,* **18**, 19.

Charwat, A. F., 1978. *Studies of the vertical mist transport process for an OTEC cycle.* DOE Contract No. EY-76-5-03-0034, Nov.

——, P. R. Hammond, and S. L. Ridgway, 1979. "The mist transport cycle: Progress in economic and experimental studies." *Proc. 6th OTEC Conf.,* Washington, D.C., June, **8c-3**, 1.

Chen, F. C., 1979. "A thermodynamic assessment of OTEC open-cycle power system." *Proc. 10th Annual Pittsburg Modeling and Simulation Conf.,* Pittsburg, Pa., April, **10**, 921.

——, and A. Golshani, 1982. "OTEC gas desorption studies." *ASME Transactions J. of Solar Energy Engineering* **104**, 35.

——, and J. W. Michael, 1980. "Development of OTEC lift cycles." *Proc. 7th Ocean Energy Conf.,* Washington, D.C., June, **2**, 16.1.

Coleman, W. H., D. F. Rodgers, D. F. Thompson, and M. I. Young, 1981. "OC-OTEC turbine design." *Proc. AIAA 2nd Terrestrial Energy Systems Conf.,* New York, N.Y., Dec., **2**, 108.

Coons, K. W., 1957. *Direct condensers.* Engineering Experiment Station report, College of Engineering, University of Alabama, Bulletin **3**.

Creare R and D, Inc., 1984. *Thermoeconomic optimization of OC-OTEC electricity and water production plants*. Creare report TM-977, Sept.

Curto, P. A., 1978. "An update of OTEC baseline design costs." *Proc. 5th Ocean Thermal Energy Conversion Conf.*, Miami Beach, Fla., Feb., II-77.

Davenport, H, 1982. *Mist lift analysis summary report*. Solar Energy Research Report No. SERI/TR-631-627.

Dugger, G. L., R. W. Henderson, E. J. Francis, and W. H. Avery, 1980. "Projected costs for electricity and products from OTEC facilities and plantships." *AIAA*, paper no. 809263.

Dukler, A. E., 1971. *Distillation plant data book*. Office of Saline Water, Jan.

Fournier, T., 1985. "OC-OTEC experimental study of flash evaporation." *Proc. 1985 Ocean Engineering and Environment Conf.*, San Diego, Cal., Nov., **2**, 1222.

Fraas, A. P., and M. N. Ozisik, 1965. *Heat exchanger design*. New York: John Wiley and Sons, Inc.

Francis, E. J., 1977, *Investment in commercial development of OTEC plantships*. Johns Hopkins University Applied Physics Laboratory, SR-77-3.

Fuks, S. N., and E. P. Zernova, 1970. "Heat and mass transfer with condensation of pure steam and of steam containing air, supplied from the side to a tube bank." *Teploenergetika*, **17**, 59.

Gauthier, M., 1991. "The pioneer OTEC operation 'La Tunisie.'" *IOA Newsletter*, **2** (1&2).

Ghiaasiaan, S. M., and A. T. Wassel, 1983. "Inverted vertical spout evaporators for OC-OTEC." *International Commun. Heat and Mass Transfer*, **10**(8), 511.

Green, H. J., D. A. Olson, D. Bharathan, and D. H. Johnson, 1981. "Measured performance of falling jet flash evaporators." *8th Ocean Energy Conf.*, Washington, D.C., June.

Hagen, A. W., 1975. *Thermal energy from the sea*. Noyes Data Corp., Park Ridge, N.Y.

Henderson, C. L., and J. M. Marchello, 1969. "Film condensation in the presence of a noncondensible gas." *ASME Transactions J. Heat Transfer*, **91**, 447.

Jahnig, C. E., 1979. "Hybrid OTEC cycle avoids indirect heat exchangers." *Proc. 6th OTEC Conf.*, Washington, D.C., June, **2**, 8D-3/1.

Johnson, D. H., 1978. "The energy of the ocean thermal resource and analysis of second law efficiencies of idealized OTEC power cycles." *J. of Energy*, 927.

Kay, M. I., 1979. "Description and status report of a program to define sea-water surface structure interaction in relation to the foam system." *Proc. 6th OTEC Conf.*, Washington, D.C., June, **1**, 9.3.

Kogan, A., D. H. Johnson, H. J. Green, and D. A. Olson, 1980. "OC-OTEC system with falling jet evaporator and condenser." *7th Ocean Energy Conf.*, Washington, D.C., **2**, 12.5.

Kreith, F. K., and D. Bharathan, 1987. "Heat transfer research for OTEC." *Proc. 2nd ASME/JSME Thermal Engineering Conf.*, Honolulu, Hawaii, March, **2**, 149.

Larsen-Basse, J., 1983. "Effect of biofouling and counter measures on heat transfer in surface and deep ocean Hawaiian waters—early results from the sea coast test facility." *Proc. 2nd ASME/JSME Thermal Engineering Conf.*, Honolulu, Hawaii, March, **2**, 285.

——, N. Sonwalker, and A. Seki, 1986. "Preliminary seawater experiments with open-cycle OTEC spout evaporation." *Proc. Ocean 1986 Conf.*, Washington, D.C., Sept., 202.

Lavi, A., 1975. "Solar sea power plants, cost and economics." *Proc. 3rd Workshop on Ocean Thermal Energy Conversion*, Houston, Tex., May, 64.

Lee, G. K. B., and S. L. Ridgway, 1983. "Vapor/droplet coupling and the mist flow (OTEC) cycle." *Trans. of the ASME J. Solar Energy Engineering*, **105**, 181.

Leninger, T. F., 1983. *Mass transfer in round turbulent water jets.* University of California at LA report, UCLA-ENG-8309.

Lewandowski, A. A., D. A. Olson, and D. H. Johnson, 1980. *OC-OTEC system performance analysis.* Solar Energy Research Institute report, SERI/TR-631-692.

——, D. A. Olson, and D. H. Johnson, 1982. *Open-Cycle OTEC System Performance Analysis,* SERI Research No. SERI/TR-631-692, Oct.

Litvin, A., and A. E. Fiorato, 1979. "Development of a light-weight concrete for OTEC cold water pipes." *Proc. 7th Ocean Energy Conf.,* Washington, D.C., June, **1**, 4B-5/1.

McDermott, Inc., 1981. *Conceptual design study: cold water pipe systems for shelf-mounted OTEC power plants.* DOE report, DOE/NOAA/OTEC-48.

Mills, A. F., C. Tan, and D. K. Chung, 1974. "Experimental study of a condensation from steam–air mixtures flowing over a horizontal tube: overall condensation rates." *Proc. 5th International Heat Transfer Conf.,* Toyko, Japan, **5**, 20.

Nauf, F. E., 1976. "Economic aspect of OTEC." *Proc. Joint Conf. on Sharing the Sun, Solar Technology in the Seventies,* Cape Canaveral, Fla., 393.

Ocean Thermal Corp., 1983. *OTEC pilot plant conceptual design study.* DOE contract, DOE-AC01-82CE30716.

Ofer, S., 1983. *Experimental study of evaporation from round turbulent water jets.* UCLA report, UCLA-ENG-8308.

Olson, D. A., 1981. *Water flows from slotted pipes.* DOE contract, AC02-77CH00178.

ORI, Inc., 1983. *Analysis and summary of OTEC cold water pipe costs.* DOE report, DOE/NOAA/OTEC-42.

Panchal, C. B., and K. J. Bell, 1984. "Theoretical analysis of condensation in the presence of noncondensible gases as applied to OC-OTEC condensers." *ASME winter annual meeting,* New Orleans, La., Dec.

Panchal, C. B., and K. J. Bell, 1987. "Simultaneous production of desalinated water and power using a hybrid-cycle OTEC plant." *ASME Transactions J. of Solar Energy Engineering,* **109**, 156.

Panchal, C. B., and H. C. Stevens, 1986. "Apparatus for OC-OTEC heat and mass transfer experiments at the seacoast test facility." *Proc. 1986 Oceans Conf.,* Washington, D.C., Sept., **1**, 213.

Parsons, B., D. Bharathan, and J. Althof, 1984. "OC-OTEC thermal-hydraulic systems analysis and parametric studies." *Proc. 1984 Ocean Conf.,* Washington, D.C., Sept., **1**, 370.

Parsons, B. K., D. Bharathan, and J. A. Althof, 1985. *Thermodynamic analysis of OC-OTEC.* Solar Energy Research Institute report, SERI/TR-252-2234.

Penney, T. R., 1984. *Small scale OC-OTEC power systems.* Solar Energy Research Institute report, SERI/TR-252-2184.

——, 1986. "Composite turbine blade design options for Claude OC-OTEC power systems." *5th International Symp. and Exhibit on OMAE,* Tokyo, Japan, April.

——, and J. A. Althof, 1985. "Measurements of gas sorption from seawater and the influence of gas release on OC-OTEC system performance." *1985 Solar Energy—The Diverse Solution Conf.,* Montreal, Canada, June.

——, D. Bharathan, J. Althof, and B. Parsons, 1984. "OC-OTEC research: Progress summary and a design study." *ASME winter annual meeting,* 84-WA/SOL 26.

Ridgway, S. L., 1977. "The mist flow OTEC plant." *Proc. 4th Annual Conf. on Ocean Thermal Energy,* New Orleans, La., March, **VIII**, 37-4D.

——, and P. R. Hammond, 1978. *Mist flow ocean thermal energy process.* R & D Associates Report No. RDA-TA-107800-002, Sept.

——, R. F. Hammond, and C. K. B. Lee, 1980. *Mist flow ocean energy thermal process.* Oakridge National Laboratory Report No. ORNL-7613/1, Apr.

Rogers, L. J., R. J. Hays, and A. R. Trenka, 1989. "A status assessment of OTEC technology." *Proc. EEZ Resources: Technology Assessment Conf.*, Honolulu, Hawaii, Jan., 6.32.

Rosard, D. D., 1978. "Working fluids and turbines for OTEC power systems." *Proc. Fluids Engineering in Advanced Energy Systems Conf.*, San Francisco, Cal., Dec., 229.

Saha, P., 1979. "A review of condensation of steam on subcooled water jets." *NRC/RSR Condensation Workshop*, Silver Spring, Md.

Sam, R. G., and B. R. Patel, 1982. *OC-OTEC evaporator/condenser test program data report.* Creare R & D, Inc., report, TN340.

——, 1984. "An experimental investigation of OC-OTEC direct-contact condensation and evaporation processes." *ASME Transactions J. of Solar Energy Engineering,* **106**, 120.

Science Applications, Inc., 1979. *OTEC modular experiment: cold water pipe system design study.* DOE contract, DOE/NOAA/OTEC-15.

Sklover, G. G., and V. G. Grigorev, 1975. "Calculating the heat transfer coefficient in steam turbine condensers." *Teploenergetika,* **22**, 67.

Solar Energy Research Institute, 1983. *Demisters for potable water production for open-cycle OTEC systems.* Solar Energy Research Institute report, SERI/TR-252-1991.

Trimble, L. C., 1975. "OTEC systems study report." *3rd Workshop on Ocean Thermal Energy Conversion*, Houston, Tex., May.

TRW Systems and Energy Group, 1979. *Ocean thermal energy conversion cold water pipe preliminary design project.* DOE report, DOE/NOAA/OTEC-14.

Wallis, G. B., et al., 1980. "A computer model for the mist lift process in an OTEC open cycle system." 11th Annual Pittsburgh Conf. on Modeling and Simulation, Pittsburgh, Pa., May.

Wassel, A. T., and S. M. Ghiaasiaan, 1985. "Falling jet flash evaporators for OC-OTEC." *Int. Comm. Heat and Mass Transfer,* **12**, 113.

Watt, A. D., F. S. Matthew, and R. E. Hathaway, 1978. "Open cycle energy conversion— a preliminary engineering evaluation." *Proc. 5th Ocean Thermal Energy Conversion Conf.*, Miami, Fla., Feb., **VII**, 13.

——, 1977. *Open cycle energy conversion, a preliminary engineering evaluation.* Colorado School of Mines report.

Westinghouse Electric Corp., 1978. *100 MW OTEC alternative power systems.* DOE contract report, EG-77-C-03-1473.

Zener, C., and J. Fetkovich, 1975. "Foam solar sea power plant." *Science,* **180**, 294.

6

OTEC CLOSED-CYCLE ENGINEERING STATUS

Engineering analyses and component design studies during the period 1974–1977 indicated the feasibility of constructing and operating floating OTEC plants and plantships in a variety of configurations ranging in power from 40 to 500 MWe. In August 1979, an at-sea test of a complete OTEC power system (Mini-OTEC) demonstrated performance in good accord with engineering predictions and established a firm basis for scale-up to larger sizes (Owens and Trimble, 1980). Heat exchanger operation at a level equivalent to 1-MWe power generation was demonstrated 1 year later in the OTEC-1 program. In 1981, a complete land-based OTEC power plant was constructed and operated under Japanese direction at the island of Nauru on the equator in the mid-Pacific ocean.

During the period 1977–1980, a U.S. plan was developed, supported by public laws PL 96-310 and PL 96-320, to demonstrate OTEC feasibility at a 100-MWe level by 1985 and 500 MWe by 1990. Testing was to start with a pilot demonstration at 40 MWe (net). Preliminary design of baseline demonstration plants at this power level for moored operation off Punta Tuna, Puerto Rico, and for grazing operation west of equatorial Brazil with on-board ammonia production was completed in 1980 (George and Richards, 1980). Conceptual designs of larger plants and power systems for demonstration at the baseline level were also completed.

In accord with the requirements of the Congressional actions, a Program Opportunity Notice (PON) was issued in September 1980 by DOE that offered cost-sharing support for innovative OTEC systems designs that contractors believed would be commercially viable if government cost sharing were made available during development of demonstration vessels. The PON asked for proposals for a development program to design, construct, and test a 40-MWe (net) closed-cycle OTEC system, which would be conducted in six phases beginning with conceptual design and continuing to preliminary design, engineering design, construction, deployment and operation, and, finally, transfer of ownership and contractor operation. The schedule was set to be consistent with the goal established by PL-96-310 of demonstration of 100-MW OTEC operation by 1985. DOE stated its intent to fund five to eight awards for the first phase, with DOE providing $900,000 as its share of each contract awarded (Dugger et al., 1983).

Responses to the PON were received by DOE from eight contractor teams. Summaries of the proposals, which contained proprietary material, are presented in Section 6.3.5. Part of the material has been made available to the authors by the respondents. This material, combined with information from published reports, provides a good basis for judging OTEC component availabilities, costs, and delivery schedules. An important conclusion that emerges is that a large fraction of the total costs of a floating OTEC system would be for components that are commercially available and could be purchased at a firm cost. Furthermore, all construction would be possible with existing technology, with facilities adapted as required for OTEC use.

Sections 6.1 to 6.4.6 of this chapter present a condensation of technical material and costs gleaned from the available reports that show the state of U.S. closed-cycle OTEC technology when development support by DOE was terminated in 1984. The state of OTEC technology in Japan is reviewed in Section 6.4.7, in France in Section 6.4.8, and in Taiwan in Section 6.4.9.

6.1 STANDARD EQUIPMENT APPLICABLE TO OTEC

During the late 1970s, surveys were conducted in support of the DOE program to establish the availability and costs of components that could satisfy the specific requirements of closed-cycle OTEC systems and installations (Barness et al., 1978; Lockheed, 1979a and b; Olmsted et al., 1978). These studies included development of computer programs to define component dimensions and performance specifications to satisfy specific design goals. Cost estimates based on surveys of available technology were included.

The documentation of a study of OTEC power system development reported by Westinghouse (Barsness et al., 1978) is typical of the depth of the analyses performed in these programs. The Westinghouse report includes a power module steady-state analysis program (OSAP) that takes data from subprograms dealing separately with design and performance characteristics of evaporators, condensers, turbine, pumps, valves, and CWPs. Detailed discussions of engineering requirements, estimated development schedules, and estimated costs are presented for:

1. Construction and deployment of concrete hulls (Dillingham);
2. Weight and stability estimates for a complete 10-MWe OTEC plant;
3. Packing factor data for steel and concrete floating plants;
4. Machinery and equipment list;
5. Cost and performance data of heat exchanger tubing alternatives;
6. Working fluids and turbines for OTEC power systems; and
7. Hydrodynamic design of OTEC pumps for cold and warm seawater.

Similar material is in the reports by Lockheed (1979a and b) and GE (Olmsted et al., 1978) and by the PON respondents.

6.2 FACILITIES FOR OTEC CONSTRUCTION

The OTEC system requires construction of CWPs and advanced types of heat exchangers as special items in a complete assembly that incorporates commercially available turbines, generators, pumps, power distribution equipment, control equipment for power systems and sea-keeping, hotel and shop facilities, and miscellaneous items. The platform can be constructed in an existing shipyard. Some of the equipment can be installed at the drydock. Most of the remaining equipment will be installed after the hull is floated from the drydock to a nearby pier where the water is deep enough to accommodate the desired draft. A schematic of this procedure is shown in Fig. 6-1 (George and Richards, 1980). The platform is then towed to a suitable deep-water area near the operational site, where the cold-water pipe is installed. A facility for construction of a concrete or plastic CWP is built at this location to avoid the need for a long sea journey by the CWP.

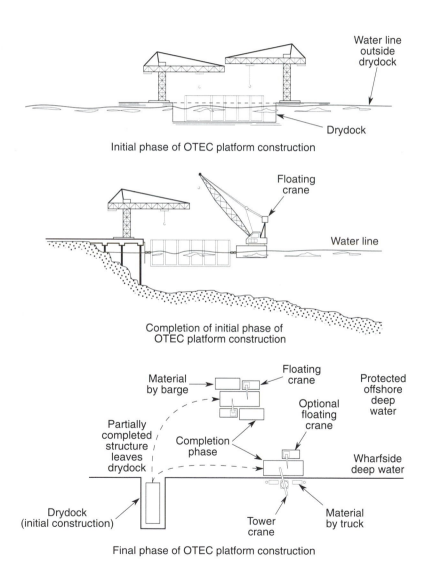

FIG. 6-1. Schematic of OTEC platform construction (George and Richards, 1980).

6.2.1 Shipyards

The discussion in this section applies only to floating OTEC platforms. Facilities for land-based plants are site specific to the local terrain and construction capabilities. Shipyards in the United States suitable for construction of OTEC vessels were surveyed in 1978 in connection with the OTEC 40-MWe baseline design program (George et al., 1979). The basic requirements defined were for a graving dock of width exceeding 50 m and length over 150 m, along with supporting facilities that would permit construction of a fully outfitted concrete platform. With heat exchanger wells closed and without the CWP, the required draft is approximately 12 m. The characteristics of shipyards throughout the world are available in the *Directory of Shipowners, Shipbuilders, and Marine Engineers* (Motor Ship Press, 1988) and in *Lloyd's Maritime Directory* (Lloyd's of London Press, 1988). The candidate U.S. shipyards that were identified will be discussed briefly here.

1. Newport News Shipbuilding Co., Newport News, Virginia: The shipyard is a large complex with multiple shipbuilding ways and drydocks. Of particular interest is a graving dock that is 490 m × 76 m × 13.4 m depth and is served by an 870-tonne gantry crane. The shipyard is equipped for steel ship construction, but the management expressed interest in expanding the facilities as needed for construction of OTEC vessels of the concrete barge design.

2. Avondale Industries, Inc., Avondale Shipyard Div., New Orleans, Louisiana: Avondale's main shipyard has three outfitting docks with supporting shops and over 1800 m (6000 ft) of pier space. The upper yard has two large positions that can accommodate vessels up to 311 m in length by 53 m beam. Part of one ship can be constructed at position #1 while a second part is being built at position #2. The parts can be moved laterally for final assembly.

3. Concrete Technology Corporation (CTC), Takoma, Washington: Concrete Technology is a world leader in construction of marine structures based on reinforced and post-tensioned concrete. Its shipyard includes a graving dock that is 156 × 9 × 3.7 m (510 × 160 × 12 ft) and an outfitting pier with water depth of 9.1–10.7 m (30–35 ft). Concrete Technology built and outfitted the ARCO LPG concrete barge, of approximately the same dimensions as the baseline 40-MWe OTEC design. The lower part of the hull of the ARCO barge was constructed in the graving dock, after which it was floated out and stationed next to the pier where construction and outfitting were completed. Construction of the OTEC demonstration vessel at this facility would involve minimal technical and financial risk.

4. National Shipbuilding (NS), San Diego, California: National Shipbuilding has a large facility with shipways of suitable size for OTEC ship construction. There is no graving dock, but construction could be done in one of the shipways that have a depth of 3.7 m (12 ft). Water depth at the outfitting dock is 8.5 m (28 ft). This might require construction in three phases.

5. Hawaii: There is no commercial graving docks or shipways in Hawaii that would be suitable for OTEC hull construction; however, there are large-scale concrete and steel construction capabilities in Hawaii that could support OTEC

platform construction if a graving dock and pier were installed. It appears that unused facilities at the U.S. Navy Pearl Harbor shipyard could be made available if Federal support for commercial development of OTEC emerged, reflecting world concerns about fossil fuel resources and global warming.

Shipyards in Germany, Spain, Portugal, Japan, and Korea that have graving docks and berths large enough for construction of vessels exceeding 500,000 dwt are listed here. These and many other smaller shipyards would be obvious candidates for construction of commercial-size OTEC platforms:

Howaldswerke-Deutsche Werft (HDW), Kiel, W. Germany
Division de Constructione Naval (DCN), Madrid, Spain
Setanave-Estaleiros Navas de Setubal, Setubal, Portugal
Hitachi-Zosen Ariake Works, Osaka, Japan
Kawasaki Heavy Industries Ltd., Kobe, Japan
Hyundai Heavy Industries Co., Ltd., Kyung-nam, S. Korea

6.2.2 Hull and CWP Construction Technology

Steel, concrete, aluminum, titanium, and reinforced plastics are all in current use in marine engineering and are candidates for OTEC application. The requirements for the platform and CWP differ and employ different selection criteria.

6.2.2.1 Platform Construction

Economic viability of OTEC requires that the construction and operating costs be low enough to permit OTEC to be competitive with alternative ways of supplying energy. This implies low construction cost and long life of the system at sea with minimum requirements for maintenance or replacement of equipment. Systems engineering leads to a preference for concrete for the platform for the following reasons:

1. Concrete is durable in marine environments without maintenance. Some concrete ships built during World War II are still in service, and concrete piers and breakwaters are used worldwide without maintenance. Steel ships must return to drydock at approximately 5-year intervals for repainting. This would require a steel OTEC vessel to be designed so that the CWP could be detached or separately supported. This would add to both construction and operating costs.

2. Preliminary estimates indicate that the cost of a concrete platform for OTEC would be roughly 30% less than the corresponding steel platform.

3. The greater weight and material volume of hulls constructed of concrete rather than of steel are not a disadvantage for OTEC vessels that graze at 0.5 knot or are fixed in place. In the preferred ship designs the buoyancy of the OTEC heat exchangers must be compensated by ballast even with concrete platform construction. The lower hull drag of a steel platform is inconsequential, since OTEC drag is determined primarily by the drag of the CWP.

4. The most cost-effective heat exchangers for OTEC will be made of aluminum, which could be incompatible with a steel platform because of the possibility

that rust particles from the hull could be trapped on the heat exchanger surfaces. This could induce pitting and perforation of the aluminum separating surfaces.
5. Aluminum and titanium are too expensive to be considered as construction materials for the OTEC platform.

6.2.2.2 OTEC Hull Construction

The extension of concrete technology from construction of small barges to a size useful for OTEC was demonstrated in 1976 in the construction of the 145.5-m-long ARCO barge, by Concrete Technology Corp. and ABAM engineers of Takoma, Wash. A feature of the design was the use of post-tensioning of the concrete to maintain the entire structure under compression at all operating conditions. Full details of the project are presented in Mast (1975) and Anderson (1977).

The vessel was designed to store liquefied propane and butane produced 3.2 km (20 miles) offshore near Jakarta, Indonesia, in an area where land is marshy and unsuitable for construction of land facilities. The complete vessel, equipped to liquefy and store 375,000 barrels of LPG for periodic transfer to tankers, was designed, constructed, and outfitted in Takoma during a 26-month period. It was then towed 15,000 km to the Indonesia site where it has been in use ever since. Final weight of the vessel was 59,000 tonnes. Nine thousand cubic meters of concrete were used in the construction. Overall dimensions of the hull are as follows: length, 140.5 m (461 ft); beam, 41.5 m (136 ft); depth, 17.2 m (56 ft) (Mast 1975). A photograph of the completed vessel en route to Indonesia is shown in Fig. 6-2 (Mast, 1975).

The ARCO barge was constructed in two stages. The first section of the hull was built in the Concrete Technology graving dock: 170 m long, 50 m wide, and 3.65 m depth. This section was then floated, moved out of the dock, and anchored next to a pier where water depth of 11 m was available. The upper sections were then completed.

The vessel design satisfies all requirements of the American Bureau of Shipping (A.B.S.) and related international agencies for safe operation on the high seas, with a safety factor of two on the ultimate load, including ability to withstand the worst stresses expected for the 100-year storm.

The design of the hull for the baseline 40-MWe OTEC demonstration vessel is based directly on the ARCO barge technology. The dimensions are nearly the same: length, 135.2 m (443.5 ft); beam, 42.7 m (140 ft); depth, 27.1 m (89 ft). Thus, construction risks will be minimal for the demonstration vessel and will provide a firm base for scale-up to sizes suitable for commercial OTEC vessels that will operate with minimum maintenance for the 30-year design life and beyond.

The concrete selected in the baseline design study for the OTEC platform is of normal weight, reinforced with steel bars of 414 MPa (60,000 psi) tensile strength [276 MPa (40,000 psi) where welding is necessary], and prestressed using 1860-MPa (270,000 psi) strands of 1.27 cm (0.5 in.) diameter, bundled in ducts. The design compressive strength is 41.4 MPa (6000 psi) (George et al., 1979).

FIG. 6-2. ARCO barge en route to Indonesia (Mast, 1975).

6.2.2.3 Land and Shelf Installations

Designs for construction of platforms for land-based and shelf-mounted OTEC installations have been prepared. The details depend on the topography and the facilities available at the selected sites. These are discussed in Section 6.4.

6.2.2.4 Concrete Pipe Construction

Concrete pipes are used worldwide in a large variety of applications for sewers, conduits, and water-distributed systems, and facilities are available for concrete pipe construction in a wide range of diameters, thicknesses, and lengths (George et al., 1979). Longitudinal reinforcing with steel bars and circumferential reinforcing with wire mesh are typical, as shown in Fig. 6-3 (Baumeister et al., 1978).

A concrete CWP was selected by JHU/APL for the 40-MWe baseline OTEC demonstration vessel after an investigation of state-of-the-art construction methods and materials that could be used to fabricate and deploy a pipe of the required dimensions and durability. The study showed a minimum cost for a concrete CWP, as well as maximum probability that the pipe could be built and deployed within the desired time schedule.

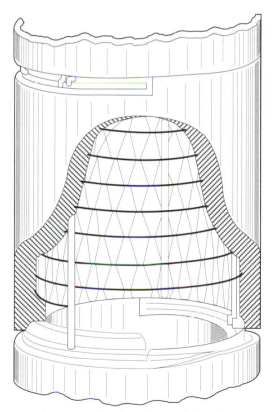

FIG. 6-3. Typical concrete pipe construction (Baumeister et al., 1978).

The use of concrete for suspended OTEC CWPs introduces several design requirements not present in standard concrete pipe structures:

1. The pipe must have longitudinal strength to withstand the weight of the pipe in seawater. This requirement is met by using concrete of lower than normal density and by post-tensioning the concrete to 41 MPa (6000 psi).

2. It is desirable to select a pipe weight that will ensure that the angle between the pipe and the platform will not exceed allowable values in storm or grazing conditions, and will also be compatible with the maximum load specifications of the pivot joint. These requirements can be met by making the pipe of low-density concrete weighing about 1350 kg/m^3 (85 lb/ft^3). (The density of seawater is 1025 kg/m^3.) The submerged weight is then 325 kg/m^3. Concrete with this density is made by using a low-density granular material to replace the sand used in standard concrete. Flyash or various proprietary materials are suitable (O'Connor, 1981).

3. As shown in Section 4.2, the bending moments would be excessive for a concrete CWP without flexible joints along the length of the pipe. Systems engineering indicates that a pipe made of segments connected by flexible joints at approximately 15-m intervals would be optimal for on-land construction, transport to the floating platform, and deployment of the (oversized) 9.1-m-diameter concrete CWP for the 40-MWe baseline OTEC plantship (Paulling, 1980).

These considerations led to selection of the pipe design shown in Fig. 6-4 for the baseline 40-MWe OTEC vessel (George and Richards, 1980). This design is suitable for OTEC power plants of power levels up to 60–80 MWe (net). The design is firmly based on experience in the construction of concrete structures by ABAM Engineers, Inc., Concrete Technology, and the Portland Cement Institute (Mast, 1975; Cichanski and Mast, 1979; O'Connor, 1981).

The post-tensioned concrete technology developed for the ARCO barge is directly applicable to the construction of the 9.1-m-diameter CWP chosen for the 40-MWe baseline OTEC barge and for engineering design of larger sizes. The feasibility of using post-tensioned concrete to satisfy the specific design requirements of the joined OTEC design was demonstrated by construction and test to failure of a 1/3-scale section of a CWP including a flexible joint of the proposed design. The test section was 3 m (10 ft) in diameter and 5 m (16.4 ft) long, including the flexible joint. Failure occurred at 140% of the predicted maximum stress for the 100-year storm. The test section was designed by ABAM and was constructed on time and within the projected cost by Concrete Technology production employees using their standard procedures. Details of the test are given in O'Connor (1980). The test arrangement is shown in Fig. 6-5.

The pipe is made by casting 9.1-m-diameter by 15.2-m-long concrete sections, incorporating structures at the two ends for the bayonet-type locking mechanisms and bearing pads for the flexible joint. Details of a typical CWP segment are shown in Fig. 6-6. After curing and outfitting, the sections are placed on a barge for delivery to the OTEC platform where they are progressively joined together and lowered to form the complete CWP.

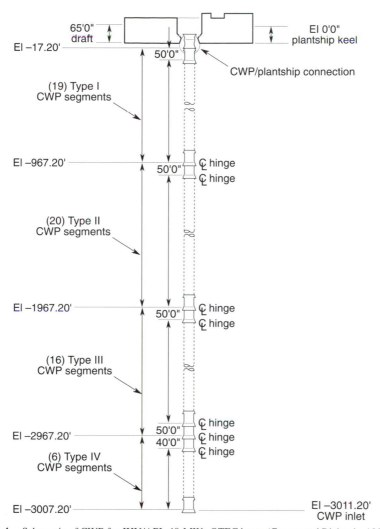

65'0" draft
El −17.20'
El 0'0"
plantship keel
50'0"
CWP/plantship connection

(19) Type I
CWP segments

El −967.20'
50'0"
₵ hinge
₵ hinge

(20) Type II
CWP segments

El −1967.20'
50'0"
₵ hinge
₵ hinge

(16) Type III
CWP segments

El −2967.20'
50'0"
₵ hinge
₵ hinge
40'0"
₵ hinge

(6) Type IV
CWP segments

El −3007.20'
El −3011.20'
CWP inlet

FIG. 6-4. Schematic of CWP for JHU/APL 40-MWe OTEC barge (George and Richards, 1980).

6.2.3 Reinforced Plastics CWP Manufacture

Plastic structures reinforced with fiberglass (FRP) are widely used in industry and marine construction, and offer an attractive option for the construction of OTEC cold-water pipes. This possibility received careful consideration in studies of various designs and materials conducted under DOE sponsorship by TRW (Griffin and Mortaloni, 1980) and SAI (Science Applications, Inc., 1979). Both organizations concluded that FRP pipes were the best choice for satisfying design requirements at low fabrication and deployment cost.

Cylindrical FRP storage vessels have been made in diameters up to 25 m (80 ft) by the method shown in Fig. 6-7 (TRW, 1979). The circumferential windings are

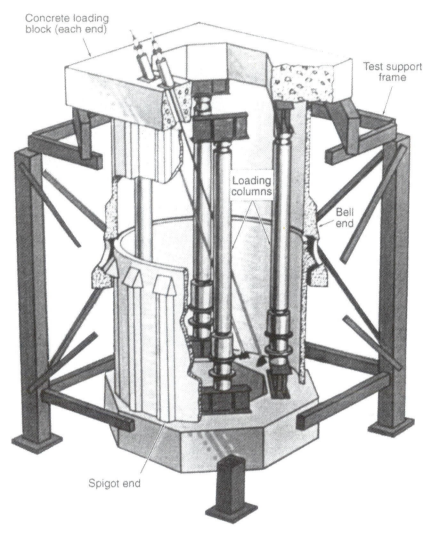

Fig. 6-5. Setup design for test of 1/3-scale concrete CWP (O'Connor, 1980).

supplemented by longitudinal fibers incorporated in strips that are bonded under tension. These are needed in the CWP to maintain the FRP integrity under the tension loads in the pipe. TRW made a comprehensive study of the suitability of an FRP CWP for the OTEC operating regime, including design of a double-walled structure to withstand buckling loads, a study of effects of hydrostatic pressure and long-time immersion on structural characteristics, and deployment scenarios. Details are given in the TRW report by Griffin and Mortaloni (1980). It was concluded that facilities could readily be developed that would make it feasible to manufacture and deploy an FRP pipe of 9.1 m diameter and 1000 m length for the baseline vessel. Construction and deployment time was estimated to be 24 months.

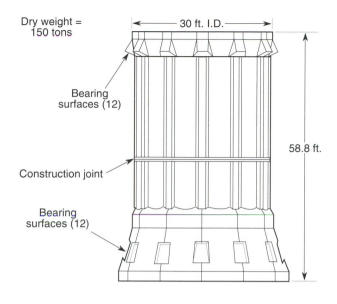

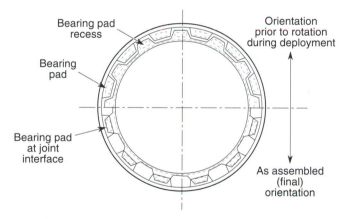

FIG. 6-6. Design of section of segmented pipe for JHU/APL OTEC CWP (George and Richards, 1980).

The estimated costs for manufacturing, deploying, and installing the FRP pipe on the OTEC platform were approximately 20% higher than for the concrete CWP, but could offer advantages in deployment procedures.

The TRW and SAI studies also examined elastomer and polyethylene CWP constructions and concluded that they would be feasible to build and deploy. The elastomer option was estimated to be 50% higher in cost than FRP but was judged worthy of further consideration as an alternative to the FRP design because of possible advantages in deployment. The cost of polyethylene appears to rule out this option. Further details and additional references are available in TRW (1979) (see also Gershunov et al., 1981).

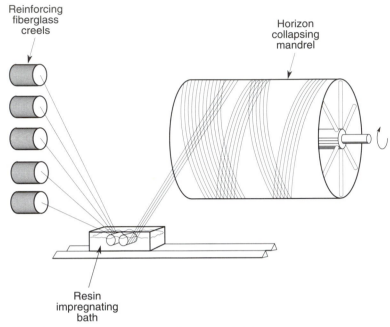

FIG. 6-7.   Schematic of construction of FRP pipe (TRW, 1979).

6.2.4 Heat Exchanger Manufacture

The economic promise of OTEC depends strongly on the use of heat exchangers that offer significant gains in performance and cost in comparison with conventional designs. Five types that offered promise to meet these criteria were demonstrated in the test programs conducted by the Argonne National Laboratory (ANL) during 1979–1981, as discussed in Section 4.1. A brief assessment of the facilities that would need to be developed for manufacture of these heat exchangers is given in Table 6-1. None of the types requires more than modest extension of current technology for use in OTEC. The total requirements for new facilities, fabrication, and predeployment testing are minimal for the Trane and folded-tube types. Further information for specific development plans is given in Sections 6.4 and 6.5.

6.3 PILOT-SCALE OTEC SYSTEMS TESTS

6.3.1 U.S. Mini-OTEC Program

A complete OTEC system was constructed, deployed, and operated successfully at sea in the Mini-OTEC program. The test provided the first demonstration of net OTEC power production. The experiment was completed in November 1979 after 4 months of successful operation offshore Kailua-Kona, Hawaii (Owens and Trimble, 1980). The program was a joint effort involving contributions from the organizations shown here. There was no DOE participation.

Table 6-1. Facilities needed for heat exchanger manufacture

HX type	Existing facilities	Additions needed for OTEC	Predeployment testing needs
S&T	Many	Scale-up to ~5 MWe	Not modular Full-scale tests req.
Plate (Alfa-Laval)	Some	Scale plate to larger size	Scale tests
Compact (Trane)	Two	Assembly facilities	Module tests at 2.5-MWe size
CMU	None	New plant	Not modular Full-scale tests req. at 5–10 MWe size
Folded-tube (JHU/APL)	None	Assembly facilities for std. pipe sections	Minimal module tests at full scale completed at ANL

Lockheed Missiles and Space Div.: Systems integration. Design, build, operate, test power plant. Dynamic analysis of CWP–platform system.

State of Hawaii: Obtain, operate, maintain the platform. Provide diver service.

Dillingham Corp.: Design, modify, deploy platform. Design, fabricate, deploy CWP.

Alfa-Laval ESD: Provide HXs, instrumentation, testing.

Worthingham Pump: Provide seawater pumps.

Rotoflow Corp.: Provide turbine-generator.

Only 19 months were needed from the date of the initial go-ahead decision until the Mini-OTEC experiment produced net power at sea. The entire program was completed by using off-the-shelf equipment and materials and standard fabrication techniques. A Navy hopper barge (YGN-70) loaned to the State of Hawaii was modified to form the platform for Mini-OTEC. The general dimensions were as follows: length, 32 m (105 ft); width, 10.3 m (33.9 ft); draft, 1.3 m (4.3 ft). A photograph of the at-sea deployment is shown in Fig. 6-8 (Owens and Trimble, 1980).

The CWP was made by joining 11.6-m sections of high-density polyethylene to form a single pipe 0.56 m ID (1.84 ft), 0.61 m OD (2.00 ft), and 661 m long (2170 ft). The characteristics are listed in Table 6-2.

The CWP was suspended from a buoy, and transfer of cold water to the ship was via an underwater hose, as shown in Fig. 6-9. A special setup onshore was needed for assembling and joining the sections of PVC pipe, but the process used available materials and techniques, and proceeded as planned, on schedule.

Design characteristics of the power plant, which used Alfa-Laval plate heat exchangers, are listed in Table 6-3.

The observed performance of the complete power system agreed closely with the analytical predictions. For the nominal warm- and cold-water temperatures, the

FIG. 6-8. Mini-OTEC deployed off Kea-hole Point, Hawaii (Owens and Trimble, 1980).

Table 6-2. Mini-OTEC CWP characteristics

Parameter	Value
Length	661 m (2170 ft)
Inside diameter	0.561 m (22.1 in.)
Outside diameter	0.610 m (24.0 in.)
Mass	28,560 kgm (63,000 lbm)
Ultimate axial load	100,000 kgf (220,000 lbf)
Dynamic flexural modulus	9.84×10^7 kg/m^2 (140,000 psi)
Fatigue strength (22.8°C, 10^5 cycles)	700,000 kg/m^2 (1000 psi)
Specific gravity	0.95

predicted isentropic power production was 93 kW; however, the efficiency of the small geared-down turbine-generator was 56%, so that only 53 kW of gross power was produced. Water and ammonia pumps and auxiliaries consumed 35 kW, operating at roughly half the efficiency that would be available in larger, state-of-the-art equipment. Thus, the net power output of 18 kW was in close agreement with expectations.

The experiment demonstrated that the systems engineering requirements of OTEC power system design, construction, and operation at sea were understood at a small scale [93 kW (gross)] and that plans for construction of pilot-size OTEC vessels could proceed with confidence (Trimble and Potash, 1979; Trimble and Owens, 1980). Details of the construction, deployment, and test results are given in the references cited.

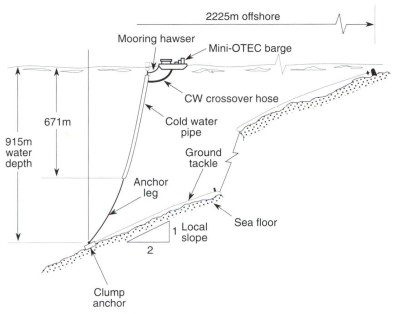

FIG. 6-9. Schematic of Mini-OTEC at-sea deployment (Owens and Trimble, 1980).

Table 6-3. Mini-OTEC power plant characteristics

Heat exchangers			Turbine-generator	
Evaporator			Turbine	
Surface area (m²)	408		Type	Radial in-flow
Material	Titanium		Speed (rpm)	28,200
Thermal load (MWt)	2.84		Enthalpy drop (kJ/kg)	4.00
Ammonia flow (kg/s)	2.31 (5.08 max)		Ammonia flow (kg/s)	2.25
Seawater flow (kg/s)	174		Output (kW)	75
Seawater temp. (°C)			Efficiency	83%
in/out	26.1/23.9			
Seawater Δp (kPa)	30.6		Generator	
Condenser			Type	Synchronous/
Surface area (m²)	408			rotating field
Material	Titanium		Speed (rpm)	3600
Thermal load (MWt)	2.75		Exciter	Shaft-mounted
Ammonia flow (kg/s)	2.31 (5.08 max)			brushless
Seawater flow (kg/s)	174		Output	50 kW, 0.8 P.F.,
Seawater temp. (°C)				120/208 V,
in/out	5.67/8.33			60 Hz, 3 phase
Seawater Δp (kPa)	30.6		Efficiency	90%
			T/G efficiency	56%

6.3.2 OTEC-1 Program

The OTEC-1 program of the DOE was established to create an at-sea test facility at a nominal 35-MWt heat loading for testing OTEC components and subsystems of moderate size that could furnish heat exchanger operational information for later demonstration of larger systems (Svenson, 1979). A deliberate decision was made by DOE not to include a turbine, but to use an expansion valve to simulate the pressure drop through the turbine (35 MWt would generate about 0.9 MWe if the system incorporated a turbo-generator). After investigation of possible platforms, a decision was made to convert a mothballed Navy T-2 tanker, *Chepachet*, to become the test vessel. Global Marine, Inc., was given responsibility for ship conversion and CWP construction and deployment, and TRW was assigned responsibility for the power system design and installation.

   Conversion of the vessel was done in the shipyard of the Northwest Marine Iron Works in Portland, Oregon. Construction was completed in June 1980, and the vessel arrived in Hawaii in July 1980. CWP construction, deployment, and installation on the ship, now named SS *Ocean Energy Converter*, were completed in December. The at-sea test began with operation of the seawater pumps on December 13, 1980. Heat exchanger testing was conducted until March 27, 1981. The planned 2-year program was terminated because the costs of supplying the pumping and auxiliary power needs via diesel-fueled generators, and of maintaining the ship's crew, could not be funded within the reduced fiscal year 1981 budget (Castellano, 1981; Castellano et al., 1983). Layouts of OTEC-1 indicating the arrangement of the installed equipment are shown in Figs. 6-10 and 6-11 (Svenson, 1979). The hull dimensions were as follows: length, 160 m (525 ft); width, 21 m (69 ft); depth, 12 m (40 ft). The displacement was 22,000 tonnes (24,200 tons). The OTEC-1 test loop is outlined in Fig. 6-12 (Lorenz et al., 1981a, 1981b).

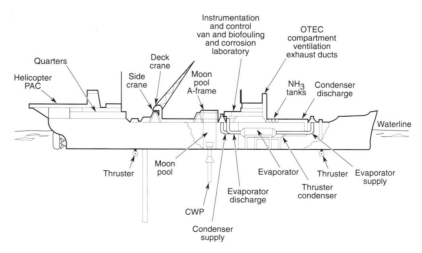

Fig. 6-10. Side profile of OTEC-1 vessel (Svenson, 1979).

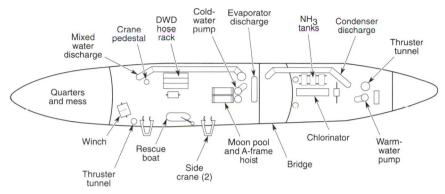

FIG. 6-11.   Layout of main deck equipment of OTEC-1 (Svenson, 1979).

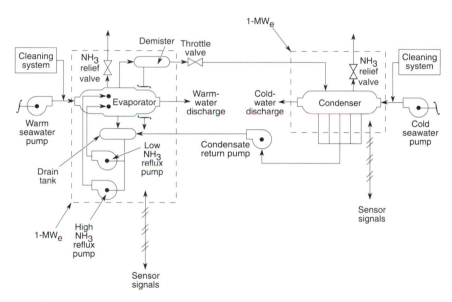

FIG. 6-12.   OTEC-1 test loop (Lorenz et al., 1981a and 1981b).

The installation included one evaporator and one condenser, with water and ammonia pumps and control equipment to permit ammonia to be circulated through the system under conditions simulating OTEC HX operation. The turbo-generator needed for OTEC power generation was replaced by the throttle valve shown in the sketch.

The evaporator was constructed in two sections that could be operated independently. The upper HX employed a bundle of plain titanium tubes, and the bottom used a bundle of enhanced tubes of the Linde design. The HX was designed to be tested in either a sprayed or flooded mode; however, time and funding limitations prevented tests of the latter type. The condenser contained only plain tubes. The system was designed so that various biofouling control methods could be evaluated, such as Amertap and chlorine. Design details of the evaporator and condenser are shown in Fig. 6-13 (Lorenz et al., 1981a, 1981b; Snyder, 1980).

Since OTEC-1 was designed only as a heat exchanger test vehicle, no attempt was made to simulate an operational CWP. Instead, a CWP was made by binding together three 1.22-m-diameter (4-ft) polyethylene pipes to form the 670-m-long

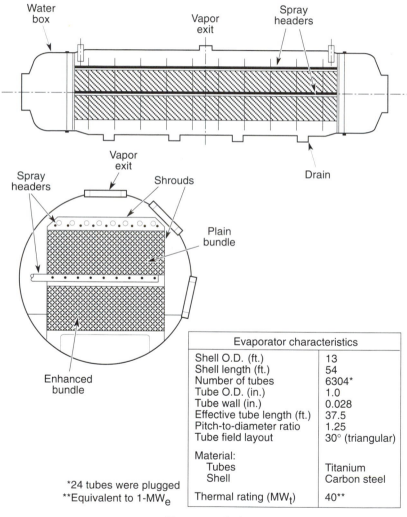

Evaporator characteristics	
Shell O.D. (ft.)	13
Shell length (ft.)	54
Number of tubes	6304*
Tube O.D. (in.)	1.0
Tube wall (in.)	0.028
Effective tube length (ft.)	37.5
Pitch-to-diameter ratio	1.25
Tube field layout	30° (triangular)
Material:	
Tubes	Titanium
Shell	Carbon steel
Thermal rating (MW$_t$)	40**

*24 tubes were plugged
**Equivalent to 1-MW$_e$

a. Evaporator design

Fig. 6-13. (a) Design of OTEC-1 evaporator (Lorenz et al., 1981a and 1981b). (b) Design of OTEC-1 condenser (Lorenz et al., 1981a and 1981b).

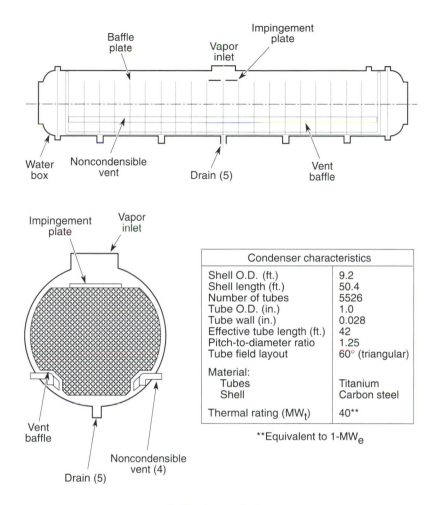

The following table appears within the figure:

Condenser characteristics	
Shell O.D. (ft.)	9.2
Shell length (ft.)	50.4
Number of tubes	5526
Tube O.D. (in.)	1.0
Tube wall (in.)	0.028
Effective tube length (ft.)	42
Pitch-to-diameter ratio	1.25
Tube field layout	60° (triangular)
Material:	
Tubes	Titanium
Shell	Carbon steel
Thermal rating (MW$_t$)	40**

**Equivalent to 1-MW$_e$

b. Condenser design

FIG. 6-13b.   (continued).

(2200-ft) CWP. Details of the construction and deployment are depicted in Fig. 6-14 (Gershonov et al., 1981). The pipe was supported at the platform by a gimbal that allowed rotation of the joint through an angle of 30 degrees. The pipe was assembled onshore and towed to the vessel mooring site off Kona, Hawaii. A thorough study was made prior to deployment to define the loads expected and the procedures necessary to avoid excessive pipe stresses or deflections. Deployment and operation of the buoyant pipe were accomplished according to plan and provided valuable insight into the requirements for deploying the larger FRP pipes proposed for actual OTEC use (Gershonov et al., 1981).

- 4 feet – outside diameter of each pipe,
- 2.4 inches – wall thickness,
- 68,000 pounds – wet weight of the lower transition and inlet screen,
- 75,000 pounds – wet weight of the stabilizing ballast at the lower end cable system,
- 2,200 feet – total length of the polyethylene pipes,
- 2,500 feet – total length of cables.

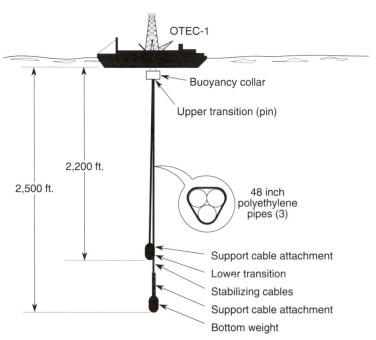

FIG. 6-14. OTEC deployment details (Gershonov et al., 1981).

Significant contributions of the OTEC-1 program included the following (Castellano, 1981):

1. Operation of the heat exchanger cycle in close accord with behavior predicted from model test results at ANL.
2. Successful deployment of the 2.2-m-diameter, 670-m-long CWP, and operation in waves and currents that caused a 30 degree excursion of the gimballed pipe, with the ship constrained by the mooring line.
3. Demonstration of biofouling control with injection of chlorine at a concentration of 0.06 ppm for 1 hour per day. This is well below FPA limits. Heat exchanger performance was in accord with expectations.
4. Successful operation of the heat exchanger test system and seawater systems during a 4-month period in which maximum wave height of 4 m, maximum subsurface currents of 3 knots, wind gusts up to 45 knots, and ship roll angle

of 18 degrees were encountered. Normal operation during drifting (grazing) operation was demonstrated. When subsurface currents caused the gimbal angle to reach 30 degrees, detaching the ship from the mooring reduced the pipe angle to 10 degrees. This demonstrated the advantage of the grazing mode in reducing CWP stresses and pivoting problems.

5. The test provided further confirmation that the marine engineering aspects of OTEC operation are understood. The test program provided no surprises.

6.3.3 Japanese Shore-Based 100-kW OTEC Pilot Plant at Nauru

In cooperation with the government of the Republic of Nauru, engineers of the Tokyo Electric Power Co. and the Toshiba Corporation in 1980 constructed a 100-kW (gross) OTEC plant at a site on the island of Nauru. Power-generating operation was performed from October 1981 to December 1981. Heat-loop operation was continued to July 1982 (Mitsui et al., 1983; Ito and Seya, 1985). The location of Nauru in the Pacific region is shown in Fig. 6-15. The inset shows the position on the island of the OTEC plant site (Uehara, 1982).

Selection of the Nauru site for the pilot program offered many advantages:

1. The combination of high average surface water temperature and cold water close to shore provided an optimal environment for the OTEC experiment.
2. Demonstration of successful operation could be followed by scale-up to 5-10

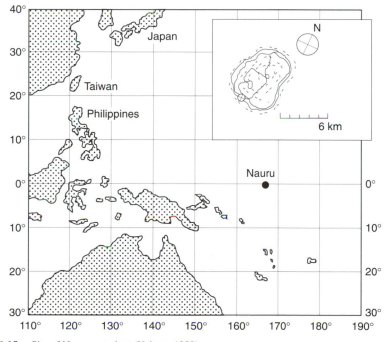

FIG. 6-15.   Site of Nauru test plant (Uehara, 1982).

MWe (net), to supply needed power at reasonable cost to the island grid, with full support by the island utility.

3. The pilot plant would supply basic information for the design and development of shore-based closed-cycle OTEC plants that could be sited at a larger number of island sites and some continental sites in the tropical oceans.

4. Specific information would be obtained that is necessary for projecting to higher power levels the feasibility and costs of CWP fabrication and deployment for land-based OTEC plants. Accurate data would be obtained on pressure losses in cold-water flow, biofouling effects and remedial procedures, and overall system performance and costs.

The island of Nauru is an elevated coral reef surrounded by a lagoon 50–150 m wide. At the plant site, the seabed is irregular from the shore to a depth of 50 m and then slopes at a nearly constant angle of 40–45 degrees to 700-m depth. The seabed is hard near the shore but then becomes porous; however, the bearing capacity is adequate for seabed emplacements.

The experimental program included bathymetric surveys of the site, development of procedures for constructing and deploying the CWP, construction of the shore-based OTEC power plant, and intermittent plant operation for 1 year while the operating parameters were varied and performance changes were documented.

6.3.3.1 Site

A plot of temperatures and currents versus depth at the site is shown in Figure 6-16 (Uehara, 1982).

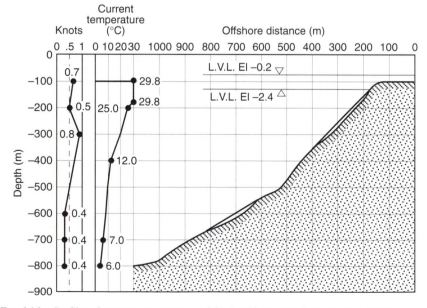

Fig. 6-16.   Profiles of temperature, current, and depth at Nauru OTEC site (Uehara, 1982).

6.3.3.2 Power Plant

The power plant used Freon-22 in a Rankine closed cycle, as depicted in Fig. 6-17 (Mochida et al., 1983). Freon was chosen for safety in handling, although analysis showed that ammonia would be preferred for commercial operation. The heat exchangers were of the shell-and-tube type with titanium tubes, but employed new surface treatments to enhance heat transfer. The evaporator tubes were coated with sprayed copper particles that enhanced nucleate boiling. The condenser was mounted vertically. It used spirally grooved tubes sealed at intervals to separator plates that stripped off the condensed film to prevent it from becoming thick enough to degrade heat transfer significantly (Kawano et al., 1983).

6.3.3.3 Cold-Water Pipe

After a review of properties of steel, FRP, PVC, and high-density polyethylene, polyethylene was selected for the CWP. The polyethylene specific gravity was slightly less than 1.0. The pipe was assembled by first joining 10-m lenths to form 18 sections 50 m long. The sections were then joined as deployment proceeded to form a single pipe 900 m long. The deployment procedure is shown in Fig. 6-18 (Mitsui et al., 1983). The slightly buoyant pipe, weighted with steel collars and supported by floats, was pulled out from the shore to its full length. The floats were then successively removed, allowing the pipe to settle gradually onto the sea floor. The weights of the metal collars were adjusted (allowing a safety factor of 1.5) to hold the pipe securely on the bottom in the presence of waves near the shore and of undersea currents along the full length of the pipe. Characteristics of the Nauru plant and its operation are listed in Table 6-4.

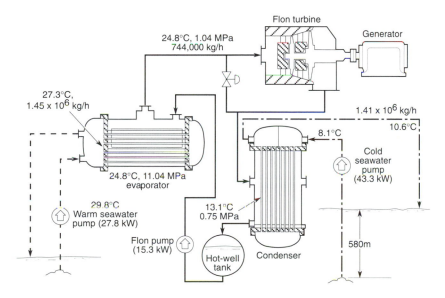

FIG. 6-17. Diagram of Japanese plant at Nauru (Monchida et al., 1983).

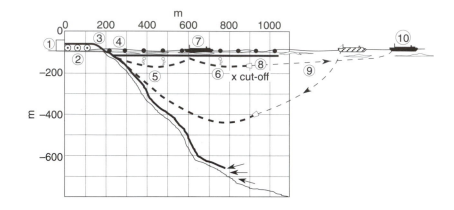

Polyethylene pipeline
Total length in the water: 945 m
Inner diameter: 700 mm

  1. Concrete foundation
  2. Temporary work stage
  3. Slope stage
  4. Buoy
  5. Counter weight
  6. Polyethylene pipeline
  7. Supporting barge
  8. Water intake
  9. Pulling wire
10. Pulling barge

FIG. 6-18.   Deployment procedure for CWP installation (Mitsui et al., 1983).

The Nauru test was designed to provide accurate experimental data on the performance of the complete power cycle, as well as to demonstrate total pilot-plant construction and operation. The goals were all successfully accomplished. The plant provided the first demonstration of land-based OTEC net power generation and established a record for total net power production, at 31.5 kW. It also was the first demonstration of OTEC power connection to a utility grid.

Details of the evaporator and condenser design and equipment arrangements, and documentation of the test results are presented in papers of Mochida et al. (1983) and Kawano et al. (1983).

6.3.4 Japanese Mini-OTEC Program

A Mini-OTEC plant was constructed for at-sea tests off the coast of Shimane Prefecture, as an initial phase of the Japanese program directed to demonstration of a 100-MWe moored OTEC system. Deployment took place on October 11, 1979 (Uehara et al., 1980).

Table 6-4. Characteristics of the Nauru 100-kW pilot power plant

Power System
Closed Rankine cycle   Freon R-22, S & T HXs, Titanium tubes
Horizontal evaporator   Vertical condenser
Evap: shell ID 1.9 m, L 8 m   870 tubes enhanced: OD 25.4 mm, t 0.7 mm, L 5.35 m
Cond: shell ID 1.5 m, L 9.1 m   914 tubes grooved: OD 25.4 mm, t (min) 0.6 mm, L 6 m

	Nominal performance	Max. performance
Gross power (kWe)	100.5	120
Net power (kWe)	14.9	31.5
Seawater ΔT (°C)	22.0	—

	Evaporator	Condenser
Seawater flow (kg/s)	402.8	391.7
Seawater flow velocity (m/s)	2.0	1.97
Inlet/outlet temp. (°C)	29.8/27.3	5.6/8.1
Pressure drop (m)	2.90	3.71
Freon flow (kg/s)	20.6	20.6
Freon temp. (°C)	24.8	13.0
Effective HX area (m²)	371.4	437.6
Heat duty (kWt)	4024	3911
Overall U (W/m² °C)	3000	2560

The organizations participating in the development included:

Saga University: Team leader. Engineering, measurement, operation of power systems.

Toboshima Corp.: Overall management. Deployment of CWP and works at sea.

Toden Sekkei: Engineering of power plant.

Toshiba: Turbo-generator.

Hisaka Works, Ltd.: Heat exchangers.

Torishima Pump Mfg. Co., Ltd.: Production of pumps for OTEC plant.

Kitazawa Valve Co., Ltd.: Production of valves for power plant.

Nippon Steel Corp.: Production and assembly of intake pipe.

Mihama Corp.: Flushing power plant system and filling with Freon-22.

Namura Shipbuilding Co.: Arranging loading area. Operation of warm- and cold-water pipes.

Saga Machine and Metal Corp.: Dismantling of testing system, conversion, reassembly, unit body testing.

The tests were conducted using as the platform a vessel *Ocean Discoverer*, owned by the Fukada Salvage and Machine Works. The characteristics are listed in Table 6-5.

The layout of the vessel including the OTEC equipment and cold-water pipe is shown in Fig. 6-19. The test included deployment of a steel CWP 190 m long made of 12-m-long tubular sections of 26.7 cm OD and 0.45 cm wall thickness, joined by flanges. The pipe was supported by a float and held in place by a steel cable attached to the pipe and anchored at a depth of 300 m. The OTEC power plant (with

Table 6-5. Characteristics of "Ocean Discoverer"

Overall length	54.0 m
Depth	5.0 m
Breadth	11.0 m
Deadweight	704 mt
Gross tonnage	669.6 mt
A-frame	60 mt
Derrick	10×15 mt-m
Power (BHP)	2100×2 sets

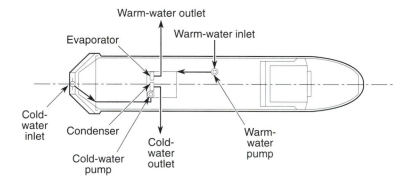

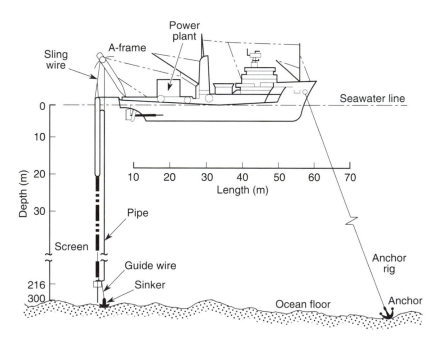

Fig. 6-19. Layout of the Japanese Mini-OTEC vessel (Uehara et al., 1980).

plate-fin heat exchangers) and water pumps were mounted on the research vessel, which was attached to the float. A flexible pipe carried the cold water from the float to the vessel. The power plant performance is documented in Table 6-6.

6.3.5 Design Studies in Response to DOE Program Opportunity Notice

A Program Opportunity Notice was announced requesting cost-sharing proposals from industry for the construction, deployment, and operation of a 40-MWe (net) OTEC power plant or plantship that would demonstrate the technical feasibility of OTEC at a scale large enough to furnish reliable engineering information for commercial development. Funding of $6.3 million was appropriated by the U. S. Congress "to support cost sharing of seven or eight conceptual design proposals." A stated purpose of the PON was to support investigation of a wide range of concepts for OTEC design, siting, management, and financing that would give a broad basis for evaluation of the potential contributions of this technology to future U.S. energy needs. Respondents were asked to propose a development program that would involve five successive cost-shared phases: I, conceptual design; II, preliminary design; III, engineering; IV, construction, deployment, and acceptance testing; V, joint operational test and evaluation; and, finally, VI, transfer of ownership to the industrial contractor. DOE announced its intent to fund five to eight Phase I awards, with the DOE share to be $900,000 for each.

Table 6-6. Performance of Japanese Mini-OTEC power plant

Parameter	Value
Evaporator	
Warm-water flow rate (kg/s)	36.1
Warm-water inlet temperature (°C)	23.6
Warm-water outlet temperature (°C)	22.4
R-22 inlet temperature (°C)	18.4
R-22 outlet temperature (°C)	18.9
R-22 flow rate (kg/s)	0.7
Log mean temperature difference (°C)	4.3
Heat transfer rate (kWt)	181
Overall heat transfer coefficient (W/m^2 °C)	5120
Heat transfer area (m^2)	8.2
Outlet power (kWe)	~1
Condenser	
Cold-water flow rate (kg/s)	38.9
Cold-water inlet temperature (°C)	9.2
Cold-water outlet temperature (°C)	10.3
R-22 inlet temperature (°C)	16.8
R-22 outlet temperature (°C)	12.1
R-22 flow rate (kg/s)	0.5
Log mean temperature difference (°C)	4.5
Heat transfer rate (kWt)	179
Overall heat transfer coefficient (W/m^2 °C)	1858
Heat transfer area (m^2)	21.6

The PON was issued in compliance with Public Laws PL 96-310 and PL 96-320, which called for demonstration of 100 MWe of OTEC power by the year 1985 and 500 MWe by 1990. Background material made available to potential contractors included the baseline 40-MWe OTEC system design prepared by JHU/APL; the power plant design studies by Westinghouse, TRW, and GE; and the platform and CWP investigations by Hydronautics, Gibbs and Cox, Lockheed, and Global Marine.

Responses to the PON were received from eight teams representing major U.S. organizations with a broad range of expertise in systems engineering, chemical engineering, equipment manufacturing, power plant design, marine engineering, and shipbuilding. Each response represented a substantial engineering design effort carried out in the anticipation that DOE would award at least five contracts if the material supplied contained the technical depth expected, displayed innovative concepts, showed reasonable management capabilities, and offered roughly 50% cost sharing.

The proposals are briefly described in the following paragraphs abstracted from Dugger et al. (1983).

1. General Electric (GE), as the system integrator, proposed a 40-MWe (net) plant on an offshore oil-rig-type tower in 100 m of water off Kahe Point, Ohau, Hawaii, with an undersea power cable to shore. Further details are given in Section 6.4.2.

2. Ocean Thermal Corporation (OTC) proposed a 40-MWe (net) plant on an artificial island in 8.5 m of water off Kahe Point, Ohau, Hawaii, with an overhead cable to shore. It would use the warm water discharged from the Hawaiian Electric Company (HECO) 600-MWe oil-fired plant, mixed with surface seawater, to enhance the OTEC warm-water inlet temperature. Further details are presented in Section 6.5.2.

3. Puerto Rico Electric Power Authority (PREPA) proposed a 40-MWe (net) OTEC power plant mounted on a steel tower in 70 m of water off Punta Tuna. Available offshore oil-rig technology would be used in the construction. Futher details are given in Section 6.4.3.

4. SOLARAMCO (The Solar Ammonia Company) proposed a cruising 40-MWe OTEC plantship sited south of Hawaii that would produce ammonia for shipment to Hawaii or other markets. It would provide test operational information for four advanced 10-MWe power modules of types found attractive in tests at the Argonne National Laboratory (ANL) of DOE and include a state-of-the-art ammonia plant for evaluation of shipboard operation. Further information is provided in Section 6.4.4.

5. Virgin Islands Water and Power Authority (VIPWA), with EBASCO as systems integrator and J. R. McDermott and TRW as subcontractors, proposed a 12.5-MWe plant on a shelf-mounted tower 1.5 km off St. Croix to deliver 10 MWe of power plus 190,000 m^3 day of fresh water. The vertical, shell-and-tube heat exchangers (with titanium tubes) would be submerged. Part of the discharged cold water would be used for mariculture experiments. The

principal features of the proposed system are described in Jones et al. (1981). The total installation cost was estimated at $141 million, and net annual receipts were estimated at $34 million from the sale of electric power, fresh water, and deep water for mariculture. The after-tax rate of return (IRR) was estimated at 15%, making the concept potentially commercially viable. Significant reductions in plant costs and increases in receipts would result from the use of aluminum plate-type heat exchangers and improved water desalination processes that have resulted from recent developments. A flow chart for the process is shown in Fig. 6-20.

6. Ocean Solar Energy Association, with Sea Solar Power as the systems integrator and General Dynamics and CEER of Puerto Rico as subcontractors, proposed an OTEC electric power plant mounted on a steel semisubmersible platform that would deliver power to shore by underwater cable from a site offshore from Punta Tuna, Puerto Rico. The platform would be maintained in position by vectoring the exhaust water plumes from the heat exchangers. The plant would use innovative designs of the platform, heat exchangers, turbines, and CWP to achieve minimal weight and cost.

7. Commonwealth of Northern Marianas Islands (CNMI), with Science Applications, Inc., as systems integrator and Dillingham/Hawaiian Dredging, Burns & Roe, and Global Marine Development, Inc., as subcontractors, proposed a 10-MWe plant on a concrete barge to be installed on a foundation 100 m off Saipan, with overhead cables to the island grid. It would use a 4.9-m-diameter CWP, with the upper 300 m made of concrete mostly trenched and buried. The last 1300 m would be of FRP installed on bottom-mounted piers.

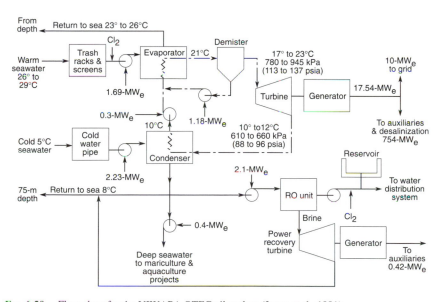

Fig. 6-20. Flow chart for the VIWAPA OTEC pilot plant (Jones et al., 1981).

8. Florida Ocean Thermal Energy Consortium (FOTEC), with Florida Solar Energy Center as systems integrator and TRW, Stone and Webster, and Santa Fe International as subcontractors, proposed an OTEC power plant mounted on a steel ship that would be anchored 48 km off Key West using a 2700-m-long single-point moor. Power would be transmitted to shore through a buoy-supported riser cable and buried bottom cable. The power plant would have titanium shell-and-tube heat exchangers. A 9-m-diameter, 793-m-long FRP CWP would be supported by a gimbal on the platform.

With the change of Administration in 1981, no funding was requested to continue the program. Of the appropriated funds, $1.8 million were used to support only the two conceptual design proposals of GE and OTC. The remaining funds were then deferred to be used for a single follow-on preliminary design program. DOE support of the program was then terminated.

The technical and financial data supplied to DOE in the unfunded PON responses were subsequently destroyed on the grounds that DOE could not protect the proprietary data they contained. Information from the unfunded proposals by PREPA and SOLARAMCO has been supplied to the authors by the PON respondents and is included in Section 6.4.

6.4   CONCEPTUAL DESIGNS OF OTEC SYSTEMS

6.4.1 Ocean Thermal Corporation's Land-Based Hawaii Site

At completion of the Phase I work, this program was funded to continue to Phase II, preliminary design. It is discussed in Section 6.5.2.

6.4.2 General Electric Corporation's Shelf-Mounted Hawaii Site

The following discussion is based on the final report on the GE Phase I contract (General Electric, 1983).

Firms participating in the Phase I contract included:

General Electric Company: Systems engineering and direction.
Brown and Root Development, Inc. : Tower, CWP, seawater piping.
Trane Corp.: HX and manifolding.
Worthington Co.: Pumps.
Hawaiian Electric Co. (HECO): Land-based facilities.

The GE OTEC design is based on the use of a tower-mounted OTEC plant installed 2200 m offshore at a depth of 100 m, at Kahe Point, Oahu, Hawaii. Power produced at the OTEC platform is carried by underwater cable to the Hawaiian Electric Company (HECO) plant for distribution to its customers.

An artist's sketch of the tower and a schematic drawing are shown in Fig. 6-21. Cold water is supplied through a pipe made of steel pipe sections 8.5 m in diameter, 150 m long, connected end to end and extending along the sloping sea bottom for 2000 m, to a depth of 670 m. The pipe-laying sequence is depicted in Fig. 6-22. The GE concept involves tradeoff and optimization studies of siting options,

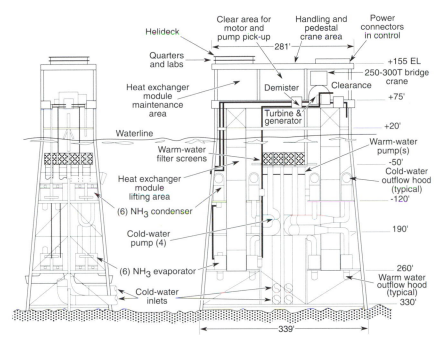

FIG. 6-21.   GE tower design for OTEC offshore plant (General Electric, 1983).

tower designs, CWP design and deployment, heat exchanger selection, turbo-generator and water pump sizing, and power generation and distribution. Proce-dures for biofouling control and component maintenance and replacement were investigated, and optimal arrangements were defined. Only the final concept is discussed here.

The GE concept calls for the tower to be constructed onshore and then barged to the offshore site, where it is erected by using procedures and equipment in common use for emplacement of offshore oil-drilling rigs. The tower is designed for convenient transfer of the power plant components to and from their mounting positions on the tower, via built-in elevators, semi-automated underwater transfer equipment, and derricks. Thus, pumps and heat exchanger modules may be blocked off, moved to the main deck for servicing, and returned without total plant shutdown. Cold water from the 8.5-m-diameter CWP enters a manifold near the base of the tower, where it transfers to four 5.9-m-diameter pipes before being routed to the condenser modules. Each pipe incorporates a pump operated by an external motor sealed for underwater operation. Each pipe is routed to serve four condenser modules. Warm water is drawn from a manifold mounted at a depth of 15 m equipped with 12 stainless steel screens, each 9 m × 12 m. Eight 4.7-m-diameter pipes conduct the water to 8 pairs of evaporator modules mounted at the 79-m level. Elevation views of the piping systems for the evaporators and condens-ers are shown in Figs. 6-23 and 6-24.

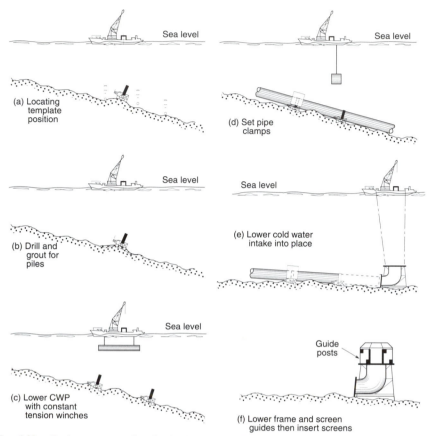

FIG. 6-22.   Deployment procedure for GE OTEC plant (General Electric, 1983).

A survey of advanced heat exchanger designs examined three candidates in some detail:

1. A brazed-aluminum compact design by the Trane Co.,
2. A plate design of the Alfa-Laval type, and
3. A new compact design by GE.

The Trane design was given the highest rating and was selected for the final system design (Ashworth et al., 1980; Foust, 1981). A cutaway view of the Trane heat exchanger construction is shown in Fig. 6-25.

The principal features of the GE concept are listed in Table 6-7. Discussion of cost estimates for this and other concepts is deferred to Chapter 7.

6.4.3 Puerto Rico Electric Power Authority

PREPA proposed a tower-mounted 40-MWe (net) OTEC plant to be sited at the edge of the continental shelf at 75-m depth approximately 2007 m offshore from Punta Tuna, Puerto Rico (Sanchez, 1980).

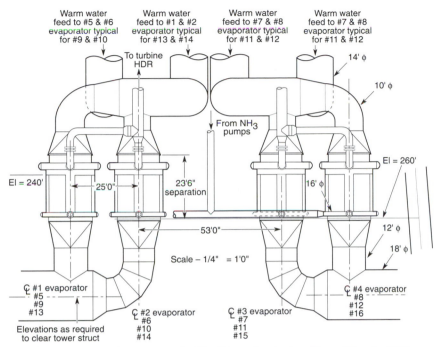

FIG. 6-23. Warm-water duct geometry for GE offshore electric plant (General Electric, 1983).

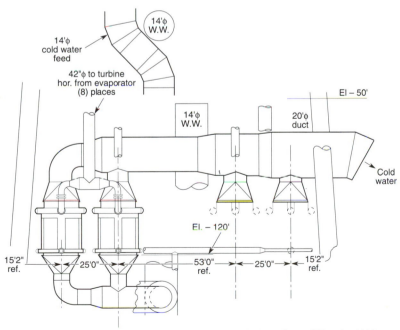

FIG. 6-24. Cold-water duct geometry for GE offshore OTEC plant (General Electric, 1983).

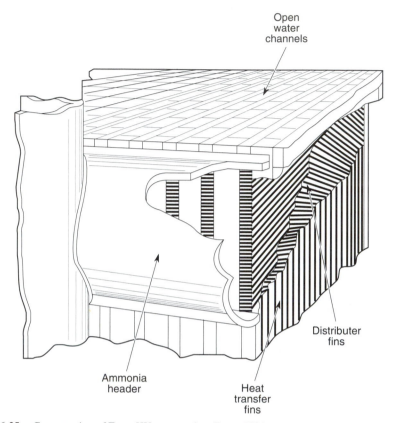

FIG. 6-25. Cutaway view of Trane HX construction (Foust, 1981).

The platform concept is shown in Fig. 6-26. Electric power is transmitted by underwater cable to the PREPA distribution system. Cold water is drawn from a depth of 1000 m at a distance of approximately 1500 m from the tower. The topography at the site is shown in Fig. 6-27. The following organizations experienced in systems engineering, offshore construction, power plant siting, and environmental assessments cooperated in defining a baseline configuration for the PREPA Phase I proposal:

Puerto Rico Electric Power Authority (PREPA)
Brown and Root Development, Inc. (BARDI)
Westinghouse
United Engineers and Constructors (UE&C)

Assistance on special topics would be provided by:

Consultores Tecnicos Asociados (CTA): Financing.
McClelland Engineers Marine: Geotechnical services.
Capacete Martina & Assoc. (CMA): Geotechnical services.
Cora's Industrial Fiber Corp.: Fiberglass-reinforced plastic.

Table 6-7. GE tower-mounted 40-MWe (net) OTEC plant

Tower dimension
  Top platform: W 30 m, L 86 m; height to top platform 149 m
  Base: W 44 m, L 103 m; height to WL 100 m
Power plant
  Closed Rankine cycle; ammonia working fluid
  Trane Corp. brazed aluminum heat exchangers
  Modular power units
    20 Evaporators: L 6.6 m, W 6.1 m, H 7.7 m; evaporators at 79 m depth
    16 Condensers: L 5.5 m, W 6.1 m, H 7.7 m; condensers at 36 m depth
Operating characteristics
  Nominal ΔT (˚C) 20.9⁺
  Seasonal ΔT (˚C): May 20.5, Aug 22.1, Nov 21.5, Jan 20.4, Mar 19.9
Power output (MWe)
  Gross turbine-generator output   54.95
  Warm-water pump power   4.61
  Cold-water pump power   9.89
  Hotel and misc. power   0.41
  Net power output   40.00
  Seasonal net power output: May 39.5, Aug 46.3, Nov 44.0, Jan 38.6, Mar 36.4

	Evaporator	Condenser
Heat load (MWt)	1915	1871
Water flow (kg/s)	212,600	184,300
Ammonia flow (kg/s)	1533	1533
Overall heat trans. coeff. (W/m² K)	4528	4506
Fouling coeff. (W/m² K)	3.5×10^5	1.8×10^5
Water inlet/outlet temp. (˚C)	25.56/23.33	4.94/7.49
Ammonia inlet/outlet temp. (˚C)	10.33/21.23	10.30/10.62
Seawater Δp (m)		
Inlet pipe	0.51	2.61
HX	1.10	1.48
Seawater velocity (m/s)	1.45	1.60
Condenser flow [m³/MW (net)]	5.3	
Net power/total HX area (kW/m²)	0.16	

The principal features of the PREPA design are described in the following subsections.

6.4.3.1 Platform

The tower mounting is selected to avoid the perceived problems of power transmission from a floating plant, and to avoid design uncertainties in the mooring and cold-water pipe installations for a floating configuration. The platform design employs steel construction and depends on design practice and technology developed for offshore oil installations. A wide range of computer codes is available for establishing static and dynamic specifications to withstand wave, wind, and current loads for extreme hurricane conditions. A survey of the topography led to selection of a sloping site 200 m offshore for platform emplacement near the edge of the continental shelf. The slope and character of the sea bottom at the site implies the

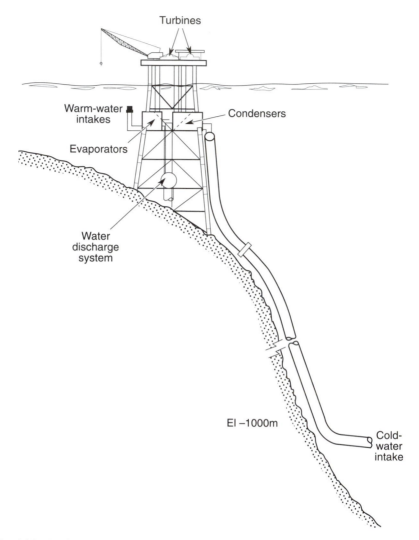

Fig. 6-26. Platform concept for PREPA OTEC plant (PREPA Phase I proposal, 1981).

need for drilling and grouting piles and some extension of existing off-shore technology, but no major problems were foreseen to cause significant risk in the installation. A scenario for platform installation is depicted in Fig. 6-28.

6.4.3.2 PREPA Power Plant

The power plant design employs four 10-MWe (net) modules that use shell-and-tube heat exchangers with ammonia as the working fluid. The power plant turbines and generators are mounted on the platform deck. The evaporators and condensers are provisionally mounted at a depth of 15 m (to the top of the heat exchangers). This

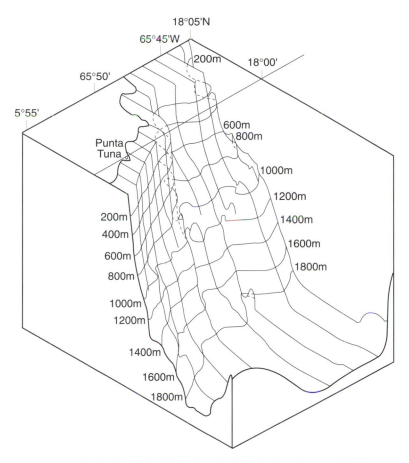

FIG. 6-27.   Topography of site for PREPA OTEC plant (PREPA Phase I proposal, 1981).

appears to be an optimal balance between selection of maximum depth to minimize wave forces and placement near the surface to maximize power output.

A schematic of the power cycle is shown in Fig. 6-29. Each heat exchanger contains three independent tube bundles. A detailed study of tube designs to enhance the heat transfer characteristics was made and analyzed to determine tradeoffs between heat transfer enhancement and pressure loss. Cost effectiveness was then estimated. This led to selection of tubes for the evaporator that are enhanced with the Linde coating on the outside and are bare on the inside. Plain tubes were found to be the best choice for the condenser. Titanium was selected for the tube material because of its proven long life in seawater applications. The cost advantages of aluminum were noted, but lack of performance data made its selection appear to involve unacceptable risk.

The preliminary layout of the power system components involving consideration of accessibility, maintenance, and replacement dictated placing the turbines,

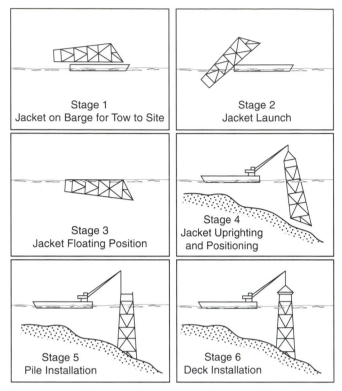

Fig. 6-28. Deployment procedure for PREPA OTEC platform (PREPA Phase I proposal, 1981).

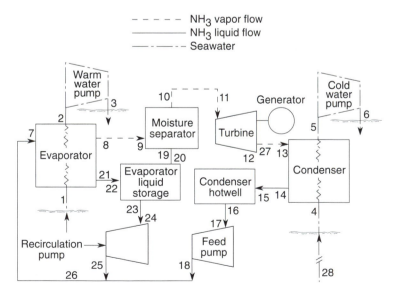

Fig. 6-29. Power cycle for PREPA OTEC power plant (PREPA Phase I proposal, 1981).

generators, cranes, power conditioning and control systems, support facilities, and personnel accommodations on the top deck.

Minimization of pressure losses required location of the water pumps and heat exchangers below the surface. For minimum pressure losses in the ammonia systems, turbine–heat exchanger distance should be as short as possible. To avoid excessive loads caused by wave action, underwater components should be placed at a depth well below the top of the maximum significant wave.

Tradeoff studies then led to the layouts for the power system components shown in Fig. 6-30.

6.4.3.3 Water Ducting

Fiberglass-reinforced plastic (FRP) was selected as the preferred material for the CWP and the discharge pipes. The 8-m-diameter CWP, approximately 2000 m long, would be deployed as shown in Fig. 6-31. Thre 8-m-diameter discharge pipes emplaced in a suitable trench on the shelf were designed to conduct the mixed effluent from the plant to a depth below the thermocline.

6.4.3.4 Energy Transfer

The design calls for power to be transmitted as 115 kV AC through an entrenched power cable joining the bus bar at the platform to the onshore utility. Cables of the required capacity are commercially available, and PREPA and Brown and Root have extensive experience in laying and operating oil-filled cables of much larger capacity. The design 10-MWe output at 13.8 kV of the four power modules would feed into a single bus and then be stepped up to 115 kV to match the voltage of the onshore substation.

The principal features of the PREPA 40-MWe (net) OTEC concept are listed in Table 6-8. Estimated costs are listed in Chapter 7.

6.4.4 SOLARAMCO Concept of OTEC 40-MWe (net) Grazing Ammonia Plantship

The SOLARAMCO group proposed a floating 40-MWe (net) OTEC ammonia plantship system, designed to graze in a region south of Hawaii indicated Fig. 6-32, where a ΔT of 22–23°C would be available (Babbitt, 1981). The plantship would use the net OTEC electric power for water electrolysis to produce hydrogen for on-board ammonia synthesis. This method efficiently stores the OTEC electric energy output in the form of internal chemical energy contained in the ammonia produced. The concept makes possible 24-hour-per-day operation of OTEC at full power and enables OTEC to utilize the total thermal energy resources of the tropical oceans. The proposal calls for the design and evaluation of four alternative power modules using alternative heat exchangers that could be incorporated as test modules in the final pilot plant. This would be narrowed by work in Phases II and III to two power systems that would be installed in the final pilot plant. An FRP CWP is proposed. It was also proposed that after 2 or 3 years of grazing OTEC ammonia production, the ship could be converted for direct power transfer to a Hawaii utility for moored

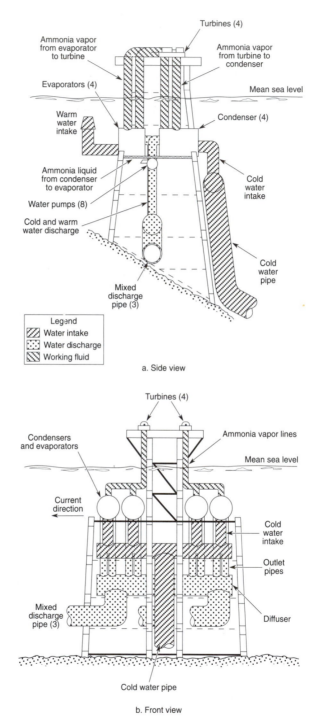

Fig. 6-30. Subsystem layout for PREPA OTEC power plant (PREPA Phase I proposal, 1981): (a) side view; (b) front view.

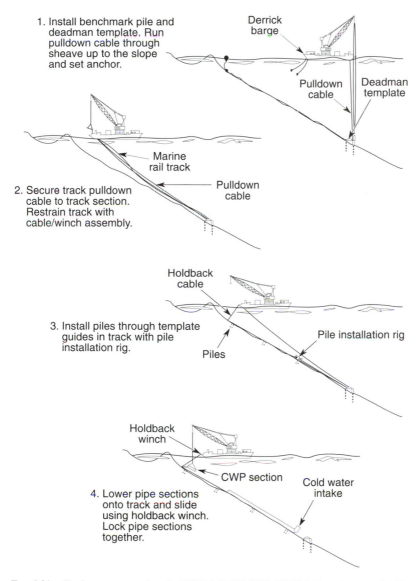

1. Install benchmark pile and deadman template. Run pulldown cable through sheave up to the slope and set anchor.

Derrick barge

Pulldown cable

Deadman template

Marine rail track

Pulldown cable

2. Secure track pulldown cable to track section. Restrain track with cable/winch assembly.

Holdback cable

3. Install piles through template guides in track with pile installation rig.

Pile installation rig

Piles

Holdback winch

CWP section

Cold water intake

4. Lower pipe sections onto track and slide using holdback winch. Lock pipe sections together.

Fɪɢ. 6-31. Deployment procedure for PREPA OTEC CWP (PREPA Phase I proposal, 1981).

operation off Kahe Point, Hawaii. In this way the same basic plant could provide engineering design information for commercial-size OTEC operation in both grazing and moored modes.

Participants in this PON response included:

SOLARAMCO: A company formed to develop, build, test, and operate a 115-ton/day [40-MWe (net at busbar)] grazing OTEC ammonia plantship, and market the ammonia product.

Table 6-8. PREPA 40-MWe (net) OTEC electric power plant concept

Tower dimensions (steel construction)
    Top platform: *W* 85 m, *L* 45 m; height above SL 20 m
    Base: *W* 90 m, *L* 50 m[a]; depth (landside) 70 m, (oceanside) 100 m
Power plant
    Closed Rankine cycle; ammonia working fluid
    Shell-and-tube heat exchangers; titanium tubes OD 2.54 cm, ID 2.40 cm
    Each 10-MW HX contains 3 bundles of 14,000 tubes
    Evaporator tubes enhanced on shell side with Linde Hi Flux coating
    Four 10-MWe (net) power modules
        Evap: Diam. 11.3 m; *L* 13.3 m, 11 m[b]
        Cond.: Diam. 11.3 m; *L* 20 m, 17.7 m[b]
Evaporators and condensers at 15-m depth (top)
Nominal operating characteristics
    ΔT 22.5°C; seasonal variation ±2°C
    Gross power ~53 MWe
    Net power 40 MWe
    Warm-water inlet temp. (°C) 27 ±2
    Cold-water inlet temp. (°C) 4.5 ±0.5
    Overall U_0 (W/m²° C): evaporator 4000, condenser 2800
Water ducts
    CWP
        FRP construction 8-m-diam., 2000-m-long, intake at 1000-m depth
        Assembled by joining 10-m sections lowered from a surface ship
        Emplaced via a marine rail
    Mixed effluent: three 8-m-diam. pipes laid on shelf to 200-m depth

[a]At 70-m depth.
[b]Tube length.

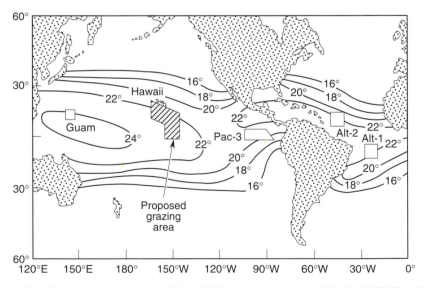

FIG. 6-32.   Proposed grazing area for SOLARAMCO ammonia plantship (SOLARAMCO Phase I proposal, 1981).

Lockheed Missiles and Space Co. (LMSC): Systems integration.
ABAM Engineering, Inc. (ABAM): Concrete structures, engineering.
State of Hawaii: Outfitting and checkout, siting.
Subcontracts:
  M. W. Kellog: Ammonia plant.
  Owens–Corning: CWP.
  ABAM: Hull.

The general features of the SOLARAMCO PON concept will be discussed here.

6.4.4.1 Platform

The plantship uses a barge-type hull made of post-tensioned concrete. The lower section of the hull is constructed in a graving dock and is then floated to a nearby deep-water pier where final hull construction and outfitting are completed. A schematic of the hull construction and outfitting showing the layout and principal dimensions is given in Fig. 6-33. Rotatable thrusters below the hull provide force for seakeeping and grazing at an average speed of 0.4 knot. The platform construction and deployment sequence is depicted in Fig. 6-34.

6.4.4.2 Power Plant

The baseline power system includes four different 10-MWe (net) power modules, each a complete closed Rankine cycle using ammonia as the working fluid,

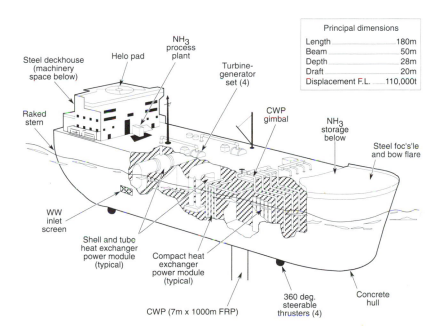

FIG. 6-33. Design of SOLARAMCO ammonia plantship (SOLARAMCO Phase I proposal, 1981).

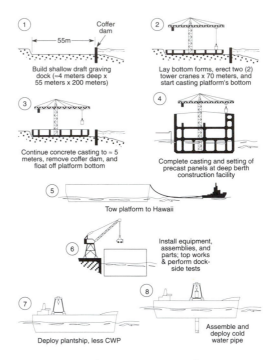

Fig. 6-34. Construction and deployment sequence for SOLARAMCO plantship (SOLARAMCO Phase I proposal, 1981).

equipped with a turbine-generator, an ammonia system, seawater pumps, electric systems, control equipment, biofouling control, and auxiliary equipment. The modules differ in being optimized for operation with different types of heat exchangers. Three are "compact" types: Alfa-Laval plate-fin, Trane finned-plate, and APL folded-tube. These draw water by gravity flow from overhead ponds. The fourth is an advanced shell-and-tube design. Schematics depicting the heat exchanger types, subsystem layouts, and power cycles for the first three HX types are shown Fig. 6-35. The power cycle diagram for the shell-and-tube HX system is shown in Fig. 6-36.

6.4.4.3 Water Ducting

Warm water enters the plantship through screened openings at the sides of the platform, passes through the warm-water pump, and enters the head ponds above the evaporators, or is pumped to the shell-and-tube evaporator. Cold water, pumped from the central moon pool, is channeled to head ponds above the condensers, or is pumped to the shell-and-tube condenser.

6.4.4.4 Cold-Water Pipe

The proposed CWP is a 7-m-diameter tube 1000 m long made in a double-wall construction, with a syntactic foam or balsa wood inner layer 51 mm thick

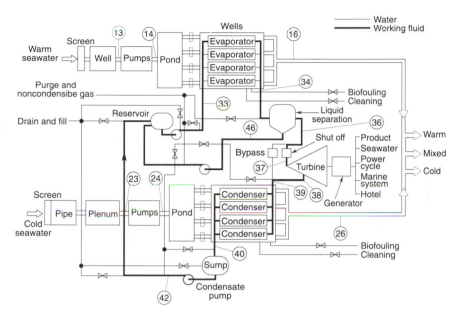

FIG. 6-35. Diagram of power cycle for Alfa-Laval, Trane, and folded-tube HX systems (SOLARAMCO Phase I proposal, 1981).

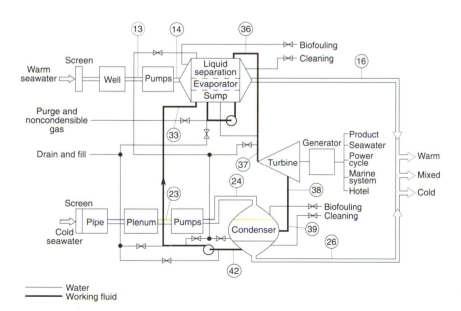

FIG. 6-36. Diagram of power cycle for shell-and-tube HX (SOLARAMCO Phase I proposal, 1981).

separating walls of 25-mm-thick FRP. The pipe is made in sections 10–20 m long that are barged to the platform stationed at the final assembly site off Kahe Point, Hawaii. Here the pipe is assembled by progressively joining and lowering the 10- to 20-m sections through a moon pool in the hull. Pipe joint options are shown in Fig. 6-37. A two-axis, in-plane gimbal (Fig. 6-38) provides a flexible connection between the pipe and platform.

6.4.4.5 Exhaust Ducting

Warm- and cold-water effluents are discharged, unmixed, at the bottom of the heat exchanger bays. Forward movement of the plantship prevents reinjection of the discharged plumes into the warm-water inlets.

6.4.4.6 Energy Transfer

The net power available from the OTEC turbine-generators is used in water electrolysis to produce hydrogen that is combined with nitrogen to form ammonia. The ammonia is liquefied and stored on-board the grazing plantship for a nominal 1-month period and is then transferred to an ammonia tanker for shipment to port. The SOLARAMCO members then market the ammonia through their normal commercial channels.

The principal characteristics of the SOLARAMCO Phase I design are listed in Table 6-9. Estimated costs are summarized in Chapter 7.

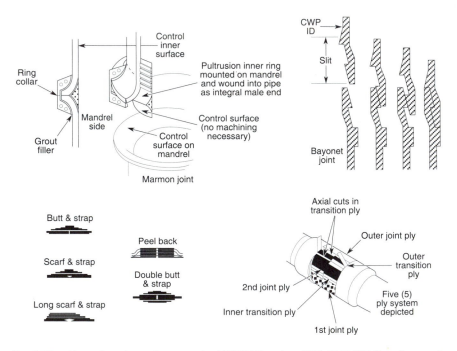

FIG. 6-37.   Alternative schemes for connecting FRP CWP sections (SOLARAMCO Phase I proposal, 1981).

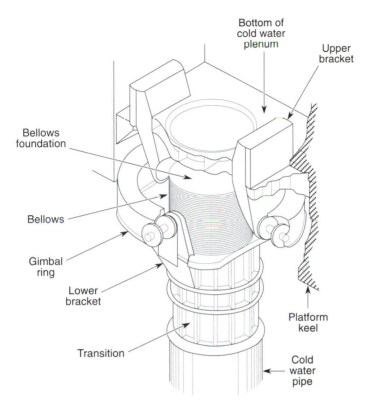

Bottom of
cold water
plenum

Upper
bracket

Bellows
foundation

Bellows

Gimbal
ring

Lower
bracket

Transition

Platform
keel

Cold
water
pipe

FIG. 6-38. Gimbal for CWP connection to the SOLARAMCO OTEC plantship (SOLARAMCO Phase I proposal, 1981).

6.4.5 Conceptual Design of OSEA 100-MWe OTEC Power Plant

In 1981 a consortium named OSEA, which included Sea Solar Power (SSP), GE, and General Dynamics, responded to the DOE PON with a proposal to build a 40-MWe floating OTEC plant of the SSP design that would supply power to the Puerto Rico Water and Power Authority. The technical details are proprietary but a summary of the estimated costs has been made available (Anderson, 1985). The following discussion is based on the published reports of the SSP 100-MWe OTEC plant conceptual designs and analyses. SSP in 1977 published a concept for a floating 100-MWe OTEC plant that incorporated many innovative features designed to maximize the thermal efficiency and minimize total system cost (Anderson and Anderson, Jr., 1977). Subsequent revisions led to the design shown in Fig. 2-3 (Anderson, 1985). The systems engineering study by the Andersons recognized the importance of many interactions between the ocean environment and the design and operation of the closed-cycle OTEC power plant that are unique to OTEC and that should be considered in achieving an optimum overall design. A successful demonstration of the SSP design features could have a major impact on estimated costs of OTEC plants. The design features include:

Table 6-9. Characteristics of SOLARAMCO 40-MWe OTEC ammonia Phase I
proposed design

Platform
  Dimensions: L 180 m, W 50 m, D 28 m, draft 20 m
  Displacement: 100,000 mt
  Thrusters: (4) 1875 kW, 360° steerable, ducted
  Annual average power demand: 2 MWe
Power plant
  10-MWe (net) modules (4) of different types employing ammonia working fluid in a closed
    Rankine cycle
  Various HX materials are candidates
  Nominal operating characteristics: ΔT (nom) 22.5°C ±0.8°C

	Avg. values	Heat exchanger type (individual values)			
		Plate	Finned plate	Folded tube	S&T
Power summary (MWe)					
Gross turb/gen power	16.45	16.47	16.42	16.46	16.48
Seawater pumps	3.1	2.79	3.18	3.33	3.19
Cycle pumps	1.2				
Thruster	1.9				
Hotel/services	0.3				
Ammonia plant					
Electrolysis	9.55				
Air plant, compress.	0.45				
Net power to NH$_3$ plant	10.0	10.08	10.14	10.24	10.14
NH$_3$ production (mt/d)	115.0				
Operating parameters					
Warm seawater flow (kg/s)		35,380	41,330	42,560	39,770
Inlet (outlet) temp. (°C)		26.5(22.7)	26.5(23.5)	26.5(23.4)	26.5(23.3)
Cold seawater flow (kg/s)		31,530	31,330	33,690	33,710
Inlet (outlet) temp. (°C)		4.0(8.2)	4.0(7.9)	4.0(7.7)	4.0(7.6)
NH$_3$ turbine flow (kg/s)		444.6	411.4	424.9	412.9
Inlet (outlet) temp. (°C)		20.5(9.9)	21.2(9.8)	20.6(9.5)	21.3(9.9)
NH$_3$ press. in (out) (kPa)		871(614)	891(611)	873(605)	893(612)
NH$_3$ quality in (out)		99.0(96.2)	99.0(96.0)	99.0(96.1)	99.0(96.0)
U_0 (W/m² K)					
Evaporator		3030	3610	2470	3580
Condenser		2870	2800	2470	2815
Specific area (m²/kW)		8.8	8.5	11.2	6.15
Total HX cost ($/kW)		1200–1500	800–1000	1000–1200	1450–1700

1. Design of the power plant components for submerged operation so that the major power plant components can be placed in the ocean outside the hull.
2. Use of turbine-driven pumps for the warm and cold water. This eliminates efficiency losses in conventional designs that result from conversion of turbine power to electric power for operation of electrically driven pump motors.
3. Use of a new heat exchanger design with projected heat transfer rate four times that of conventional heat exchanger designs.
4. Use of a four-stage vapor cycle to improve the average temperature difference between the water flows and working medium flow.

5. Placing the evaporators 25 m below the condensers allows the condensed working fluid, R-22, to flow by gravity to the evaporators, eliminating the need for a working fluid liquid pump and the associated power losses.
6. Use of a "stockade" construction of the CWP that is designed to permit CWP assembly on the OTEC platform from sections of steel pipe that would be easy to transport. With appropriate valves and seals, the hollow pipes could be used to control the buoyancy of the CWP.

Engineering study of the SSP concept has been limited so that many details of the construction, deployment, and operation are not available. Without development support for a thorough engineering evaluation of the Andersons' design the claimed performance and cost advantages of their total system compared to other OTEC systems that have received more engineering support remain in doubt.

The principal design characteristics of the OSEA concept will be discussed here.

6.4.5.1 Platform

The OTEC plant concept is illustrated in Fig. 2-3 (Anderson, 1985). The platform, of steel construction, is of the semisubmersible type. Essentially it is a framework that supports the power plant systems. Warm water enters through screens surrounding the plant at the water line and passes through turbine-driven warm-water pumps that direct the water through the large tubes shown at the ends to plenums at 63 m depth that house the evaporators. Cold water flows from the centrally located CWP through turbine-driven pumps to plenums that house the condensers. These are located at approximately 25 m depth. The above-water platform, approximately 25 m above the sea surface, supports personnel services and facilities for control, maintenance, and repair of the OTEC system equipment. The openings shown in the platform permit equipment mounted below the surface to be raised to the top platform. The turbo-generators can be drawn into the cylindrical chambers shown at about 50 m depth, to which their housings are attached, for servicing.

6.4.5.2 Power Plant

The power plant uses a closed Rankine cycle with Freon R-22 as the working fluid. The density of R-22, 1250 kg/m^3, allows it to flow without a pump under gravity from a condenser 25 m above the evaporator against the pressure head of the evaporating vapor. This eliminates the need for working fluid pumps. The power cycle is conducted in four stages, with different turbine inlet and outlet pressures, which enhances cycle efficiency at the cost of more complexity. The turbines in one stage drive the water pumps. The heat exchangers are of a compact SSP finned tube design, stated to give an overall heat transfer rate four times that of conventional shell-and-tube designs. Vanes are provided at the water exits that can be vectored to provide stationkeeping for the plant, thereby obviating the need for mooring.

6.4.5.3 Water Ducting

Cold-Water Pipe.   Cold water enters the plant through an 8.5-m-diameter central tube suspended by a spring that is supported by hydraulically controlled cables. The tube is made by binding together parallel steel tubes of about 1/3 m diameter to create a cylindrical "stockade" configuration. The length of the tubes is selected to permit assembly of pipe sections on the platform that can then be joined together and lowered to form a complete CWP of the desired total length. The hollow tubes may be sealed and filled with air or water to control the buoyancy of the assembly. The spring support is designed to isolate the CWP from platform motions. Details of the support and water seal between pipe and platform have not been published. The allowable angular motion between CWP and platform has not been stated. Since the cold water is discharged into the mixed layer, it is expected to enhance biomass production.

Warm Water.   Warm water enters the system through the screens shown in Fig. 2-3. The turbine-driven warm-water pumps are located in the vertical down-comer pipes that conduct the water to the evaporator plenums.

Exhaust Ducting.   Water from the evaporators and condensers discharges directly through vanes into the surrounding seawater. Under average conditions the water will enter the mixed layer and will be denser than the ambient water.

6.4.5.4 Energy Transfer

The design calls for electrical energy to be transferred ashore by an underwater cable connecting the floating OTEC plant terminal to the onshore utility. The plant is maintained in position by vectoring the exhaust flow to counter surface and subsurface ocean currents. Similar methods are employed for stationkeeping of semisubmersible drilling rigs; however, the thrust requirements are much larger because of the underwater areas of the evaporators and condensers. Stationkeeping must be possible under storm conditions that would cause loss of OTEC power, implying a need for a method independent of the OTEC pumped water flow. The method and costs of the standby equipment are not available.

6.4.5.5 Construction and Deployment

Construction of the OTEC plant, without the CWP, would be done in a shipyard experienced in fabrication of oil drilling rigs. Deployment would involve support-ing the OTEC structure on its side on a barge, towing the barge to the offshore site, and then floating the assembly free from the barge and maneuvering it into an upright position for assembly and installation of the CWP. Details of the deploy-ment scenario have not been published.

6.4.6 Conceptual Design of 160-MWe (nom) OTEC Methanol Plantship

As a follow-on to the preliminary design of the APL 40-MWe OTEC baseline ammonia plantship, effort was directed, under DOE support, to a conceptual study of a 160-MWe (nom) OTEC plantship designed to use OTEC electrolytic hydrogen and oxygen, reacting with coal, for on-board methanol synthesis. Pulverized coal for the process is transported to the plantship by conventional coal colliers. The methanol produced is accumulated in on-board tanks for approximately 1 month and then shipped to continental ports in conventional tankers. The process efficiently uses both oxygen and hydrogen produced by water electrolysis in combination with low-cost coal to produce a low-pollution liquid vehicle fuel that can replace gasoline for transportation at competitive costs. Unlike other synfuel processes involving coal as a feedstock, no CO_2 is generated as a byproduct of the OTEC methanol synthesis process.

The following organizations participated in the OTEC methanol conceptual design program:

The Johns Hopkins University Applied Physics Laboratory (JHU/APL): Technical direction and systems integration.

Brown and Root Development, Inc. (BARDI): Initial design of OTEC methanol process and plant space requirements based on Texaco coal gasification and Lurgi methanol synthesis. Hull sizing and seakeeping requirements for 160-MWe (net) OTEC methanol plantship.

EBASCO Services, Inc.: Advanced design of OTEC methanol process based on Rockwell coal gasification and ICI methanol synthesis.

Rockwell International: Design of coal gasification process (Kohl et al., 1980).

The principal features of the OTEC methanol conceptual design are discussed briefly in the following sections.

6.4.6.1 Platform

The platform configuration is based on scale-up of the APL 40- MWe (net) plantship to 160 MWe (nom) to provide power and space for a 1750-mt/day methanol plant. The rectangular hull, with vertical sidewalls, is constructed of post-tensioned concrete as in the baseline 40-MWe design. Wing walls at the sides and ends provide longitudinal strength to resist 100-year-storm bending loads. Minimal fairing of the bow is provided to reduce drag in grazing operation. The methanol plant is placed at the center of the platform with OTEC power modules of 80 MW (net), each equipped with a CWP, at either end. The layout is shown schematically in Fig. 6-39. An outboard profile is shown in Fig. 6-40. For a 184-MWe power output, enlargement of the baseline design of the cold-water pipes from 9.1 m diameter to 11–13 m, with accompanying enlargement of the flexible joint supports, is required. Power for seakeeping is provided by rotatable ducted thrusters beneath the hull. Details of the design and analysis of the platform motions under normal and storm conditions are provided in BARDI (1982). Platform construction and deployment follow the procedures discussed in Section 6.5.1 for the APL 40-MWe demonstration plantship, with appropriate modifications to accommodate the larger size.

6.4.6.2 Power Plant

A plan view of the power plant layouts is included in Fig. 6-39. Eight 10-MWe (net) power modules are grouped around one CWP at station 137.6 and eight around a second CWP at station 687.5. (Note that the drawing dimensions are in English units.) Each 10-MWe module includes two evaporators and two condensers, supplied with seawater by gravity flow from overhead ponds, as in the baseline

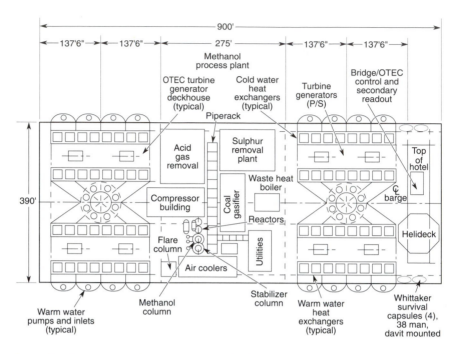

Fɪɢ. 6-39.   Plan view of OTEC methanol plantship (BARDI, 1982).

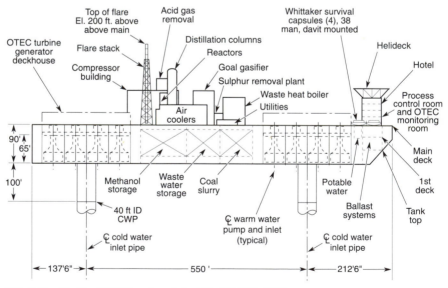

Fɪɢ. 6-40.   Profile of OTEC methanol plantship (BARDI, 1982).

40-MWe OTEC design. The BARDI study uses the APL baseline power plant performance based on the ATL-1 operational site and annual average ΔT of 23.3°C to give a nominal power output of 160 MWe (net) for their design. In the EBASCO update (EBASCO, 1984), a South Pacific site near Guam was chosen where the annual ΔT is 24.1°C. A calculation of the power output for each month of the year for this site shows that the annual average power of the nominal 160-MWe plant would be 178.4 MWe (net), with maximum output of 189.9 MWe (net) in the months of July and August. Since the turbines and generators would need to be designed for this maximum output, EBASCO proposed that a supplementary coal-fired boiler electric power system be installed that would be used to maintain constant net power input of 189 MWe to the methanol plant. This modification increases annual coal consumption by only 6%, with evident advantages in maintaining plant efficiency at maximum levels throughout the year. The details are presented in Table 6-10 (EBASCO, 1982). The added capital cost is small because coal supplies and handling equipment are already available on the plantship.

A brief study by EBASCO indicated that the average net power could be further increased to 207 MWe (net) in a thermal augmentation cycle that uses process heat to superheat the ammonia working fluid in one or two of the OTEC power modules, as shown in Fig. 6-41. The Mollier diagram for the process is shown in Fig. 6-42. This modification was incorporated in estimating the methanol plant output; however, funding was not available to permit equipment details of that process to be established.

6.4.6.3 Water Ducting

The BARDI and EBASCO conceptual design studies were not funded to determine the design requirements of the water-ducting systems for the 160-MWe (net) methanol plantship. Scale-up of the 40-MWe baseline design, with improvements suggested by the ANL experiments, indicates that two CWPs of 11–13 m diameter made of FRP or lightweight concrete and supported at the center position on the platform of each 80-MWe module by a universal joint or ball-and-socket mounting will be required. The general layout is indicated in Figs. 6-39 and 6-40 (Avery et al., 1983). No ducting is necessary for the warm- and cold-water effluents.

6.4.6.4 Energy Transfer

Energy transfer via production of methanol on an OTEC plantship with subsequent shipment to world ports was studied in two phases. In the initial phase the feasibility was studied of installing a methanol plant based on the Texaco coal-slurry coal gasification process and Lurgi methanol synthesis (BARDI, 1982; Avery et al., 1983). BARDI was selected for the design study because of their experience in operating a demonstration of the Texaco technology at Muscle Shoals, Mississippi, and the availability to them of a proprietary preliminary engineering design of a barge-mounted plant for offshore production of methanol from natural gas. Both the hull design requirements and methanol plant installation and operation were considered. Further information was available from reports on the status of floating vessels for industrial plants in a comprehensive study conducted by the U.S.

Table 6-10. Characteristics of 1750-mt/d OTEC–methanol plantship

Platform
    Rectangular barge; post-tensioned concrete construction
    L 275 m, W 108 m, depth 27.4 m; bow fairing
    Launching draft 10.8 m; operational draft 19.8 m

Dry weight at launching (mt)	290,100
Outfitting	4,355
OTEC machinery	31,000
Heat exchangers	12,650
Process plant	16,350
Hotel and superstructure	2,720
Cold-water pipes	16,350

Ballast to bring the total operating displacement to 514,000 mt at 19.8-m draft:

Pond/ballast	21,800
Misc. fluids	10,000
Process/ballast	73,000
Ballast	39,700

Power plant
    Closed Rankine cycle; ammonia working fluid
    Folded tube heat exchangers in 5-MWe (net) units
    Two complete power plants, including water ducting and CWP, each having 8 10-MWe (nom)
        power modules
    Operating characteristics
        $\Delta T(°C)$
            Annual average 24.1
            3-mo. average: Feb 23.6, May 23.9, Aug 24.7, Nov 24.4
        Electric power summary (MWe)

Gross OTEC power	234.0
Warm-water pump power	18.3
Cold-water pump power	25.9
NH_3 pump power	7.0
Hotel and misc.	1.3
Net OTEC power	181.5
Net OTEC plant power[a]	190.0
Total power with thermal augmentation	207

    Methanol process characteristics (nom)
    Input

Electrolytic hydrogen (kg/s)	1.14
Electrolytic oxygen (kg/s)	9.05
Coal (e.g., Utah Sunnyside) (mt/d)	1360
Sodium carbonate (mt/d)	39

    Output

Methanol (mt/d)	1750
Liquid sulfur (mt/d)	4.8
Ash (mt/d) (shipped for disposal)	35.4

[a]Includes coal-power supplement.

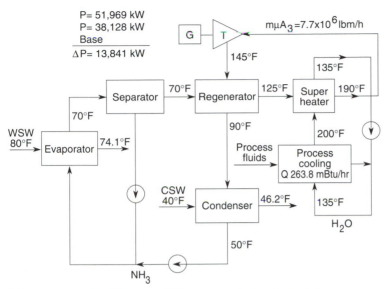

FIG. 6-41.   Augmented 20-MWe (nom) OTEC power cycle (EBASCO, 1982).

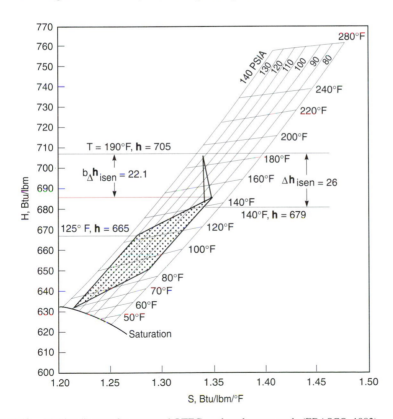

FIG. 6-42.   Mollier diagram for proposed OTEC methanol power cycle (EBASCO, 1982).

Maritime Commission (MARAD) (MARAD, 1980; Global Marine Development, Inc., 1981; International Maritime Associates, 1981). The study showed the feasibility of building a seaworthy OTEC plantship that would produce 1000 mt/d of methanol, with an OTEC power plant designed to furnish 160 MWe (nom) power to the electrolysis cells (BARDI, 1982). In the BARDI OTEC–methanol study, current technology was used as the basis for plantship design.

In the follow-on by Ebasco Services Inc. (EBASCO, 1984), methanol plantship design was based on the use of a more efficient and compact coal gasification process (Rockgas) developed by Rockwell International Corporation. This process, which uses molten sodium carbonate as a coal gasification medium, had been tested at a 10-mt/h coal rate. Thus, data on systems requirements and gas compositions were available. A drawing of the gasifier is shown in Fig. 6-43 (Richards et al., 1984). EBASCO and Rockwell cooperated in the OTEC–methanol process design. A schematic of the process is shown in Fig. 6-44 (Avery et al., 1985). The revised design also includes process improvements in the use of heat generated in the gasification and methanol synthesis steps. With these changes the EBASCO–BARDI–Rockwell system will produce 1750 mt/d of methanol within the same space and platform of the BARDI design, and with the same OTEC power system, sited at Guam rather than ATL-1.

The principal features of the OTEC–methanol plantship design are listed in Table 6-10. Estimated capital costs are summarized in Chapter 7.

6.4.7 Conceptual Design of Japanese 100-MWe Floating OTEC Plants

During the period 1974–1977 a program was established in Japan to explore the feasibility and cost of supplying power or energy-intensive products to Japan via OTEC. A presentation by Kamogawa to the Sixth General Meeting of the Pacific Basin Economic Council in 1970 that suggested OTEC as a potential means of supplying power and promoting mariculture led to the establishment in 1974 of a committee of the Ministry of International Trade and Industry (MITI) to investigate the feasibility of OTEC. The program was named the "Sunshine Project." In the same year experimental work was begun at the Electrotechnical Laboratory of the Agency of Industrial Science and Technology (AIST) in Tokyo, and at the Science and Engineering Faculty of Saga University (Kamogawa, 1972; Homma and Kamogawa, 1979). The program began with a first phase study of a 1-MWe, land-based plant and proceeded to a study of a 100-MWe floating plant. This was followed by a conceptual system design in 1975. In 1976 an improved conceptual design was developed, and in 1977 the design was further refined with emphasis on platform structure, stationkeeping, riser cable design, and heat exchanger improvements. Only the final version is discussed here.

The following organizations participated in the conceptual design studies:

Saga University
Electrotechnical Laboratory
MITI
Tokyo Electric Power Co.

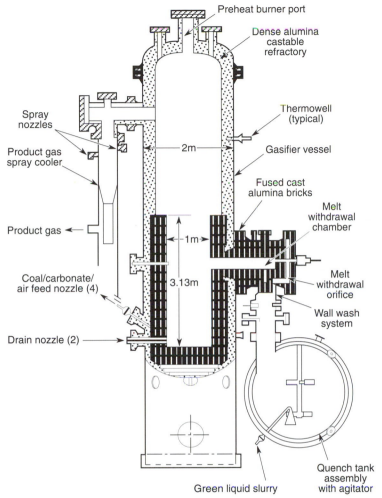

FIG. 6-43. Rockwell molten carbonate coal gasifier (Richards et al., 1984).

Japan Marine Science and Technology Center
Toshiba Corp.
Tokai University
Kyuei Co., Ltd.

The principal features of the moored 100-MWe floating OTEC plant are discussed briefly in the following sections.

6.4.7.1 Site Selection

After a preliminary survey of water temperatures, wind, and bathymetry, five representative sites for OTEC power plant installations in Japan were considered

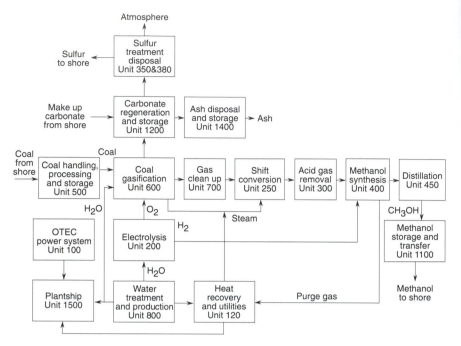

Fig. 6-44. Schematic of EBASCO–BARDI–Rockwell OTEC methanol power cycle (Avery et al., 1985).

in some detail. The sites are shown in Fig. 6-45. An estimate of the total OTEC power that could be produced from the thermal resources of the sea near Japan indicated a potential output of 3.75×10^5 GW h/y from an area 1° wide in longitude and 1° high in latitude. This output exceeds the annual electric power consumption in Japan (Kamogawa, 1980). The sites at Toyama and Osumi were selected for conceptual designs of OTEC moored plants. A summary of factors considered in the conceptual design studies is listed in Table 6-11.

6.4.7.2 Platform

Three types of platform, all of steel construction, were considered: a surface ship, a surface disc, and a submerged disc. After study of construction feasibility, mooring, CWP design, requirements to meet extreme weather conditions, and estimated capital cost, the submerged disc was chosen for evaluation at the Osumi site, and the surface ship for the Toyama site. Both platforms are designed for moored operation offshore Japan, with power transmission by underwater cable to the on-land power system.

6.4.7.3 CWP and Power Transmission

Information is not available.

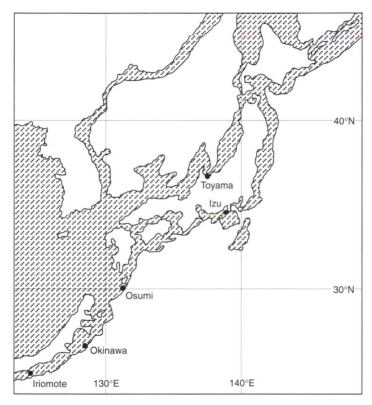

FIG. 6-45.   Sites for proposed Japanese moored OTEC power plants (Homma and Kamogawa, 1979).

Table 6-11. Site selection criteria for moored Japanese
OTEC plants

Design factor	Power output relative to a plant sited at Iriomote, with 1000-m CWP	
	At Osumi	At Toyama
CWP length (m)		
600	0.70	0.62
800	0.86	0.62
1000	0.94	
Maximum wind speed (m/s)	78.6	39.6
Significant wave height (m)	13	8.5
Shortest distance to land (km)	20	3–4
Distance to major port (km)	200	20–25
Distance to airport (km)	70	20

6.4.7.4 Power Plant

The power plant is based on the closed Rankine cycle with ammonia as the working fluid. Plate heat exchangers of the type designed by Uehara et al. (1978, 1983) are incorporated in 25-MWe power modules. The principal characteristics of the Japanese 100-MWe floating OTEC plant designs are listed in Table 6-12. Cost estimates are summarized in Chapter 7 (Homma et al., 1979).

6.4.8 Conceptual Design of French 5-MWe OTEC Plants at Tahiti

A feasibility study of shore-based and floating OTEC plants to be sited in Tahiti was conducted during 1978–1980 under sponsorship of the French Centre pour l'Exploitation des Oceans (CNEXO). It was concluded that, with estimated OTEC plant costs and projected oil prices, OTEC installations of 5–20 MWe would be competitive with oil plants. The favorable result led to a decision to proceed with the design of a land-based 5-MWe OTEC power plant, with a scheduled completion date of 1985. The Phase II program was initiated under joint sponsorship of CNEXO (renamed IFREMER) and an industrial joint venture called EUROCEAN GIE, composed of the companies that had been involved in the Phase I program. The Phase II program included detailed site studies, evaluation of closed- and open-cycle options, methods of constructing and deploying the cold-water pipe, and estimates of costs of favored alternatives.

Table 6-12. Conceptual designs of Japanese 100-MWe floating OTEC plants

Platform:	At Osumi	At Toyama
Type	Submerged disc	Surface ship
Dimensions (m)	Diameter 110	L 230, W 60
	Depth 40	Depth 27
	Draft 55	Draft 13
Displacement (m)	409,300	145,000
Water ducts		
CWP: diameter 11.2 m, length 500 m		
Discharge (2 pipes): diameter 8 m, length 100 m		
Power plant		
4 25-MWe (gross) modules; ammonia working fluid; closed Rankine cycle; titanium plate HXs of Uehara design; cross flow		
Nominal operating characteristics:	At Osumi	At Toyama
Gross power (MWe)	100	100
Net power (MWe)	78.8	83.1
Warm-water temp. (°C)	28	26
Cold-water temp. (°C)	4.56	0.747
Warm-water flow (kg/s)	217,220	192,500
Cold-water flow (kg/s)	171,100	158,100
NH_3 flow rate (kg/s)	2,750	2,520
Evaporator heat transfer area (m^2)	2,096,000	1,712,000
Condenser heat transfer area (m^2)	2,352,000	1,896,000

6.4.8.1 Site Selection

Preliminary surveys led to selection of a site at the port of Papeete for the projected plant. A map showing the plant location on the island of Tahiti is presented in Fig. 6-46. A research and development program was then conducted that included detailed bathymetry near the site as well as:

1. Evaluation of the thermal resource and its seasonal variations;
2. Measurements of waves and currents; and
3. Measurement of winds and sea states and the physico-chemical and biological composition of the seawater.

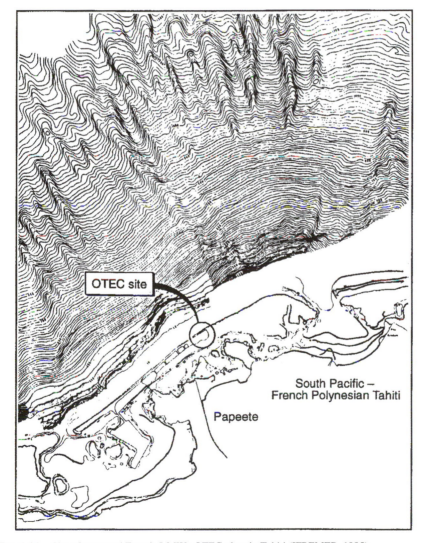

FIG. 6-46.   Site of proposed French 5-MWe OTEC plant in Tahiti (IFREMER, 1985).

The survey shows that a ΔT of $23.3 \pm 1.5°C$ is available at a depth of 1000 m within 3 km from shore and that the bottom slopes gradually from near shore to the 1000-m depth. The maximum wave is estimated to be 12 m and the wave period 12 s.

6.4.8.2 Platform

The design does not call for a platform or containment building. Instead the power plant components are mounted separately. Plant layouts for closed-cycle and open-cycle systems are illustrated in Figs. 6-47 and 6-48. A pond adjoining the power plant installation receives nutrient-rich water leaving the condensers that is to be used for mariculture programs. The water effluent streams are finally directed to a lagoon from which the water flows to the ocean.

6.4.8.3 Power Plant

Conceptual designs of both closed- and open-cycle OTEC installations have been prepared, based on selection of the most attractive of a wide range of alternative designs considered in Phase I. Investment comparisons indicate that the costs of the two systems are within 5% of each other. The closed-cycle plant uses horizontal shell-and-tube heat exchangers with smooth titanium tubes in both evaporators and condensers. A cost study showed that the performance gains with enhanced surfaces would not compensate for the added cost of enhancement of the condenser tubes. Dependent on industrial capabilities, enhancement of the evaporator tubes could be cost effective (Marchand, 1985). A decision not to use aluminum was based on the lack of data on the long-term durability of aluminum. [Durability has

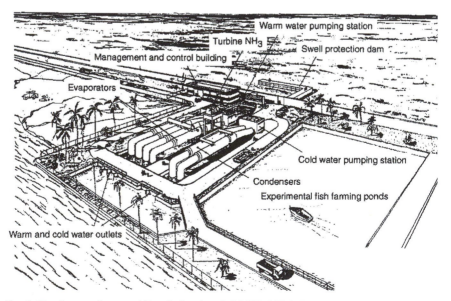

FIG. 6-47. Layout of proposed French closed-cycle 5-MWe OTEC plant in Tahiti (IFREMER, 1985).

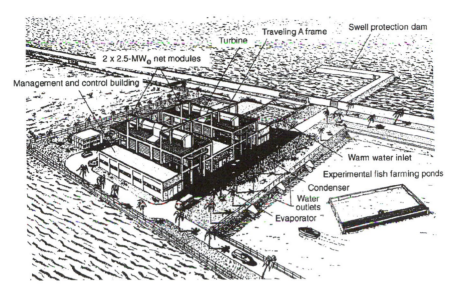

FIG. 6-48.   Layout of proposed French open-cycle 5-MWe OTEC plant in Tahiti (IFREMER, 1985).

since been demonstrated (Larsen-Basse, 1983); therefore, use of aluminum heat exchangers would make a significant reduction in the plant investment.] Ammonia was chosen as the working fluid.

The open-cycle design uses spout evaporators and direct-contact condensation of the steam with the cold seawater. A 2.5-MWe evaporator module requires 16,000 spout tubes arranged in a tray of 110 m^2 area. The condenser is of a patented new design for which details are not available. Performance data for the design were obtained in experiments with a 30-kW module in the laboratories of the Alsthom Atlantique Neyrtec Co. in Grenoble. The evaporator and the two condenser modules, of the dimensions shown in Fig. 6-49, are enclosed in a steel vacuum vessel. The floor of the vessel is 9.5 m above sea level.

6.4.8.4 Water Ducting

A variety of cold-water pipe designs was evaluated. Initial studies showed that one type of structure would be needed to support the pipe in the shallow section extending from near the knee of the escarpment through the surf zone into the onshore plant, and a second support design would be required for the deep-water section extending from the shallow zone a distance of approximately 2.5 km to tap the 4°C water at a depth of 950 m. The preferred overall design that emerged has the following features:

1. In the shallow zone the pipe is enclosed in a concrete trench 15 m deep at the shore that extends from the base of the shore plant to a point approximately 100 m from the "knee," where the sea floor drops sharply. From this point to the knee the pipe is rigidly supported by fixed guide rings set on metal structures

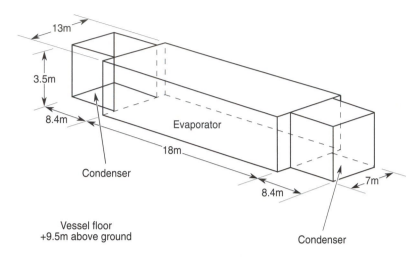

FIG. 6-49.   Dimensions of HX compartments for 5-MWe OTEC open-cycle plant in Tahiti (IFREMER, 1985).

held at the base by concrete footings. At the knee the pipe connects through a ball joint to the deep-water section.

2. The deep-water pipe is supported above the sea floor in a trajectory defined by "hawses" anchored by piles in the sea floor and supported by buoys designed to float at 50 m depth. The neutrally buoyant CWP, 3 m in diameter, assembled on shore as deployment proceeds, is deployed by pulling it through the hawses. This arrangement was selected after studies of the alternatives depicted in Fig. 6-50 (IFREMER, 1985; Geothermies Actualities, 1985).

3. Reviews of materials and methods for construction of the CWP led to selection of a double-walled design with fiber-reinforced plastic (FRP) side walls and a syntactic foam spacer.

6.4.8.5 Energy Transfer

Direct transmission of electric power, preparation of fresh water, and mariculture may all be combined to make operation of the 5-MWe Tahiti plant of commercial interest. Cost estimates for the last option have not been obtained.

The principal design features of the 5-MWe (net) Tahiti OTEC plants are listed in Table 6-13. Cost information is presented in Chapter 7.

6.4.9 Conceptual Design of Taiwan 50-MWe (net) Shore-Based OTEC Plants

In 1984 the Taiwan Power Company initiated a program to determine the feasibility and to prepare conceptual designs of shore-based 50-MWe OTEC plants in Taiwan (Liao et al., 1986). Three sites on the east coast of Taiwan were selected for evaluation: Ho-Ping, Shi-Ti-Ping, and Chang-Yuan. Studies of the bathymetry at the sites, water temperatures and currents, and availability of supply and construction facilities led to the conclusion that shore-based plants at Ho-Ping and Chang-

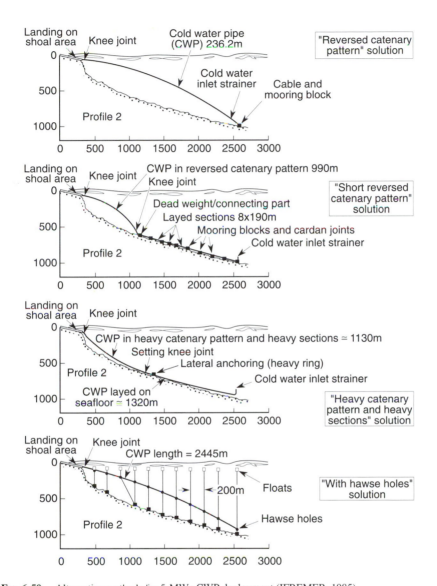

FIG. 6-50. Alternative methods for 5-MWe CWP deployment (IFREMER, 1985).

Yuan sites should be selected for conceptual design. The locations are shown in Fig. 6-51. The design features are reviewed briefly in the following sections.

6.4.9.1 Platform

After reviewing estimated costs, it was decided that the platform (containment structure) should be constructed in a shipyard, where facilities for installing heavy equipment would be available. The platform would then be floated and towed as a unit to the plant site on the beach. Here it would be lowered by flooding the sumps

Table 6-13. Conceptual designs of French 5-MWe (net) OTEC plants at Tahiti

	Closed-cycle plant	Open-cycle plant
Platform	Shore-based not enclosed	Shore-based not enclosed
Power plant	Closed Rankine cycle	Open cycle
Working fluid	Ammonia	Water vapor
	S & T HX with smooth	
	titanium tubes	Spout evaporators
	Tube diam. 25 mm,	
	thickness 0.7 mm	Direct contact cond.
Number of tubes	30,400	18,200
Tube length (total) (km)	400	320
Nominal operating characteristics		
Gross power (MWe)	6.80	6.99
Warm-water pumping (MWe)	0.60	0.34
Cold-water pumping (MWe)	0.94	0.67
Other power needs (MWe)	0.27	0.78
Net power (MWe)	4.99	5.20

	Evaporator	Condenser
U_0 (unenhanced) (W/m^2 °C)	2,000	2,200
HX area (m^2)	32,000	50,500
Heat loading (MWt)	260	252

to rest on a prepared bottom foundation behind an appropriate breakwater. The general arrangement of the containment structure is shown in Fig. 6-52.

6.4.9.2 Power Plant

The power plant employs four 12.5-MWe (net) power modules that use ammonia as the working fluid in horizontal shell-and-tube heat exchangers of conventional overall design, as depicted in Fig. 6-52. A sketch of a power module is also shown in Fig. 6-53. Titanium tubes are employed. The design permits use of either chlorine or Amertap for biofouling control. The seawater flow arrangement is shown in Fig. 6-53. Seawater enters a bay (sump) at the base of the plant from which it is pumped by a vertical, axial pump into the heat exchanger shell. After transit through the HX tubes it discharges into a second sump, where it mixes with effluent from the other heat exchangers and is then discharged through pipes that exhaust below the mixed layer.

6.4.9.3 Water Ducting

Warm and cold water enter the platform through pipes that pass through the breakwater and emerge into sumps below the platform floor. Water is forced through the heat exchangers by vertical pumps that draw water from the sumps. The heat exchanger effluents discharge into a common sump from which the water flows out by gravity.

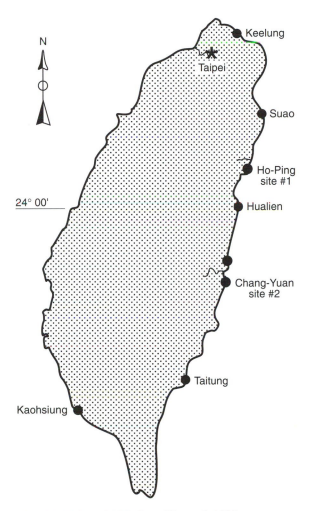

FIG. 6-51. Sites selected for Taiwan OTEC plants (Liao et al., 1986).

Cold-Water Pipe. A single CWP of 10 m diameter, made of lightweight concrete, provides cold water for all four power modules. A systems engineering study led to a conclusion that, with existing technology, the pipe diameter should be restricted to 10 m to ensure reasonable freedom from risk in the construction and deployment of the pipe. This constraint limited the total power that could be generated by the plant. Studies of the variation of net power versus water flow velocity then showed that maximum net power of 50 MWe would be possible with the 10-m pipe diameter. Consideration of various methods of pipe construction and deployment led to the conclusion that the pipe should be constructed in sections, 10 m in diameter and 15 m long, with elastomeric bell and spigot configurations at the end, as shown in Fig. 6-54. The pipe sections will be barged to the plant site, where

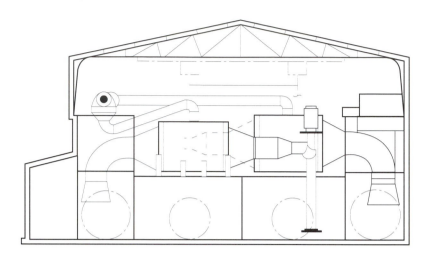

Section looking to sea

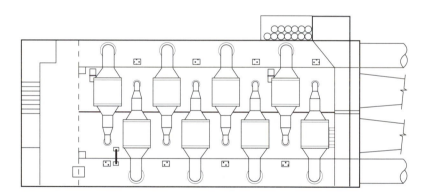

Plan at lower containment structure

Fig. 6-52. Layout for Taiwan 50-MWe OTEC power plant (Liao et al., 1986).

they will be joined end to end to form the cold-water pipe. Pipe deployment would involve the sequence of operations depicted in Fig. 6-55.

Warm Water.   Details of the warm-water and mixed effluent deployment are not available.

6.4.9.4 Energy Transfer

The installation of the power plant on the beach allows the generated power to be transferred directly to the onshore utility. Details of the power conversion and control equipment are not available.

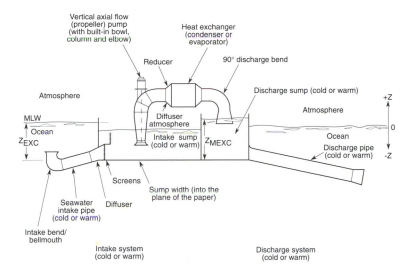

FIG. 6-53. Schematic design of Taiwan 50-MWe OTEC power plant (Liao et al., 1986).

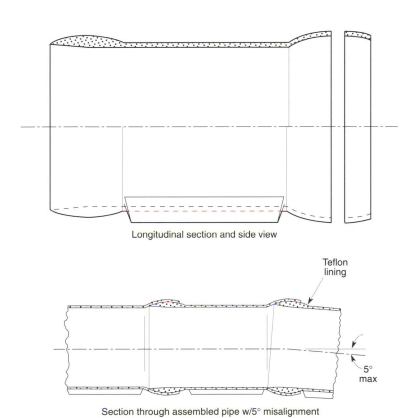

Longitudinal section and side view

Section through assembled pipe w/5° misalignment

FIG. 6-54. Design of CWP joints for Taiwan 50-MWe OTEC power plant (Liao et al., 1986).

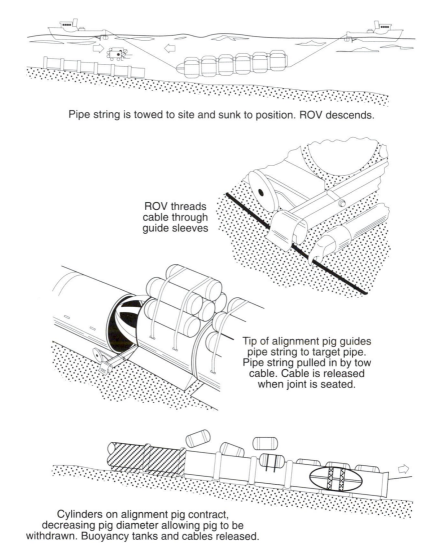

Pipe string is towed to site and sunk to position. ROV descends.

ROV threads cable through guide sleeves

Tip of alignment pig guides pipe string to target pipe. Pipe string pulled in by tow cable. Cable is released when joint is seated.

Cylinders on alignment pig contract, decreasing pig diameter allowing pig to be withdrawn. Buoyancy tanks and cables released.

FIG. 6-55.   Deployment sequence for Taiwan 50-MWe OTEC CWP (Liao et al., 1986).

The design features and operating characteristics of the Taiwan conceptual designs are listed in Table 6-14. Cost information is presented in Chapter 7.

Table 6-14. Design features of the Taiwan Power Co. 50-MWe OTEC plants

Feature	At Ho-Ping		At Chang-Huan	
Closed Rankine cycle; NH_3 working fluid; shell-and-tube HX; titanium tubes; containment system L 170 m, W 77 m, H 40 m, depth below SL 15 m				
Average ΔT (°C)	20.35		21.72	
Power plant summary	Evap.	Cond.	Evap.	Cond.
HX internal shell diam. (m)	14	12	14	12.25
Tube OD (mm)	25.4	25.4	25.4	25.4
Effective tube length (m)	14	14	14	14
No. of tubes	79,954	62,860	79,954	79,604
Water flow (kg/s)	51,000	32,780	51,000	42,140
Re no. (H_2O)	42,291	20,312	42,291	20,312
Pr no. (H_2O)	6.437	11.512	6.437	11.512
Water vel. in tubes (m/s)	1.85	1.5	1.85	1.5
H_2O inlet/outlet temp. (°C)	24.35/21.49	4.0/8.3	25.95/23.07	4.0/8.3
Water pressure drop (m)	2.45	1.93	2.45	1.93
NH_3 flow (kg/s)	543	494	733	667
NH_3 temp. (°C)	20.0	10.0	20.0	10.0
Operating pressure (kPa)	858	615	858	615
LMTD (°C)	2.74	3.41	3.77	3.41
U_0 (W/m² °C)	2,959	3,092	2,945	3,092
h_w (W/m² °C)	6,992	4,571	6,913	4,571
h fouling (W/m² °C)	16,667	16,667	16,667	16,667
h wall (W/m² °C)	10,652	10,506	10,652	10,506
h NH_3 (W/m² °C)	24,365	28,501	24,365	28,501

6.5   PRELIMINARY DESIGN OF OTEC PLANTS

6.5.1 Baseline 40-MWe (net) Moored and Grazing OTEC Plants

In 1977, The Applied Physics Laboratory of The Johns Hopkins University (JHU/APL) was tasked by DOE to prepare baseline preliminary engineering designs of moored and grazing 40-MWe (net) OTEC plants that would furnish basic data on systems requirements, performance, and costs. The information was required to provide informed specifications for projected DOE industrial contracts for OTEC plant construction. The baseline designs resulting from this task incorporate the results of a number of earlier experimental and analytical studies of OTEC systems requirements, including the conceptual design of a 100-MWe plant, design and tests of a folded-tube heat exchanger module, design and test at 1/3 scale of a jointed section of a 9.1-m-diameter segmented CWP that would use lightweight concrete, at-sea tests of a 1/6-scale jointed CWP which validated theoretical predictions of CWP dynamics, tests of ultrasonic and mechanical systems for removal of biofouling from the heat exchanger tubes, and analyses of OTEC commercialization potential (George and Richards, 1980).

The JHU/APL 40-MWe baseline designs update and augment a 1978 APL design of a 10- to 20-MWe pilot plantship, sponsored jointly by DOE and the U.S. Maritime Administration (MARAD) (George et al., 1979). The earlier design used systems engineering to achieve minimal cost of a total OTEC system. Features of the design directed to this goal included a concrete barge-type hull, a segmented CWP of lightweight concrete, an aluminum folded-tube HX, and plantship siting for grazing on the tropical oceans near the equator to maximize power output.

In the subsequent development of the 40-MWe baseline, DOE asked APL to expand the design to include the following:

1. Power increase to 40 MWe (nominal).
2. Provision of a 20-MWe power module based on the Lockheed power plant design and a 20-MWe module of the APL design.
3. Alternative plant designs to include a moored configuration supplying electric power to the island of Puerto Rico via underwater power cable, and a grazing plantship to produce and store 125 mt/d of ammonia for bimonthly shipment in ammonia tankers to onshore markets.
4. Performance and design improvements indicated by results of other OTEC research and development programs.

The baseline design program involved a team effort headed by APL in which the following organizations participated:

JHU/APL: Overall program direction. JHU/APL HX design, test, and analysis. Overall systems engineering and cost evaluations.

ABAM Engineers, Inc.: Structural analysis, design, and cost estimates for the hull and CWP. Development of the segmented CWP concept. Coordination of all platform design activities.

L. R. Glosten & Associates: Naval architecture and marine engineering. Seakeeping and stability analysis. Space layouts, outfitting, platform support systems.

Tokola Offshore, Inc.: CWP deployment and attachment. Construction and deployment scenarios for platform, CWP, discharge pipes, including costs. Selection of platform heavy lift systems.

Seward Associates: Electrical engineering design of OTEC power system, including ship services and substation for power transmission to shore.

J. R. Paulling, Inc.: Analysis of CWP and discharge pipes dynamics, including development of computer codes and comparison of theoretical performance with results of at-sea tests.

Trane Co.: Manufacturing study of folded-tube HX construction and costs, including projections for large-scale production.

The following organizations contributed data, expertise, and constructive and critical reviews of all aspects of the baseline designs:

NOAA
Argonne National Laboratory
TRW
SAl, Inc.
BARDI
M. Rosenblatt & Sons
LMSC
Simplex Wire and Cable

Det Norske Veritas
Giannotti and Associates
Gibbs and Cox, Inc.
Value Systems Engineering.

A brief review of the technical status of the subsystems in the 40-MWe baseline configurations will be presented in the following. Complete details, including engineering drawings, are available in George and Richards (1980).

6.5.1.1 Platform Systems

The studies discussed in Section 4.5 led to the selection of a barge-type hull made of post-tensioned concrete for the baseline OTEC grazing and moored system designs because of suitability for both missions, ease of installation and replacement of subsystems and components, minimal cost, compatibility with aluminum heat exchangers, and long-term durability. Assessment of the packaging alternatives for heat exchangers, water ducting including the CWP, pumps, energy delivery systems, stationkeeping systems, support facilities, and personnel accommodations resulted in the configuration shown in Fig. 6-56. The platform length is 135 m. A side view showing the subsystems is presented in Fig. 6-57.

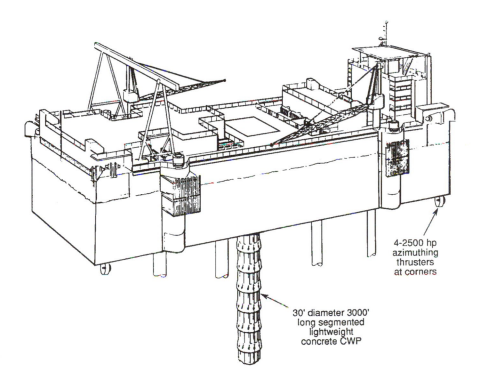

4-2500 hp
azimuthing
thrusters
at corners

30' diameter 3000'
long segmented
lightweight
concrete CWP

Fig. 6-56. Sketch of JHU/APL 40-MWe baseline OTEC ammonia plantship (George and Richards, 1980).

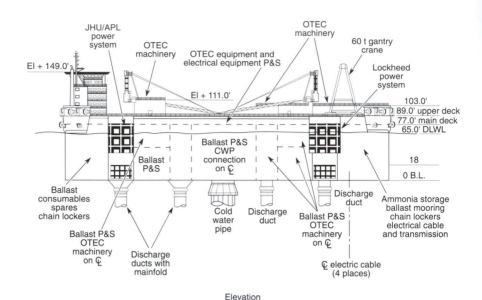

Elevation

FIG. 6-57. Subsystems layout of baseline 40-MWe OTEC barge (George and Richards, 1980).

Analysis of the maximum loads that would be imposed by the 100-year storm defined the structural design and the post-tensioning and reinforcing steel requirements to maintain the structure in compression under the most extreme predicted loads, with a safety margin of 100% (Mast, 1975; George and Richards, 1980). The hull openings for the heat exchangers and CWP require that longitudinal bending strength be provided by wing wall girders that extend the full length of the platform. The analysis included consideration of the survival requirements for grazing plantships stationed in the equatorial Atlantic Ocean [significant wave height (H_s 8 m, period (T_0) 18.0 s] and for moored plants stationed near Puerto Rico (H_s 10.9 m, T_0 13.1 s) or near the island of Oahu in Hawaii (H_s 8.4 m, T_0 11.7 s).

6.5.1.2 Power Systems

Details of the baseline 40-MWE OTEC HX design are provided in Section 4.5. The OTEC barge design accommodates alternative 20-MWe power system modules, of configurations proposed by Lockheed and JHU/APL, as shown in Fig. 6.57. The dimensions and weight of the two systems are nearly the same. Thus, the hull configuration is suitable for 40-MWe (net) power systems of either design. Other compact power systems using the Trane or Alfa-Laval HX designs could also be installed without major changes in the hull dimensions.

APL Power System. The APL power system was designed with two major objectives:

1. Minimize HX cost by using low-cost, roll-welded aluminum tubing in a compact folded-tube configuration with internal ammonia flow. This elimi-

nates the containment structure needed in conventional shell-and-tube HX designs, reducing packaging, space, and cost.

2. Minimize platform cost by designing the water ducting structures to be compact, integral, load-bearing elements of the hull structure.

The power system employs two 5-MWe (net) evaporator and condenser modules combined with a single turbine to form a complete 10-MWe (net) power module. Each 5-MWe (net) unit is housed in a separate bay with vertical sides that form an integral part of the hull structure. Two 10-MWe (net) modules, each having two evaporators and two condensers, are located at the aft end of the baseline platform. A 20-MWe power system of Lockheed design is installed ahead of the mid-section.

A schematic view of a 5-MWe (net) condenser module of the JHU/APL design is shown in Fig. 6-58. Further details of this power system design are presented in Section 4.1.2.

A thorough study of the fabrication, assembly, deployment, and cost of heat exchangers of this type was conducted by the Trane Corporation, who verified the suitability of the JHU/APL HX design for low-cost, mass production (Foust, 1979).

Lockheed Power System.   The 40-MWe (net) baseline design program called for two 20-MWe power modules of different types to be provided in the demonstra-

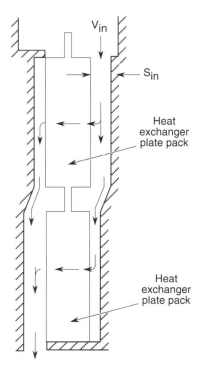

Fig. 6-58.   Schematic of Lockheed 5-MWe HX design for the 40-MWe OTEC baseline barge (George and Richards, 1980).

tion plant. In consultation with DOE, the design proposed by Lockheed for a power module employing flat plate heat exchangers of the Alfa-Laval type was selected as the second power module. A schematic view of the Lockheed HX configuration is shown in Fig. 6-58 (Lockheed, 1978).

A study showed that the platform space and HX bays allotted to the APL design were well suited to packaging the Lockheed module; thus no modifications of the baseline hull were indicated. The plate heat exchangers are mounted in two tiers in a heat exchanger bay. Seawater from the overhead pond enters the heat exchangers at the sides, flows horizontally across the heat exchanger plates, and then exits at the bottom of the HX bay. Liquid ammonia enters the evaporator from below and flows vertically upward as it vaporizes. Vapor leaving the turbines enters the condensers at the top and flows vertically downward to the sump from which it is pumped to the evaporators. Analysis by Lockheed showed that this cross-flow arrangement was superior to one using parallel flow.

6.5.1.3 Water Ducting Systems

Cold Water. Cold water is supplied to the OTEC power systems through a 9.1-m-diameter, 930-m-long, segmented concrete pipe (Cichanski and Mast, 1979). The pipe is supported by a ball-and-socket structure incorporated in the hull at approximately the center of gravity. The joint allows the pipe to pivot freely through an angle of 18 degrees to the vertical. It also allows the pipe to rotate about the vertical axis (see Fig. 4-75) .The dynamics of the coupled system have been analyzed in depth by Paulling (1980) and others, as discussed in Section 4.2 and in George and Richards (1980). The analyses show that the coupled pipe–hull system will survive the most extreme conditions of the 100-year storm for grazing operation at the Atlantic-1 site, with a good safety margin, and for moored operation at the Hawaii site. The analysis predicts that the design is marginal for the extreme hurricane condition at the Puerto Rico site, where the combination of extreme values of surface currents and waves leads to a predicted maximum pipe angle of 19.7 degrees (sum of extreme values due to hull deflections and surface currents). The calculation does not include mitigating effects of mooring forces on the platform motions or potential benefits of bow fairing of the rectangular platform. Further study is required to indicate detail changes in design that might be required for moored operation at this site. The cost of manufacture, assembly, and deployment of CWPs of this type was estimated in combined studies by ABAM, Concrete Technology, Inc., Glosten Associates, and Tokola Offshore, Inc. The CWP manufacturing cost was independently verified by a professional estimating concern.

Warm Water. Warm water enters the evaporators through screens mounted at the sides of the hull at a mean depth of 16. 6 m. The screen area is about 125 m^2 for each 10-MWe evaporator module.

Mixed Effluent. In the grazing mode of operation, the warm- and cold-water effluents from the heat exchangers are discharged at the bottom of the platform, where their higher density than the surrounding water causes them to sink. Since the grazing platform is always moving into fresh water, there is no recirculation of the discharged water into the warm-water inlets. The moored baseline plant is designed with discharge pipes to carry the effluent to a depth below the thermocline, to prevent reingestion and possible impacts on the near-shore environment. Eight pipes 4.6 m in diameter and 76 m long are used to carry the mixed effluent below the thermocline, where it is discharged and allowed to sink. A detailed structural analysis of these pipes was not funded but will be necessary if they are to be retained in the design. The OTEC-1 experiment indicated that these pipes may be unnecessary because the mixed plume descends rapidly and may not pose any reingestion problem; however, termination of the program did not permit a conclusion to be reached.

6.5.1.4 Energy Transfer Systems

The OTEC on-board electric power generation and distribution system is based on the use of standard commercial equipment for primary and secondary applications and is designed to comply with the requirements of IEEE standard No. 45 and relevant regulations of the U.S. Coast Guard, ABS, NEMA, NFPA, and UL. There are four 16-MWe (gross) ammonia turbine generators, each having its own set of auxiliaries and switch gear. Ammonia storage, liquid nitrogen, circulating cooling water, and generator loading systems are used in common. All of the generators may be used in parallel, and in normal operation all power would be derived from them. During start up or when the OTEC plant is shut down during storms, reserve power is provided by diesel–generator sets. Tie feeders between the OTEC generators and the reserve power units allow this interchange. The on-board power utilization systems differ for the grazing and moored OTEC options; however, the systems are the same up to the final phase. Figure 6-59 presents a line diagram of the basic system. In the grazing plant, step-down/DC rectifiers are installed to furnish DC current to the electrolyzer cells. In the moored plant, transformers step up the voltage to 115 kV for cable transmission to shore.

   Grazing Configuration. In the OTEC plantship, electric power generated on-board is used to produce a chemical product that stores the energy for later transport to shore. Water electrolysis provides an efficient method for converting electrical energy to chemical energy and is the basic step in making all of the surface waters of the tropical oceans a source of energy that can be transported to land sites.

   In the baseline 40-MWe (net) OTEC plantship the power produced by the turbo-generator is used to produce ammonia (NH_3). The on-board ammonia plant includes water electrolyzers to generate gaseous hydrogen (along with gaseous oxygen that is vented to the air), an air separation plant to produce pure nitrogen, a catalytic converter to combine the hydrogen and nitrogen to form ammonia, ammonia liquefaction and storage equipment, and facilities for transferring the liquid ammonia to tankers for shipment to world ports. Details of the ammonia synthesis process are described in Section 4.3.

   Commercial ammonia plants typically are sized to produce 1000 tons/d (909 mt/d) of liquid ammonia. The output of the 40-MWe (net) baseline plantship would be one-eighth of this amount.

   The feasibility and practicality of installing an ammonia plant on an OTEC platform were investigated in detail in the baseline study, including possible effects

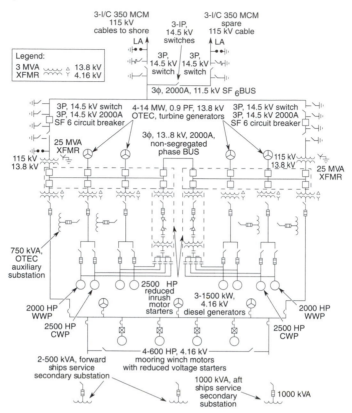

FIG. 6-59. Proposed wiring diagram for the 40-MWe grazing and moored OTEC systems (George and Richards, 1980).

of ship motions on the process operations, with favorable results. Smaller, less efficient, process components would be required for the baseline vessel in comparison with commercial plants; however, all of the equipment apart from the electrolysis system is commercially available. The process design would be similar to the commercial ammonia plant designs except for the simplification resulting from replacement by water electrolysis of the expensive steam-reforming process for hydrogen production. A flow diagram of the process is shown in Fig. 6-60. An installation diagram of a 113-ton/d ammonia plant on the OTEC platform is shown in Fig. 6-61.

Moored Configuration. It was recognized early in the OTEC program that at favorable sites moored OTEC plants could deliver OTEC power directly to onshore facilities via underwater power cables (Winer, 1977). The requirements for this mode of OTEC operation were also determined in the 40-MWe (net) baseline design study. The site selected for detailed study was near Punte Tuna, Puerto Rico.

Power transfer requires on-board equipment to convert the OTEC bus bar output to suitable voltage and frequency for transmission through the selected power cable and equipment to support the power cable that must connect to the

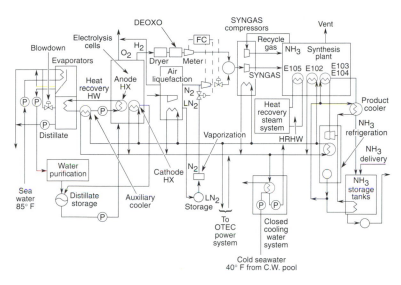

FIG. 6-60. Flow diagram for OTEC ammonia plant (George and Richards, 1980).

OTEC platform through a flexible junction. A proposed arrangement for the aft cable installation on the platform is shown in Fig. 6-62. Experiments with power cables under simulated OTEC conditions indicated satisfactory durability could be achieved (Morello, 1981; Pieroni et al., 1979; Pieroni, 1980; Soden et al., 1981, 1982); however, a complete design of the power cable system had not been completed when the program was terminated. A diagram of the electrical system, which satisfies the total requirements of both moored and grazing OTEC plants, is shown in Fig. 6-59.

The principal features of the APL 40-MWe (nom) OTEC baseline plants are listed in Table 6-15.

6.5.2 Preliminary Design of Shelf-Mounted OTC Demonstration OTEC Power Plant for Hawaii

At the completion of the Phase I conceptual design competition, a contract was awarded to the Ocean Thermal Corporation for preparation of a preliminary design of their proposed shelf-mounted plant (Yaffo et al., 1982, Paoloni, 1984; OTC, 1984). The plant is designed to be sited on the continental shelf 600 m offshore from the Hawaiian Electric Company (HECO) oil-fired power plant at Kahe Point on the island of Oahu. A feature of the OTC concept is the use of the warm-water outflow from the HECO plant, mixed with the warm-water input to the OTEC evaporator, to raise the temperature of the OTEC inlet water from 25.6 to 27.9°C, with a significant increase in OTEC performance.

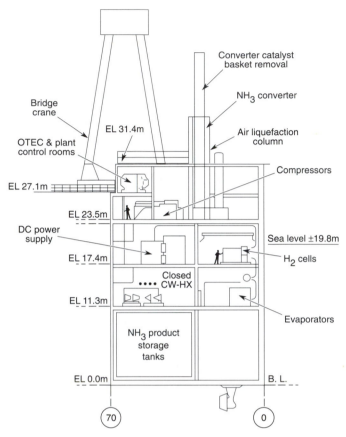

Fig. 6-61. Schematic diagram of 40-MWe OTEC ammonia plant installation (George and Richards, 1980).

The goal of the OTC contract was to develop, with government cost sharing, an engineering and economic evaluation of an OTEC power plant design that would provide a firm basis for detailed engineering and construction of a 40-MWe shore-based OTEC demonstration plant. In the conduct of the contract, the following topics were evaluated in some depth:

System performance and operational requirements
Government regulations and operational requirements
OTC performance and operational requirements
Codes, standards, and recommended practices
HECO requirements
Possible impact on military operations near Ohau
Aesthetic considerations
Site-specific design conditions
Site description

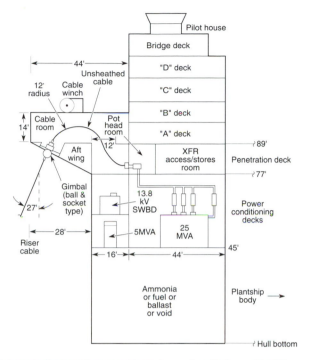

FIG. 6-62. Schematic diagram of power cable equipment installation on 40-MWe moored OTEC plant (George and Richards, 1980).

Wind, wave design criteria
Tsunamis
Tides
Currents
Seawater temperatures
HECO condenser outflow conditions
Seismicity
Climatology
Biofouling and corrosion
Preliminary design system engineering
OTEC plant system characteristics
OTEC plant system overall layout
Land-based containment system (LCBS) design
Power system design
Water ducting design
Asset acquisition
Project schedule
OTEC plant system deployment
LBCS deployment
Pipeline deployment

Table 6-15. JHU/APL preliminary design of baseline 40-MWe (nom) OTEC plants

Hull
  Rectangular barge
  Dimensions m (ft): L 135 (443.5), W 42.7 (140), D 31.4 (103)
  Hull material: post-tensioned concrete
Power plant
  Closed Rankine cycle; ammonia working fluid
  Folded tube HXs of JHU/APL design with nested 0.072-m (3-in.) tubes,
    or plate HXs of Lockheed design using titanium plates
  Internal ammonia flow; 4 10-MWe (net) modules
  Nominal operating characteristics
      Nominal $\Delta T(^{\circ}C)$
        Annual average 23.65
        Seasonal $\Delta T(^{\circ}C)$ (ATL-1): Feb 23.6, May 24.6, Aug 23.4, Nov 23.0

Power [MWE (nom)] (optimized for ATL-1)	JHU/APL power modules	Lockheed power modules
Gross turbo-gen output	52.0	52.9
Cold-water pump	5.76	9.66 (both pumps)
Warm-water pump	4.08	
Ammonia pumps	1.51	1.18
Noncondensible purge	0.16	
Ship services	0.17	2.07
Net busbar power	40.3	40.0

	JHU/APL		Lockheed	
	Evap.	Cond.	Evap.	Cond.
Heat load (MWt)	1,747	1,689	1,680	1,629
Water flow (kg/s)	132,300	136,720	105,600	102,900
NH_3 turbine flow (kg/s)	2,824	2,824	1,722	1,722
U_0 (W/m² K)	2,537	2,463	3,259	3,118
Foul. coeff. (W/m² K)	3.5×10^5	3.5×10^5	5×10^5	5×10^5
H_2O in/out temp. (°C)	27.89/25.27	4.56/7.46	27.78/23.81	4.44/8.36
NH_3 in/outlet temp. (°C)	23.94/21.94	9.50/10.00	13.37/21.71	10.37/10.47
Seawater Δp (m)	2.58	2.95	3.21	3.44
Inlet duct	0.51	0.88	0.51	0.88
HX	2.07	2.07	2.70	2.56
Seawater velocity (m/s)				
Inlet duct	0.3	2.0	n.a.	n.a.
HX	0.76	0.76	n.a.	n.a.
Seawater flow [m³/s/MWe (net)]	3.24	3.33	2.59	2.51

Inspection, maintenance, and repair
Safety analysis
Risk assessment
Environmental monitoring and assessment
Cost and commercialization prospects (Hoffman and Paoloni, 1984)

Space limitations permit only a brief summary in the following sections of the reported material. Firms associated with OTC in the 40-MWe preliminary design were:

TRW Energy Development Group
Hawaiian Dredging and Construction Co.
R. J. Brown and Associates
Burns and Roe
Hydro Research Science
Edward K. Noda and Associates
O. I. Consultants
Harding Lawson Associates
Offshore Investigations
Makai Ocean Engineering
Alfred A. Yee Division of Leo A. Daly
Tokola Offshore

6.5.2.1 Platform (Land-Based Containment System)

This comprises a concrete building 71.8 m wide (236.5 ft), 93.3 m long (306 ft), and 33.5 m high (110 ft) (10 m above sea level) mounted on the sea bottom at a depth of 22.2 m (73 ft), at 550 m (1800 ft) offshore. The land-based containment system (LBCS) houses the OTEC power plant, water and ammonia pumps, biofouling control equipment, and power conversion equipment. A breakwater structure extending from the bottom to just above sea level is attached on the seaward end. Figure 6-63 is an artist's sketch of the proposed site and a schematic of the above-

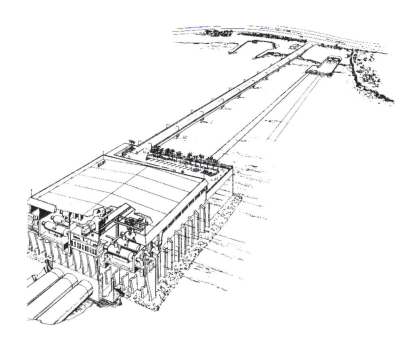

FIG. 6-63. Artist's sketch of the OTC 50-MWe OTEC power plant at Kea-hole Point, Hawaii (OTC, 1984).

water platform layout. A sketch of the subsystems layout is shown in Fig. 6-64. A schematic of the underwater structure of the OTC plant is shown in Fig. 6-65. The generated power is carried to the HECO plant via power cables mounted on the trestle shown in Fig. 6-63 that provides access to the LBCS from the shore.

6.5.2.2 Water Ducting

Design of the CWP system required an extensive study of the physical and hydrological features of the Kahe Point site. The studies included detailed bathymetric surveys with a resolution of 50 m covering approximately 5.5 km² bordering Kahe Point. Bottom core samples and subsurface data representative of the area were also acquired. With this information it was possible to define the pipe structural requirements and to lay out a pipe trajectory that would minimize risks in deployment and long-term operation and would be the most cost effective. The bathymetry and topography of the surveyed area are shown in Fig. 6-66.

In the final preliminary design of the CWP, water enters the LBCS through a bottom-mounted pipe 3660 m long that draws water from an offshore depth of 670 m (2200 ft). The CWP is divided into two major segments. The seaward section of the pipe is of FRP. It is 2560 m long (8400 ft), of 6 m diameter (19.8 ft), and 0.072 m wall thickness (2.83 in.). It is composed of segments 75–90 m long, joined by flexible bellows, and supported on a steel support structure that rests on the bottom. This structure protects the pipe from bottom erosion, relieves tension loads on the

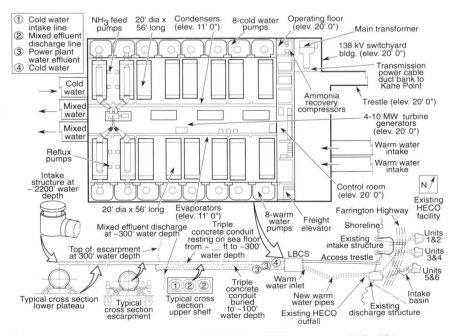

Fig. 6-64. Layout of subsystems for the OTC 50-MWe OTEC power plant (OTC, 1984).

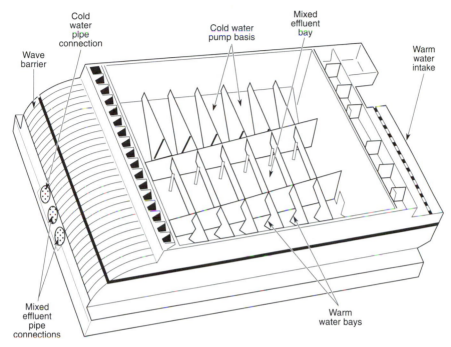

FIG. 6-65. Sketch of the underwater structure of the OTC 50-MWe OTEC power plant (OTC, 1984).

FRP, provides weight to ensure stability against hydrodynamic loads, and allows flexibility to accommodate seismic loads, bottom scour, and differential settling. This CWP section extends 2560 m (8400 ft) from the deep-water inlet to the top of the escarpment at a depth of 82 m (270 ft). A profile of the FRP pipe and a schematic of the ducting construction and support structure are shown in Fig. 6-67.

At the top of the escarpment, the FRP pipe connects to a triple-pipe concrete structure that combines a pipe to conduct the cold water to the LBCS with two parallel pipes that carry the mixed effluent (flowing in the opposite direction) from the OTEC plant to the discharge point. A profile of this section and a cross section of the concrete triple-pipe structure are shown in Fig. 6-67.

Warm near-surface water enters the LBCS through a structure on the landward side, where it is mixed with warm water conducted from the HECO condenser discharge. After passing through the OTEC power system, the exhaust streams from the evaporators and condensers are mixed and then conducted through the bottom-mounted concrete ducting structure to the edge of the escarpment, where the combined flow is discharged at a depth of approximately 115 m (380 ft).

6.5.2.3 Support Facilities

All support facilities not directly involved in plant power production are sited onshore. These include offices, laboratories, maintenance facilities, ammonia and equipment storage, and related equipment.

FIG. 6-66. Topography of the offshore area of the OTC 50-MWe OTEC power plant (OTC, 1986).

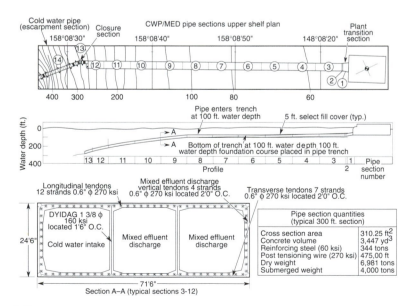

FIG. 6-67. Water ducting arrangement for the 50-MWe OTC OTEC power plant (OTC, 1984).

The principal features of the OTC design are listed in Table 6-16. Cost information is listed in Chapter 7.

Table 6-16. OTC design of a shore-mounted 40-MWe OTEC power plant

LBCS dimensions m (ft)
  Above water: *L* 93 (306), *W* 71 (234), *H* 11.3 (37)
  Under water: *L* 116 (380), *W* 71 (234), *H* 22.2 (73)
Power plant
  Closed Rankine cycle; ammonia working fluid; shell-and-tube HXs with titanium tubes
    0.01905 m OD, 0.01765 m ID; 4 10-MWe (net) modules
Nominal operating characteristics
  Nominal ΔT (K) (mixed seawater and HECO flow)   26.83
  Power (MWe)

	Gross turbine-generator power	53.8
	Warm-water pump power	4.1
	Cold-water pump power	8.4
	Ammonia pumps power	0.9
	Other	0.4
	Net power output to busbar	40.0

	Evaporator	Condenser
Heat load (MWt)	1,765	1,711
Water flow (kg/s)	13,747	11,510
Ammonia flow (kg/s)	1,434	1,434
Number of HX tubes	40,128	37,671
Tube length (m)	13.65	15.03
HX area (m²)	321,024	301,368
Water side	243,197	251,239
Ammonia side	262,442	271,121
Overall heat trans. coeff. (W/m² K)	2,516	2,601
Fouling coeff. (m² K/W) × 10⁵	1.76	1.76
Water inlet (outlet) temp. (°C)	26.8 (22.8)	5.0 (9.7)
Ammonia inlet (outlet) temp. (°C)	(21.44)	10.66
Log mean temp. diff. (LMTD) (+°C)	2.95	2.69
Seawater Δp (m)	2.39	2.25
Sum of inlet & outlet Δp	0.60	2.91
Heat exchanger Δp	2.25	2.39
Seawater velocity (m/s)	1.37	1.22
CWP flow (m³/[s MWe (net)])		2.25
Net power/total HX area (kWe/m²)	0.081	

6.6 CONCLUSIONS

The review in this chapter of the OTEC programs conducted in the United States and abroad shows that the technology required for high-confidence pilot plant demonstration of sea-based and land-based OTEC systems is now available. Satisfactory solutions have been demonstrated at reasonable scale for all of the technical problems that were identified as major barriers to OTEC commercial operation in early reviews of OTEC:

Heat exchanger biofouling
Aluminum HX durability
OTEC heat engine performance
Cold-water pipe fabrication
Cold-water pipe deployment
OTEC ocean durability
OTEC environmental impacts

A final perceived barrier, OTEC cost, is discussed in detail in Chapters 7, 8, and 9, where OTEC costs are compared on a uniform basis with other proposed fuel and power options that would reduce U. S. dependence on foreign oil and reduce environmental and safety concerns of present systems.

REFERENCES

Anderson, A. R., 1977. "World's largest prestressed LPG vessel." *J. Prestressed Concrete Institute,* **22**, 1.

Anderson, J. H., 1985. "Ocean thermal power–the coming energy revolution." *Solar & Wind Technology,* **2**, 25.

——, 1986. *Sea solar power.* Sea Solar Power Inc., Oct.

—— and J. H. Anderson, Jr., 1977. "Compact heat exchangers for sea thermal power plants." *Proc. 4th Annual Conf. on Ocean Thermal Energy Conversion,* New Orleans, La., March, **6**, 3.

Ashworth, D. J., and G. Slebodnick, 1980. "Cost and configuration of a closed cycle OTEC power system using brazed aluminum plate and fin heat exchangers." *Proc. 7th Ocean Energy Conf.,* Washington, D.C., June, **II,** 7.1.

Avery, W. H., D. Richards, and G. L. Dugger, 1985. "Hydrogen generation by OTEC electrolysis, and economical energy transfer to world markets via ammonia and methanol."*J. Hydrogen Energy,* **10**, 727.

——, D. R. Richards, W. G. Niemeyer, and J. D. Shoemaker, 1983. "OTEC energy via methanol production." *Proc. 18th Intersociety Energy Conversion Engineering Conf.,* Orlando, Fla., August, **1**, 346.

Babbitt, J. F., 1981. *Closed cycle ocean thermal energy conversion pilot plant–grazing ammonia plantship.* SOLARAMCO, Tulsa, Okla.

BARDI, 1982. *Coal-to-methanol study using OTEC technology.* For Johns Hopkins University Applied Physics Laboratory. Brown and Root Development, Inc., Houston, Tex.

Barsness, E. J., et al., 1978. *Conceptual design of an OTEC power system using modular heat exchangers.* Westinghouse, March.

Baumeister, T., E. A. Avallone, and T. Baumeister III, 1978. *Marks' standard handbook for mechanical engineers.* 8th Ed. New York: McGraw-Hill.

Castellano, C. C., 1981. "Overall OTEC-1 status and accomplishments." *Proc. 8th Ocean Energy Conf.,* Washington, D.C., June, **2**, 971.

——, E. A. Midboe, and G. M. Hagerman, 1983. "The U.S. Department of Energy's 40 MWe ocean thermal energy conversion pilot plant program: A status report." *Proc. Oceans '83* **2**, 728.

Cichanski, W. J., and R. F. Mast, 1979. " Design of a concrete cold water pipe for ocean thermal energy (OTEC) systems." *11th Annual Offshore Technology Conf.,* Houston, Tex., 1653.

Denton, J. W., P. Bakstad, and K. McIlroy, 1979. "Test article and a 10-MWe power module." *Proc. 6th OTEC Conf.* Washington, D. C., June, **1**, 8.3-1.

Dugger, G. L., J. E. Snyder, and F. Naef, 1983. "Ocean thermal energy conversion: Historical highlights, status, and forecast." *J. Energy* **7**, 293.

EBASCO, 1984. *Final report. Coal-to-methanol OTEC plantship study using the Rockwell molten carbonate gasification system.* EBASCO Services Corp., Los Angeles, Cal.

Foust, H. D., 1979. *U. S. Patent No. 4,276,927 for plate type heat exchanger.* U. S. Patent Office, June.

——, 1981. "Constructability of extended surface heat exchangers for OTEC." *Proc. 8th Ocean Energy Conf.,* Washington, D. C., June, **2**, 717.

General Electric, 1983. *Closed cycle OTEC power plant final report.* General Electric Co., Schenectady, N. Y.

George, J. F., 1979. "System design considerations for a floating OTEC modular experiment platform." *Proc. 6th OTEC Conf.,* Washington, D. C., June, 5.5-1.

——, and D. Richards, 1980. *Baseline designs of moored and grazing 40-MWe OTEC pilot plants.* Johns Hopkins Univ. Applied Physics Lab., June, SR-80A&B.

——, D. Richards, and L. L. Perini, 1979. *A baseline design of an OTEC plantship.* Johns Hopkins Univ. Applied Physics Lab., SR-78-3A.

Geothermie Actualites, 1985. *L'Energie thermique des mers.* Geothermie Actualites.

Gershunov, E. M., Y. H. Ozudogru, and J. M. Betts, 1981. "Analysis of the up-ending problem of the OTEC-1 cold water pipe." *Proc. 8th Ocean Energy Conf.,* Washington, D. C., June, **2**, 9-1.

Global Marine Development, Inc., 1981. *Floating vessels for industrial plants.* U. S. Department of Commerce.

Griffin, A. B., and M. P. Bianchi, 1979. "Cold water pipe preliminary design studies." *Technical Workshop on Ocean Thermal Energy Conversion,* Washington, D. C., Jan.

Griffin, A. B., and L. R. Mortaloni, 1980. "Baseline designs for three OTEC cold water pipes." *Proc. 7th Ocean Energy Conf.,* Washington, D. C., June, **1**, 3.2.

Hoffman, B., and E. C. Paoloni, 1984. "Commercialization of OTEC: A 40-MWe OTEC power plant for Hawaii." *Oceans '84.* Washington, D. C., Sept.

Homma, T., and H. Kamogawa, 1979. "An overview of Japanese OTEC development." *Proc. 6th OTEC Conf.,* Washington, D. C., **1**, 3.31.

IFREMER, 1985. *Ocean thermal energy conversion, the French program, a pilot electric power plant for French Polynesia.* IFREMER, March.

International Maritime Associates, 1981. *The development of waterborne alcohol fuel plants.* U. S. Dept. of Commerce, **1**, 3.

Ito, F., and Y. Seya, 1985. Operation experience of the 100-kW OTEC and its subsequent research." *Japanese–French Symp. on Ocean Development,* Tokyo, Sept.

Jones, M., V. Thiagarahan, and K. Sathjanarayna, 1981. "OTEC aluminum integration studies." *Proc. 8th Ocean Energy Conf.,* Washington, D. C., **1**, 643.

Kamogawa, H., 1972. "Equatorial marine–industrial complex. A report for National Resources Development Committee of the Pacific Basin Economic Council at the 5th general meeting in New Zealand." *Proc. Second International Ocean Development Conf.* Tokyo, October, 1.

——, 1980. *OTEC research in Japan.* Pergamon Press Ltd., London, **5**, 481.

Kawano, S., T. Takahata, and M. Miyoshi, 1983. "Performance tests of a condenser for a 100-kW (gross) OTEC plant." *Proc. ASME–JSME Thermal Eng. Joint Conf.,* Honolulu, March, **2**, 247.

Kohl, A. L., M. H. Slater, and P. R. Hsia, 1980. "Operation of the molten salt coal gasification process." *7th Energy Technology Conf. and Exposition,* Washington, D. C., March, 1.

Larsen-Basse, J., 1983. "Effect of biofouling and countermeasures on heat transfer in surface and deep ocean Hawaiian waters–early results." *ASME-JSME Thermal Eng. Joint Conf. Proc.,* Honolulu, **2**, 285.

Leitner, M. I., 1979. "Design of a 10-MWe power system and heat exchanger test articles using plate heat exchangers." *Proc. 6th OTEC Conf.,* Washington, D. C., **1**, 8.4-1.

Liao, T., J. G. Giannotti, and P. R. Mater, 1986. *Feasibility and concept design studies for OTEC plants along the east coast of Taiwan, Republic of China.* Taiwan Power Co., April.

Lloyd's Maritime Directory, 1988. Lloyd's of London Press.

Lockheed, 1978. *OTEC platform and configuration study.* Lockheed Missiles and Space Co. Inc., 4 volumes.

——, 1979a. *Preliminary design report for OTEC station-keeping (SKSS).* Lockheed Missiles and Space Co. Inc.

——, 1979b. *PSD-II final design briefing.* Lockheed Missiles and Space Co. Inc., June, Task 2.

Lorenz, J. J., D. Yung, P. A. Howard, C. B. Panchal, and F. W. Poucher, 1981a. *OTEC-1 power system test program: performance of one-megawatt exchangers.* Argonne National Lab., OTEC-PS-10.

Lorenz, J. J., D. Yung, P. A. Howard, C. B. Panchal, and F. W. Poucher, 1981b. "Heat exchanger test results." *Proc. 8th Ocean Energy Conf.,* Washington, D. C., June, **2**, 977.

MARAD, 1980. *The development of waterborne alcohol fuel plants.* U. S. Dept. of Commerce (three vol.).

Marchand, P., 1985. "French thermal energy conversion program." *Proc. French–Japanese Symp. on Ocean Development,* Tokyo, Sept., 1.

Mast, R. F., 1975. "The ARCO LPG vessel." *Conf. on Concrete Floating Ships and Terminal Vessels,* 1.

Mitsui, T., F. Ito, Y. Seya, and Y. Nakamoto, 1983. "Outline of the 100-kW OTEC pilot plant in the Republic of Nauru." *IEEE/PES 1983 Winter Meeting,* New York, Jan., 212.

Mochida, Y., T. Takahata, and M. Miyoshi, 1983. "Performance tests of an evaporator for a 100-kW (gross) OTEC plant." *Proc. ASME–JSME Thermal Eng. Joint Conf.,* Honolulu, March, **2**, 241.

Morello, A., 1981. "Constructability of submarine bottom cables for OTEC power transmission." *Proc. 8th Ocean Energy Conf.,* Washington, D. C., June, **2**, 747.

Motor Ship Press, 1988. *Directory of shipowners, shipbuilders, and marine engineers.*

Ocean Thermal Corporation, 1984. *Deployment of the land-based containment system for the 40 megawatt Ocean Thermal Corporation OTEC plant.* Makai Ocean Engineering, Inc.

O'Connor, J. S., 1981. *Light weight concrete cold water pipe tests: Phase II.* Johns Hopkins University Applied Physics Lab., SR-80-5A.

——, 1980. "The development of a lightweight concrete for ocean thermal energy conversion (OTEC) systems." *International Conf. on Performance of Concrete Marine Environment,* August.

Olmsted, M. G., M. J. Mann, and C. S. Yang, 1979. "Optimizing plant design for minimum cost per kilowatt with Refrigerant-22 working fluid." *Ocean Thermal Energy for the 80's,* I, 8.51.

Owens, W. L., and L. C. Trimble, 1980. "Mini-OTEC operational results." *Proc. 7th Ocean Energy Conf.,* Washington, D. C., June, **2**, 14.1-1.

Paoloni, E. C., 1984. "Optional strategy for financing a 40-MWe OTEC power plant." *Proc. Pacific Congress on Marine Technology (PACON 84),* MRM 8/7.

Paulling, J. R., 1980. *Theory and users manual for OTEC cold water pipe programs, ROTECF and SEGPIP.* J. R. Paulling Assoc., Berkeley, Cal.

Pieroni, C. A., 1980. "Material evaluation and testing program for OTEC riser cable." *Proc. 15th Intersociety Energy Conversion Engineering Conf.,* Seattle, Wash., **1**, 360.

——, R. T. Traut, and D. O. Libby, 1979. "The development of riser cable systems for OTEC plants." *Proc. 6th OTEC Conf.,* Washington, D. C., June, **I**, 7.2-1.

Richards, D., W. H. Avery, E. J. Francis, M. Jones, B. Heyer, A. L. Kohl, and J. K. Rosemary, 1984. *OTEC methanol from coal plantship studies.* American Society of Mechanical Engineers, New York, N. Y., 84-WA/SOL-32.

Sanchez, J. A., 1980. *OTEC alternative for Puerto Rico vs. coal-oil-nuclear. Phase II.* Clean Energy Research Institute, Miami, Fla., Dec., 415.

Science Applications, Inc., 1979. *Design of a 1/3 scale fiberglass OTEC cold water pipe.* Science Applications, Inc.

Snyder, J. E., 1978. "1-MWe heat exchanger for OTEC. Status report." *Proc. 5th Ocean Thermal Energy Conversion Conf.,* Miami Beach, Fla., **VI**, 20.

——, 1980. *1-MWe heat exchangers for OTEC. Final report.* TRW Systems Group.

Soden, J. E., R. Eaton, and J. P. Walsh, 1982. "Progress in the development and testing of OTEC riser cables." *Oceans 82 Conf. Records (Marine Technology Society),* Washington, D. C., Sept., **E**, 587.

——, D. O. Libby, and J. R. Spiller, 1981. "A study of ocean cable ships applied to submarine power cable installation." *Proc. 8th Ocean Energy Conf.,* Washington, D. C., June, **2**, 617.

Svenson, N. A., 1979. "Overview of the OTEC-1 design." *Proc. 6th OTEC Conf.,* Washington, D. C., **1**, 5.3.

Trimble, L. C., and W.L. Owens, 1980. "Review of Mini-OTEC performance," *15th Intersociety Energy Conversion Conf.* Seattle, Wash., Aug., **I**, 1331.

——, and R. L. Potash, 1979. "OTEC goes to sea. A review of Mini-OTEC." *Proc. 6th OTEC Conf.,* Washington, D. C., June, **I**, 5.2-1.

TRW, 1979. *Ocean thermal energy conversion cold water pipe preliminary design final report.* TRW Energy Systems Group.

Uehara, H., 1982. "Research and development on ocean thermal energy conversion in Japan." *Proc. 17th IECEC Conf.,* **3**, 1454.

——, H. Kusuda, M. Monda, T. Nakaoka, and H. Sumitomo, 1983. "Shell-and-plate type heat exchangers for an OTEC plant." *Proc. ASME–JSME Thermal Eng. Joint Conf.,* Honolulu, Mar., **2**, 253.

——, S. Nagasaki, and H. Yokohama, 1980. "Deployment of cold water pipe in the Japan Sea." *Proc. 7th Ocean Energy Conf.,* Washington, D. C., June, **1**, 14.4.

——, K. Masutani, and M. Miyoshi, 1978. "Heat transfer coefficients of condensation on vertical fluted tubes." *Proc. 5th Ocean Thermal Energy Conversion Conf.,* Miami Beach, Fla., Feb., **VI**, 261.

——, and H. Kusuda, 1978. "Plate type evaporator and condenser for ocean thermal energy conversion plant." *Proc. 5th Ocean Thermal Energy Conversion Conf.,* Miami Beach, Fla., Feb., **VI**, 261.

Vincent, S. P., and C.H. Kostors, 1979. "Performance optimization of an OTEC turbine." *Proc. 6th OTEC Conf.,* Washington, D. C., **I**, 8.7-1.

Winer, B., 1977. "Electrical energy transmission for OTEC power plants." *Proc. 4th Annual Conf. on Ocean Thermal Energy Conversion,* New Orleans, La., March.

Yaffo, J. E., A. F. Butler, and A. B. Griffin, 1982. "OTEC power for Hawaii." *Marine Technology Workshop,* Washington, D. C., 14.

7

OTEC CLOSED-CYCLE SYSTEMS COST EVALUATION

7.1 COST UNCERTAINTIES AND PERCEIVED RISKS VERSUS ENGINEERING STATUS

Innovative technologies such as OTEC achieve commercial development when potential investors decide that the return on the investment will repay the estimated development costs plus a profit, with an acceptably low risk of cost overruns. Industrial experience shows that the estimated cost to complete development of a new technology generally increases as development proceeds from the conceptual design through pilot development, demonstration, field testing, and final commercial manufacture (Merrow et al., 1981). The ratio between final cost and initial design estimate is strongly dependent on the extent to which the manufacturing process employs already developed equipment, procedures, and facilities. New projects that require "high technology" for their success, such as jet engines or nuclear power plants, have been characterized by large underestimates of the final costs, whereas the costs of projects that are firmly based on existing technology, such as the development of "supertankers," have been accomplished well within the usual industrial uncertainty margin of ± 15 to 20%. The accuracy of the estimate is also strongly dependent on the thoroughness of the systems engineering evaluation that is done before development proceeds.

Commercial applications of OTEC have been proposed in three principal categories. The first includes OTEC power plants mounted on floating platforms that would generate 50- to 400 MWe (net) of onboard electric power. The need to minimize plant size makes it mandatory to use closed-cycle OTEC for these applications. The second category includes land-based or shelf-mounted plants designed to supply power in the 50- to 400-MWe range to municipal utilities. Either open- or closed-cycle systems could be suitable. The third category comprises small (5- to 20-MWe) land-based or shelf-mounted OTEC plants designed for island applications where electric power generation, mariculture, fresh-water production, supply of cold water for air-conditioning systems, and fuel production could be combined to offer an economically attractive OTEC system despite the relatively high cost of power for small OTEC installations. Open-cycle OTEC plants may be the preferred choice for the third category.

The estimated investment costs of installed complete OTEC systems, measured in dollars per kilowatt of net OTEC electric power generated, differ significantly among the three categories. Large cost benefits are gained by siting OTEC where ΔT is a maximum, by scaling to optimum size, by large volume production of subsystems and components, and by integrated system design. Small land-based systems are relatively expensive in dollars per kilowatt, but the cost can be offset in part by using the cold water for air conditioning, mariculture, and fresh-water production. Each category is considered separately in the following cost assessments; however, since the importance of OTEC as a renewable energy source will depend on its ability to furnish significant quantities of vehicle fuels that will be economically and environmentally attractive, the major emphasis is placed on the OTEC plantship options.

In the OTEC research and development programs sponsored by DOE, cost estimates were made of many alternative designs of subsystems and components. To provide a systematic method for describing subprograms and comparing costs, participants in a DOE program were asked to document their designs in terms of a "work breakdown structure" (WBS) that required the system developer to list costs first by designated subsystems (for example, "power system") and then for each subsystem by designated subsystems or components (for example, "heat exchangers," "turbine"). This method enables the system manager to rate alternative design options on a uniform basis. Furthermore, it provides a means for assessing the total cost uncertainty of a development program in terms of the cost uncertainties of the individual subsystems and components. The DOE work breakdown structure prepared for the APL 40-MWe (net) demonstration baseline vessels has been used in describing and assessing costs for the programs discussed in Chapter 6 and is adopted for the cost assessments described here and in Chapters 8 and 9.

In the United States, the most detailed OTEC development cost studies were conducted in the APL and OTC preliminary design programs, which drew heavily on industrial experience and work by other DOE contractors. These studies have been used as the primary basis for the estimates listed in the tables presented in later sections of this chapter. Cost studies in less detail have also been reported in other American studies, and by Japanese, French, and Taiwanese investigators. Some of these are also listed. Because of unknown differences in estimating procedures, the results may not be strictly comparable but are deemed useful to include.

The following sections of this chapter are designed to yield reliable estimates of OTEC system costs through division of the cost estimates for total systems development into WBS sections for each of which the cost uncertainty is estimated to be low, moderate, or high. A weighted average uncertainty is then calculated for the total system cost. The cost uncertainty estimates are based on the estimated costs developed in the DOE-funded preliminary designs of the 40-MWe baseline floating OTEC systems and the 40-MWe land- and shore-based 40-MWe systems, described in Chapter 6.

A low uncertainty of ±10–20% is assigned to cost estimates for components or construction technologies that are in standard industrial use and were based on firm quotes. The fraction of the total cost in this category ranges from 45% of the total for the ammonia plantship to 70% for the methanol plantship. Moderate uncertainty of ±20–35% is assigned to components or construction technologies that are commercially available at smaller scale or under different working conditions but must be modified for OTEC use. The corresponding fractions of the total cost in this category are 41% and 23% for the two cases. High cost uncertainty of −35 to +100% is assigned to fabrication and deployment of the cold-water pipe. The corresponding percentages of total cost are 13% and 7%. The weighted average risks of cost underestimates or overruns in constructing and deploying OTEC systems are thus judged to be between 20% and 30%. Scaling factors and experience factors[1] are used in deriving the final estimates.

It should be noted that all complete OTEC installations will incorporate:

1. A platform system or land-based containment system to support and house the OTEC subsystems, including water-pumping and distribution, platform stationkeeping, and auxiliary platform services.
2. An OTEC power plant including heat exchangers, turbines, electric generators, water and working-fluid pumps, and control equipment.
3. Water ducting equipment.
4. An energy transfer system. This may include a process plant and related equipment for fuel production, storage, and transfer; equipment for routing electric power to consumers, mariculture installations, fresh-water production, and use of the cold-water effluent for air conditioning.
5. Support facilities and equipment, including shops, personnel accommodations, etc.

7.2   ESTIMATED COSTS OF OTEC SYSTEMS

7.2.1 *Costs of Floating OTEC Systems*

7.2.1.1 Baseline 40-MWe (Nom) OTEC Grazing Plantship and Moored Plant

A cost assessment of floating OTEC systems at the preliminary design level has been completed only for the baseline 40-MWe (nom) grazing ammonia plantship and moored electric power plant designs (George and Richards, 1980). Optimization studies by Pandolfini (1985) show that the grazing plantship stationed near the equator where the water average annual temperature difference is 23.3°C will deliver 46 MWe (net) at the onboard busbar. The moored plant at Punta Tuna, Puerto Rico, with the same power system but with inlet water ΔT of 40.2°C, will deliver 40 MWe (net). These values of the generated net power are used in Table 7-1.

7.2.1.2   SOLARAMCO 40-MWe (Nom) OTEC Demonstration Plantship

As discussed in Chapter 6, the Solar Ammonia Company (SOLARAMCO) prepared a conceptual design of a 40-MWe (nom) ammonia plantship that would produce 115 mt/d of liquid ammonia. The estimated cost of this system is $250 M (1980$). A detailed cost breakdown is not available.

Table 7-1. Cost estimates for floating OTEC subsystems, components, facilities, and procedures ($M1990)[a]

Subsystem	46 MWe (net) Grazing plantship Mid-1980$	Mid-1990$	Est. cost uncertainty − %	+ %	40 MWe (net) Moored plant Mid-1980$	Mid-1990$	Est. cost uncertainty − %	+ %
Platform	47.3	64.8	13	13	66.6	91.2	14	28
Hull	20.4	27.9	15	20	20.4	27.9	15	20
Water pumps	9.7	13.3	15	15	9.7	13.3	15	15
Position control	7.6	10.4	10	10	24.2	33.1	15	50
Other	9.6	13.1	10	10	12.3	16.8	10	10
Power systems	42.0	57.5	24	17	42.0	57.5	24	18
Heat exchangers	28.3	38.7	30	20	28.3	38.7	30	20
NH$_3$ systems	3.5	4.8	15	15	3.5	4.8	15	15
Power generation	7.6	10.4	15	15	7.6	10.4	15	15
Other	2.6	3.6	10	10	2.6	3.6	10	10
Water ducts	9.4	12.9	25	99	13.0	17.8	22	76
WW & disch. pipes	0	0.0	15	15	3.5	4.8	15	15
CWP	6.6	9.0	25	100	6.7	9.2	25	100
CWP/hull joint	2.7	3.7	25	100	2.7	3.7	25	100
Other	0.1	0.1	15	15	0.1	0.1	15	15
Energy transfer								
NH$_3$ production	27.7	37.9	23	25				
Liquid N$_2$ plant	3.2	4.4	15	15				
Electrolysis	12.0	16.4	30	35				
Plant NH$_3$ synthesis	8.7	11.9	20	20				
Other	3.8	5.2	15	15				
Elec. power to shore					23.1	31.6	25	76
Control and bus					2.2	3.0	15	15
Conditioning					1.4	1.9	15	15
Riser cable					9.0	12.3	30	100
Bottom cable					4.1	5.6	30	100
Buoys					2.4	3.3	25	25
Misc. hardware					2.4	3.3	15	15
Potheads, gimbals, disconnects					1.6	2.2	15	15
Deployment	11.6	15.9	31	84	20.7	28.3	34	90
Platform	1.3	1.8	15	15	0.9	1.2	15	15
Power system	1.1	1.6	20	40	0.7	1.0	20	40
CWP	9.2	12.5	35	100	9.4	12.9	35	100
Discharge pipe					1.7	2.3	35	50
Electric cable					8.0	11.0	35	100
Acceptance, ind. fac., E&A	6.4	8.8	20	20	8.4	11.5	20	20
Weighted cost uncertainty (%)			22	30			26	54
Direct cost ($M)	144.4	197.7			170.2	233.0		
Interest dur. constr. (15% of direct cost)	22.1	30.2			26.0	35.6		
Tot. investment ($M)								
Nominal		228				269		
Minimum		182				199		
Maximum		296				414		

[a] 1990$ = 1980$ × 1.369 = 1983$ × 1.128 (Avery et al., 1985).

7.2.1.3 Commercial-Sized OTEC Grazing Plantships

Conceptual design studies have been completed for a 160-MWe (nom), 200-MWe (optimized), methanol plantship (Avery et al., 1985). These data have been used also for an estimate of the cost of a 320-MWe (nom) (368-MWe des. est.) ammonia plantship. The data provide a basis for estimating the cost and cost uncertainties of OTEC systems that would enter commercial production. Cost information available from conceptual design studies of other OTEC systems and preliminary designs of subsystems, as well as prices of standard items and components, have been used in these estimates.

The cost data are based on industrial estimates made in the period 1978–1982. Termination of OTEC development funding has prevented the acquisition of comparable estimates of current costs. Therefore, the quoted estimates of total system costs in 1990$ simply represent an updating of earlier data to account for inflation in the costs of industrial equipment and construction. The inflation factors are taken from the periodic surveys of construction costs published by *Chemical Engineering* magazine.

Estimated costs for the conceptual design of commercial-size OTEC methanol and ammonia plantships are listed in Table 7-2.

7.2.1.4 Sea Solar Power, Inc. 100-MWe Floating OTEC Power Plant

Costs estimated for the innovative 100-MWe dynamically positioned OTEC power plant, of the conceptual design published by Sea Solar Power, Inc. (SSP), are $250M (Anderson, 1985). A cost breakdown is not available.

7.2.1.5 Japanese Floating OTEC Plants

Conceptual designs of moored OTEC plants sited in the Sea of Japan were discussed in Section 6.4. The costs estimated by the Japanese engineers are listed in Table 7-3.

7.2.2 *Estimated Costs of Shore-Based OTEC systems*

7.2.2.1   OTC Shore-Based Power Plant

During 1981–1983 a preliminary design of a 40-MWe (nom) shelf-mounted OTEC power plant to be sited at Kahe Point on the island of Oahu, Hawaii, was completed by the Ocean Thermal Corporation (OTC) under DOE auspices (Ocean Thermal Corp., 1984). The program is described in Section 6.5. Details of the system design costs are proprietary, but the estimated costs of the major subsystems are available and are presented in Table 7-4. Performance analysis by OTC in 1983 showed that the nominal 40-MWe plant would produce a design power output of 50 MWe (net), with state-of-the-art components. The cost uncertainties listed in the table are those of the authors of this book, based on a review of the OTC preliminary design report and comparison with other relevant information. The much larger system costs, compared with those of the floating baseline moored power plant, are caused by the use of titanium heat exchangers instead of aluminum in the OTC design, by the expensive construction and deployment of the 3-km-long, bottom-mounted CWP,

Table 7-2. Cost estimates for grazing OTEC methanol
and ammonia plantships ($M1990)[a]

Subsystem	First 200-MWE methanol plantship	Subsystem cost uncertainty − %	+ %	First 368-MWe ammonia plantship	Subsystem cost uncertainty − %	+ %
Platform	190.7	14.0	16.9	247	13.8	16.3
Hull	109.7	15	20	124	15	20
Water pumps	42.7	15	15	65	15	15
Position control	20.7	10	10	31	10	10
Other	17.6	10	10	27	10	10
Power systems	180.4	23.4	17.6	273	24.5	17.6
Heat exchangers	124.3	30	20	188	30	20
NH$_3$ system	12.2	15	15	18	15	15
Power generation	30.5	10	10	46	10	10
Other	13.4	15	15	20	15	15
Water ducts	30.8	24.9	99.0	47	24.9	99.0
WW & disch. pipes	0.0	0				
CWP	21.9	25	100	33	25	100
CWP/hull joint	8.5	25	100	13	25	100
Other	0.4	15	15	1	15	15
Energy transfer						
NH$_3$ production				175	18.7	18.7
Liquid N$_2$ plant				23	15	15
Electrolysis plant				86	15	15
NH$_3$ synthesis				62	20	20
Other				27	15	15
CH$_3$OH production	311.2	16.5	16.5			
Coal prep. molten Na$_2$CO$_3$ gasifier	47.0	25	25.0			
Gas clean-up, CO shift, sulfur recov.	104.6	15	15.0			
CH$_3$OH synthesis	68.0	15	15.0			
Steam syst.	25.3	15	15.0			
H$_2$O, waste treatment Na$_2$CO$_3$ regen.	27.1	15	15.0			
Serv. & maint.	39.2	15	15.0			
Electrolysis	58.4	30	35.0			
Deployment	39.0	30.6	81.4	59	30.6	81.4
Platform	4.9	15	15	7	15	15
Power system	3.7	15	15	6	15	15
CWP	30.5	35	100	46	35	100
Discharge pipe	0.0					
Electric cable	0.0					
Acceptance, ind. fac., E&A	45.1	20	20	68	20	20
Weighted cost uncertainty (%)		13	24		20	26
Direct cost ($M)	855.6			869		
Interest dur. constr. (15% of direct cost)	131			133		
Tot. investment ($M)						
Nominal	986			1002		
Minimum	789			802		
Maximum	1292			1313		

[a]1990$ = 1980$ × 1.369 = 1983$ × 1.128.

Table 7-3. Estimated costs of 100-MWe (net) Japanese moored
OTEC power plants (1980$)

	At Osumi	At Toyama
Type of platform	Submerged disc	Surface ship
Net power output (MWe)	100	100
Unit construction cost (yen/kW)	78×10^5	59×10^5
($/kW)	3550	2690
Unit power cost at busbar (yen/kW h)	12.21	12.36
($/kW h)	0.0555	0.0562
Power transmission cost (yen/kW h)	0.66 (AC)	0.66 (AC)
($/kW h)	0.003	0.003

Table 7-4. Estimated costs of OTC 40-MWe (nom) land-based
OTEC power plant [50-MWe (net) final design] ($M 1985)

Land-based containment system	60.3	Energy transfer system	5.0
Structure	38.1	Power control & bus	1.4
Pumps and motors	12.1	Power conditioning	0.5
Facility support	7.7	Cable system	1.0
Outfit and furnishings	2.4	Shore-grid station	1.8
Cold-water pipe system	73.1	Biofouling corrosion control	0.3
Pipe	67.7	Deployment services	54.2
Intake	0.1	LBCS deployment	6.8
CWP/LBCS transition	0.7	CWP deployment	27.4
CWP embedment/anchoring	1.7	WWP deployment	6.8
Biofouling corrosion control	2.9	Effluent pipe deployment	11.2
Warm-water pipe system	3.7	Power system deployment	2.0
Pipe	3.4	Energy transfer syst. deployment	0.1
WWP/HECO basin transition	0.2	Industrial facilities	12.0
WWP/LBCS transition	0.2	Construction facilities	11.5
Mixed effluent pipe system	5.0	Logistic support facilities	0.5
Pipe	2.3	Acquisition, acceptance, testing	2.0
Effluent pipe/LBCS transition	1.1	Engineering & detail design	5.2
Outfall & discharge channel	0.5		
Embedment/anchoring	1.1	Total direct cost	327.8
Power system	107.2	Construction interest	50.1
Evaporators	39.7	Plant investment (M1984$)	378
Condensors	37.1	Plant investment (M1990$)	419
Turbine/generators	15.4		
Working fluid loop	4.8		
Nitrogen purge	0.3		
Instrument & control	5.4		
Biofouling corrosion control	3.0		
Auxiliary systems	1.6		

and by the requirement of a platform that will withstand the wave action in the near-shore zone.

7.2.2.2 General Electric Conceptual Design of a Tower-Mounted OTEC Power Plant for Hawaii

Costs estimated in the DOE supported conceptual design study, discussed in Section 6.4, are listed in Table 7-5.

Table 7-5. Estimated costs of GE 40-MWe (nom) tower-mounted OTEC plant

System cost estimates	$M(1982$)
GE tower-mounted design	
Tower system	90
Fabrication	63
Transportation	3
Installation	17
Engineering	4
Insurance	4
Cold-water pipe system	64
Pipe material	49
Templates	1
Deployment	12
Weather allowance	2
Power plant	166
Heat exchangers	34
Ammonia charge	1
Water pumps	6
Turbine-generators	20
Deploy, install, misc.	105
Engineering, misc.	8
Total system cost	320
$/kWe (net)	8000

7.2.2.3 Puerto Rico Electric Power Co. Conceptual Design of a 40-MWe (Nom) Tower-Mounted OTEC Plant

The costs quoted for the PREPA installation, described in Chapter 6, are presented in Table 7-6.

7.2.2.4 French Designs of 5-MWe OTEC Land-Based Plant for Tahiti

French plans for a 5- to 10-MWe OTEC demonstration plant to be sited in Tahiti were discussed in Chapter 6. Preliminary designs of a 5-MWe closed-cycle plant and of an alternative 5-MWe open-cycle plant were completed, including equipment for the production of fresh water in each system design. The cost assessment led to essentially the same totals for the two options, as shown in Table 7-7 (Gauthier, 1986).

7.2.2.5 Taiwan 100-MWe (Gross) Land-Based OTEC Power Plant

Characteristics of land-based OTEC plants designed for sites on the island of Taiwan were presented in Chapter 6. The estimated costs for a 74-MWe (net) plant are listed in Table 7-8 (Liao et al., 1986).

7.3 OTEC CAPITAL INVESTMENTS AND POTENTIAL SALES PRICES FOR PRODUCTS

The capital investment data for OTEC systems, presented in Section 7.2, provide a basis for estimating the delivered cost of electric power and energy products from

Table 7-6. Estimated phase IV costs of PREPA 40-MWe (net) pilot plant ($M1982)

Tower system	21
Fabrication	13.9
Transportation	0.6
Installation	6.5
Power plant system	117.4
Heat exchangers	62.8
Seawater systems	12.9
Turbine-generator	10.0
Biocontrol	7.4
Ammonia systems	5.0
Startup/standby power	7.0
Control and elect.	10.6
Miscellaneous	1.5
Water ducting	22.2
CWP incl. deployment	18.6
Warm and mixed systems	3.6
Energy transfer system	66.0
Acceptance testing	2.0
Deployment services	20.0
Industrial facilities	2.0
Eng. and detail design	2.0
Total	259.4
$/kW	6485

Table 7-7. Estimated costs of French land-based 5-MWe (net)
OTEC electric power and fresh-water plants for Tahiti (MFF 1985)

	Closed cycle	Open cycle
General	23	23
Power system	144	167
Land structures		
(incl. land part of CWP and civilian work)	221	221
CWP	135	136
Total	523	547
FF/kW	105,200	106,900

land-based and floating plants and plantships. However, many development, financing, marketing, and siting alternatives are possible that lead to large variations in the estimated costs and revenues from sales of power or energy products, which will determine the profitability of future OTEC installations. Analyses and cost-estimating programs have been published by American investigators and by French, Japanese, and other authors (Francis and Richards, 1983; Dugger et al., 1982; Avery et al., 1985; Gritton et al., 1981; Reid, 1978; Chan, 1981; Curto and Cohen, 1978; Gauthier, 1986; Homma and Kamogawa, 1978; Kamogawa and Nakamoto, 1981; Liao et al., 1986; Mossman, 1989).

The primary factor that determines the cost of delivered OTEC energy is the financing cost to amortize the capital investment. Tax benefits and federal subsidies available to investors can reduce the needed capital investment or annual costs. The

Table 7-8. Power and costs of Taiwan 50-MWe (nom) land-based OTEC plants (1980$)

Site	Cold-water pump	Warm-water pump	Module pump power	Total pump power	Gross turb/gen power	Optimized plant Net power output (nom)	Power output (ann.avg)	Investment (M$)
Ho-Ping (long CWP)	1.54	3.07	4.62	18.46	68.46	50.0	53	301.5 ($5690/kW)
Ho-Ping (short CWP)	1.50	3.07	4.57	18.27	68.27	50.0		
Chang-Yuan (long CWP)	1.71	3.13	4.84	19.37	69.37	50.0	74	364.5 ($4920/kW)
Chang-Yuan (short CWP)	1.60	3.13	4.73	18.93	68.93	50.0		

sales price of the product is determined by the market. The profitability of the investment is then the difference between the two values over the life of the plant. Although the amortization costs will be known before construction begins, the estimate of profitability must include a forecast of the course of inflation or deflation of energy and fuel prices as a function of plant life. Fossil fuel prices are expected to rise as U.S. reserves are depleted, but are subject to manipulation by the Arab nations, who possess the largest share of the world's oil reserves. Thus, a decision by private investors to proceed with OTEC involves an assessment of risk that affects the rate of return needed to make the venture of interest. Government cost sharing of construction of OTEC demonstration vessels in the 50- to 100-MWe range, which was a provision of the 1980 public laws PL-310 and PL-320, and government-supported loan guarantees such as the U.S. Maritime Administration (MARAD) Title XI program or the Japanese Sunshine program may be essential for large-scale OTEC commercialization.

Project financing in general requires two parties: (1) the project sponsors who build and operate the OTEC plant, sell the product, and receive the profits, and (2) lenders who provide the funds for construction and implementation and receive interest on the money lent. The first group are "venture capitalists," the second bankers and other financial institutions. Thus the total investment consists of equity and debt. Loan guarantees available from MARAD for ship construction can be arranged to cover 87.5% of the total investment, but 75-80% is more common. Such loans substantially reduce the risk to the investor because loan approval, after in-depth scrutiny by MARAD, carries a guarantee backed by the government. The project will have a negative cash flow in the early years. For investors who have profits in other operations, OTEC losses can be deducted from the total income, thereby reducing the investor's total taxes, in effect providing income from the operating losses via the tax benefit. Other sponsors will invest because, if profitable operation is attained, the receipts from the product sales can lead to a large rate of return on their investment. For example, a return of 10% on total investment corresponds to a 50% rate of return on equity for an investor who has supplied 20%

of the plant investment from private funds and borrowed the remaining 80%. Thus, the total financial package may take many forms depending on the ingenuity and resources of the investors.

OTEC plants and plantships can be profitable if conditions can be created that will enable commercial production to be established. The requirements are documented in the following sections of this chapter and in Chapter 8.

7.3.1   OTEC Methanol Financing

The following paragraphs present the methodology and a listing of the parameters involved in a recent financial evaluation by Mossman (1989) of a proposed first-of-a-kind 160-MWe (nom) [200 MWe (des. est.)] OTEC methanol plantship designed to produce 1750 metric tons per day of high-octane methanol fuel for automobiles. Two years of financed construction followed by an operating life of 25 years are assumed, for the nominal investment (see Table 7-2). The project is assumed to be undertaken by a firm or consortium that will use all possible tax shields provided by depreciation and interest payments. The analysis calculates a quantity termed "the adjusted present value" (APV), which measures the profitability of a potential venture. If the APV has a positive value, the venture will return a profit sufficiently high to attract entrepreneurs to invest in it rather than to select alternative profit-making opportunities. If the APV has a negative value, the project will lose money or be too risky to be considered. The Mossman APV methodology calculates the sales price of product needed to achieve a positive APV by following the procedures listed here, condensed from Mossman. Full details are presented by Mossman (1989).

1.   The project is valued as an all-equity financed firm. Cash flows for a selected product sales price over the life of the project are identified for each period and are then discounted using the standard formula to find a net present value (NPV). The cash flows are established by entering data in a spreadsheet that lists values over the life of the project for the terms calculated from the following function:

   a.   List revenues from product sales;
   b.   Subtract operating expenses;
   c.   Subtract depreciation;
   d.   Equals earnings before taxes (EBT);
   e.   Subtract taxes on EBT;
   f.   Equals earnings after taxes (EAT);
   g.   Add-back depreciation;
   h.   Subtract additional working capital;
   i.   Subtract net capital expenditures;
   j.   Equals unlevered [sic] free cash flows.

2.   The unlevered free cash flows are then discounted using a required rate of return based on the expected rate of return offered to investors by equivalent-

risk investments traded in capital markets. This is the net present value (NPV) of the investment.

3. Side effects of financing, primarily associated with tax shields provided by interest payments on debt, are calculated.

4. The present value of the side effects is calculated, using the interest rate of the debt as the discount rate, and then added to the net present value of the unlevered free cash flows. This defines the adjusted present value (APV).

5. The decision rule is used that the APV must be greater than zero dollars. If the APV is negative, the project will either provide a negative rate of return or a positive rate of return that is inadequate to compensate investors for their opportunity cost relative to other possible investments of equivalent risk.

The following assumptions were made in assessing the OTEC methanol base case model:

1. *Rate of return.* For risk-free debt this is assumed to be equal to the yield on 30-year U.S. Treasury notes, in 1989 around 8.00%.

2. *Market premium.* Six percent is assumed.

3. *Project beta.* This is a factor that is used to define the interest premium on the borrowed capital relative to U.S. Treasury notes, that is paid by investors in a particular industry. If Treasury notes pay 8%, the market premium is 6%, and the value of beta is 0.83 for the OTEC project, then the interest charge will be $8\% + 6\% \times 0.83 = 12.98\%$. The beta value of 0.83 is based on 1988 equity betas of Dow Chemical, DuPont, Hercules, Monsanto, Olin, Rohm and Haas, and Union Carbide. Beta is a measure of the expected price stability of methanol or ammonia relative to the average of all commodities listed in the consumer price index. A value less than 1.0 indicates more stability than the average.

4. *Permanent financing interest rate.* A value of 13% was chosen based on the yield of high-interest-rate corporate bonds (BB/Ba-C) as of November 1988.

5. *Construction financing.* An interest rate of 12% was chosen, reflecting 1990 interest rates for short-term construction loans plus a premium to allow for uncertainties in construction time.

6. *Equity.* A value of 20% was selected. Debt as percentage of total capital was assumed to be 80% equal to the value of the debt for a methanol plant of similar size completed in 1987 in Punta Arenas, Chile.

7. *Methanol price.* The amortized cost of methanol from a new U.S. plant using natural gas at the market price in 1989 was estimated to be in the range of $0.55–0.60 /gal. In 1990 the sales price of methanol was $0.70/gal.

8. *Methanol production.* The design methanol production rate per OTEC plantship is 199 million gallons per year. The annual coal cost is 23.5 million dollars ($23.5M) (470,000 tons/year at $50/ton). Coal and methanol shipping costs are $20/ton each. (These values are taken from Avery et al., 1985.)

9.   *Marginal tax rate.*   The 1989 top marginal tax rate for corporations was
        34%. This value is used.
10.   *Depreciation schedule.*   This is based on the assumption that accelerated
        depreciation will be available.
11.   *Inflation rate.*   The calculation is based on zero rate of inflation.

With the assumptions listed, a computer program was prepared following the
Mossman rationale. It relates the profitability of an OTEC venture to the values of
the plant investment, interest rates, operating and maintenance charges, and
methanol sales prices.

For the first-of-a-kind 200-MWe OTEC methanol plant, the nominal plant
investment, taken from Table 7-2, is estimated to be 960 million dollars ($960M,
in 1990$).[2]

The fixed cost estimate will be reduced and the financial prospects enhanced
by the following considerations:

1. The APL plant investment is based on costs of construction in U.S. shipyards.
    Construction costs, including outfitting, of large ships in Japan in 1986 were
    one-half those in U.S. shipyards and, in Korea, were only one-third (Bunch,
    1987, 1988).
2. The sales price of methanol is assumed constant over the plant life; that is, there
    is no allowance for inflation in the calculated revenues of $78.0M/year.
    Operating and maintenance costs are also assumed to be constant at $50.9M/
    year. Inflation acting on the difference would increase the calculated present
    value of the net receipts.
3. U.S. Navy experience with large orders for destroyers of the same specifica-
    tions showed that the delivered cost of the second vessel was 0.8 times that of
    the first and an experience factor[3] of 0.93 applied to the remaining orders of
    20+ vessels. Applying these values of OTEC methanol plantships, the cost for
    the 8th is 70% of the cost of the first, the cost is 60% for the 32nd, and so on
    (Fink, 1987).
4. A government commitment to provide incentives for the production of low-
    polluting vehicle fuels will lead to demands for large increases in methanol
    production that will cause methanol prices to increase toward parity with
    premium gasoline on a miles per gallon (mpg) basis.

The graph shown in Fig. 7-1 summarizes the results of the financial analysis.
The lines of the graph show the APV as a function of methanol sales price for
estimated plant investment values. The lines represent the following selection of
values for the estimated plant investment, starting from the bottom line:

1. Plant investment 30% above the nominal value of $960M, reflecting the upper
    estimate of cost uncertainty for the first plant constructed in a U.S. shipyard at
    1988 estimated costs.
2. Plant investment equals the nominal value.
3. Plant investment 20% below the nominal value, reflecting the lower estimate
    of cost uncertainty for U.S. construction.

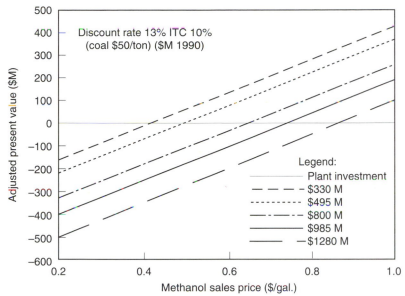

FIG. 7-1. Mossman profitability diagram for 200-MWe OTEC methanol fuel production.

4. Plant investment one-half the nominal value, representing estimated construction cost in Japan.
5. Plant investment one-third the nominal value, representing estimated construction cost in Korea.

An investment tax credit (ITC) of 10% has been used in the calculation of these values.

The graph shows that to attract financing for the first plant at the nominal estimated investment value of $960M (i.e., an APV greater than 0), the methanol sales price would need to be $0.73/gal or higher. For a plant investment of $495M, the project is profitable with a methanol sales price of $0.50/gal. With a plant investment of $330M, the investment would be profitable at a methanol sales price of $0.43/gal.

As shown in Chapter 9, OTEC methanol will be competitive in price with gasoline on a dollars per mile basis in the 1990s if the price of gasoline includes the externality costs of oil imports and fossil emissions and if the commercialization program receives positive support from government and industry. Such support will include cost sharing of construction of a demonstration plant of about 50 MWe (net), sponsorship of methanol vehicle development by the automobile industry, and loan guarantees and tax incentives to stimulate construction of OTEC methanol plantships.

Construction of 20 plantships per year at an annual total cost of $7–$15 billion would provide in 10 years an inexhaustible, environmentally superior supply of vehicle fuel in an amount equivalent to 1.8 million barrels per day of premium gasoline. This would displace crude oil imports costing $14 billion/year at the 1990 price of $20/bbl. The economics are discussed in more detail in Chapter 9.

Methanol is also a favored fuel for use in fuel cells to generate electric power (Patel, 1983; Patel et al., 1982), as an alternative to electric power generation from fossil fuel or nuclear power plants. A discussion of the economics of this option is presented in Section 8.3.

7.3.2 OTEC Ammonia Commercialization

Production of ammonia on OTEC plantships can supply world needs for a practical motor vehicle fuel that produces only water and nitrogen as products of combustion. Experimental work has demonstrated that ammonia can be a practical substitute for gasoline, burning cleanly with high efficiency and a measured octane number of 111 (Avery, 1988). Use of OTEC ammonia instead of gasoline would eliminate 5 tons per year of CO_2 added to the atmosphere by an average automobile, or roughly 500 million tons per year for the total number of U.S. automobiles. Ammonia is also an attractive fuel for fuel-cell production of electric power. This option is discussed in Section 8.3.

Profitability projections for OTEC ammonia production have been made using the Mossman financial analysis. The results of five cases are presented in Fig. 7-2. Data for the estimate of plant investment and operating costs are:

1. Plantship direct cost—$975 (nom);
2. Annual production—1.7×10^8 gal/y of fuel-grade ammonia;
3. Operating and maintenance and shipping costs—$16.9M/y;
4. Rate of return on risk-free debt—8.00%;

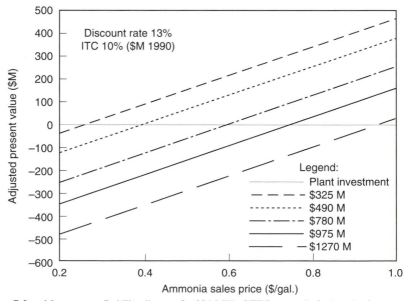

FIG. 7-2.   Mossman profitability diagram for 384-MWe OTEC ammonia fuel production.

5. Market premium—6%;
6. Project equity beta—0.83;
7. Permanent financing interest rate—13.00%;
8. Construction interest—12.00%;
9. Debt as percentage of total capital—80%;
10. Wholesale price of ammonia—$230/tonne;
11. Marginal tax rate—34%;
12. Depreciation schedule—based on the assumption that accelerated depreciation will be available;
13. Rate of inflation—assumed to be zero.

An assessment of the profitability of this option is given in Section 9.1.

7.3.3 Ocean Thermal Corporation Land-Based Containment System

A preliminary design of a nominal 40-MWe (net) land-based OTEC plant to be sited at Kahe Point, Hawaii, was completed in 1984 by the Ocean Thermal Corporation (OTC). The OTC plan called for power in the amount of 320,000 kW h/y to be purchased by the Hawaiian Electric Co. (HECO) under a long-term contract at a fixed initial price of $0.125/kW h with an (undisclosed) allowance for inflation. The drop in oil prices in late 1984 made OTC and HECO consider the plan unprofitable.

It was stated by OTC in 1984 that improvements in design had been made that would increase the net output of the plant to 50 MWe. This implies that operation would be profitable with the nominal estimated plant investment.

A Mossman analysis using 1990s conditions supports the conclusion that the venture would be profitable with the projected receipts. A graph showing the adjusted present value of the project as a function of capital cost and sales price of power is presented in Fig. 7-3. The plant output is taken to be 50 MWe (net), and sale of the power produced in 8000 h per year operation is assumed. The results indicate that with the nominal value of the estimated plant investment and 10% investment tax credit, operation would be profitable at a sales price of electric power of $0.125/kW h.

7.3.4 Cost of OTEC Power in Japan

The Japanese OTEC program was discussed in Section 6.4. Estimates of the costs of moored OTEC plants in Japan and projected cost of delivered power are presented in Table 7-3 (Homma and Kamogawa, 1978). The benefits of lower construction cost compared with the United States are apparent. The results of the financial assessment are presented in Fig. 7-4.

7.3.5 Cost of OTEC Power and Fresh-Water Production in Tahiti

French plans for an OTEC installation on the island of Tahiti and cost estimates were discussed in Section 6.4 (Gauthier, 1986). Estimates of the profitability of these plants using the Mossman analysis are presented in Fig. 7-5.

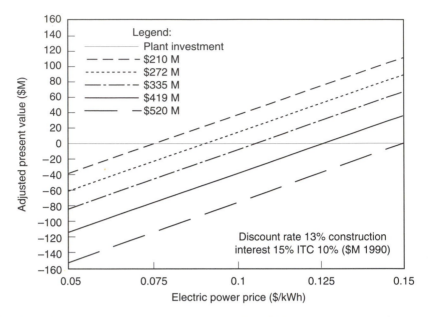

FIG. 7-3. Mossman profitability diagram for OTC 50-MWe OTEC electric power production.

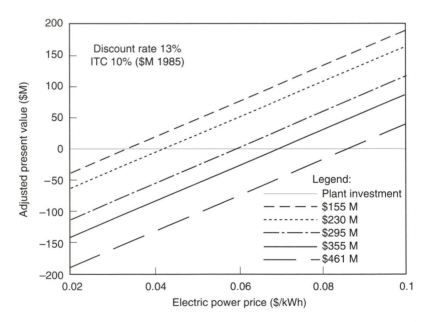

FIG. 7-4. Mossman profitability diagram for Japanese 100-MWe OTEC electric power production.

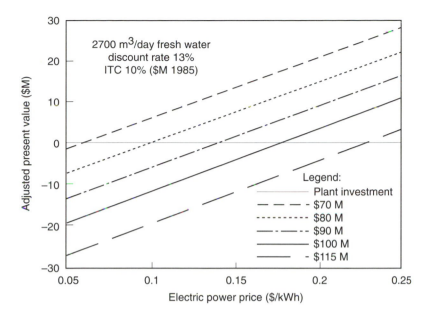

FIG. 7-5. Mossman profitability diagram for French 5-MWe OTEC electric power and fresh-water production.

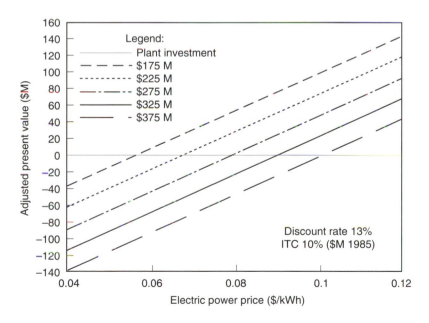

FIG. 7-6. Mossman profitability diagram for Taiwan 100-MWe OTEC electric power production.

7.3.6 *Cost of OTEC Power at Sites on the Island of Taiwan*

Design studies for land-based OTEC power plants in Taiwan were discussed in Section 6.4 (Liao et al., 1986). Cost data are listed in Table 7-8. Profitability estimates for the proposed installation at Chang-Yuan with the long cold-water pipe are shown in Fig. 7-6.

The economic potential of OTEC fuel and electric power production options is compared with other existing and proposed alternative sources in Chapter 8.

NOTES

1. Industrial cost surveys show that unit costs of items in commercial production tend to decrease by a constant factor with each doubling of the number produced; that is, the second unit cost x times the cost of the first, the 4th x times the cost of the second, and so on. The factor x is called the "experience factor."
2. In the text, tables, and graphs "$M" signifies "millions of dollars," not "thousands of dollars," as is conventional in banking terminology.
3. The cost is reduced by the factor 0.93 for each doubling of the number produced.

REFERENCES

Anderson, J. H., 1985. "Ocean thermal power—the coming energy revolution." *Solar and Wind Technology* **2**, 25.

Avery, W. H., 1988. "A role for ammonia in the hydrogen economy." *Int. J. Hydrogen Energy* **13**, 761.

——, D. Richards, and G. L. Dugger, 1985. "Hydrogen generation by OTEC electrolysis and economical transfer to world markets via ammonia and methanol." *Int. J. Hydrogen Energy* **10**, 727.

Bunch, H. M., 1987. "Comparison of the construction planning and manpower schedules for building the PD214 general mobilization ship in a U.S. shipyard and in a Japanese shipyard." *J. of Ship Production* **1**, 25.

——, 1988. "Testimony on ship costs, to Commission on Merchant Marine and Defense, July 18, 1988."

Chang, G. L., 1981. "An introductory study of factors favorable to an OTEC electric power generating station in Saipan." *Proc. 8th Ocean Energy Conf.*, Washington, D.C., June, 941.

Curto, P. L., and R. Cohen, 1978. "Producer incentives for OTEC commercialization." *Proc. 7th OTEC Conf.*, Washington, D.C., June, **II**, 77.

Dugger, G. L., J. E. Snyder, and F. Naef, 1982. "Ocean thermal energy conversion: Historical highlights, status and forecasts." *J. of Energy* **5**, 231.

Fink, L., 1987. "Costs of Navy ship construction." Personal communication to W. H. Avery.

Francis, E. J., and D. Richards, 1983. "OTEC—Status and potential of private funding." *'82 MTS-IEEE Conf. and Exposition*, Washington D.C., Sept., 100.

Gauthier, M., 1986. *The economic context of the utilization of the ocean thermal temperature gradient.* IFREMER, Brest, France, May.

George, J. F., and D. Richards, 1980. *Baseline designs of moored and grazing 40-MW OTEC pilot plants. Detailed report.* Johns Hopkins University Applied Physics Lab. SR-80A.

Gritton, E. C., R. Y. Pei, and J. Aroesty, 1981. "A quantitative evaluation of closed-cycle OTEC technology in central station applications." *Proc. 8th Ocean Energy Conf.*, Washington, D.C., 873.

Homma, T., and H. Kamogawa, 1978. "Conceptual design and economic evaluation of OTEC power in Japan." *Proc. 5th Ocean Thermal Energy Conversion Conf.*, Miami Beach, Fla., **2**, 161.

Kamogawa, H., and Y. Nakamoto, 1981. "A feasibility study of the OTEC complex for island applications." *Proc. 8th Ocean Energy Conf.*, Washington, D.C., June, 161.

Liao, T., J. G. Giannnotti, P. R. Van Mater, Jr., and R. A. Lindman, 1986. "Feasibility and concept design studies for OTEC plants along the east coast of Taiwan, Republic of China." *Proc. International Offshore Mechanics and Arctic Engineering Symp.*, **2**, 618.

Merrow, E. W., K. E. Phillips, and C. M. Myers, 1981. *Understanding cost growth and performance shortfalls in pioneer process plants*. RAND Corp., RAND report #2569, Sept.

Mossman, B. J., 1989. "OTEC methanol plantship proposal: A financial evaluation." Based on a project in fulfillment of the requirements for the M.B.A. degree from the Wharton School of the University of Pennsylvania. Document available from the author.

Ocean Thermal Corp., 1984. *40-MWe OTEC power plant. Preliminary design engineering report.* Ocean Thermal Corp., Nov.

Pandolfini, P. P., 1985. *OTEC system analysis. Final report*. Contract 51352402. Johns Hopkins Univ. Applied Physics Lab.

Patel, P. S., 1983. *Assessment of a 6500-Btu/kWh heat rate dispersed cell generator*. Energy Research Corp., Danbury, Conn., EM-3307

——, H. C. Naru, and B. S. Baker, 1982. "Direct molten carbonate fuel cell." *National Fuel Cell Seminar, 1982*, 88.

Reid, A. F., 1978. "Ocean thermal energy conversion (OTEC): Temperature increase system (TIS) assisted." *Proc. 5th Ocean Thermal Energy Conversion Conf.*, Miami Beach, Fla., Feb., 59.

SOLARAMCO, 1981. *Response to DOE program opportunity notice for 40-MWe OTEC system concept*. Solar Ammonia Company, Inc., Tulsa, Okla.

8

OTEC ECONOMICS

8.1 INTRODUCTION

The economic contribution that OTEC will make to the solution of the nation's energy problems depends on its perceived merits relative to existing and alternative sources of energy and fuels. The previous chapters have shown that OTEC technology is ready for large-scale demonstrations that will provide a firm basis for commercial development. OTEC can have a large impact on U.S. energy needs by supplying liquid fuels for direct use in transportation, or for electric power production via fuel cells. Its commercial development will depend finally on political and other factors that cannot be assessed quantitatively. The national security and environmental impacts of continuing dependence on oil should receive major emphasis in decisions to implement new processes for fuel production. In this chapter we review the estimated sales prices of fuels and electric power from existing and proposed sources and compare them with OTEC prices. Actual manufacturing *costs* are generally unavailable and are highly dependent on the financing methods and resources of the individual producer. However, an objective comparison of the *sales prices* of fuels produced by proposed processes can be made by using the Mossman financial analysis method to estimate the sales price of fuel or electric power that must be charged for profitable operation. This requires only information on the plant investment, input fuel costs, and operation and mainte-nance costs. The sensitivity of the product costs to changes in the estimates in plant investment can then be displayed in a suitable graph. With this procedure the alternatives can be equitably compared from a uniform point of view.

8.2 FACTORS THAT DETERMINE THE COMMERCIAL VIABILITY OF ALTERNATIVE ENERGY OPTIONS

8.2.1 *Capital Investment*

This includes costs of related facilities as well as the plant investment.

8.2.2 *Annual Financing Costs*

See the discussion in Section 7.3.

8.2.3 *Fuel Costs*

These costs vary with time in an unpredictable manner. Past forecasts have been in error by large factors.

8.2.4 *Operating and Maintenance Costs*

These costs are a small part of the total, typically a few percent of plant investment for capital-intensive projects. Environmental impact costs have usually been ignored.

8.2.5 *Federal, State, and Local Taxes, Licensing Fees, Insurance, and Incidentals*

These items are a few percent of plant investment. State and local property taxes will be zero for sea-based OTEC systems.

8.2.6 *Demand for the Energy Option, and Potential Revenues*

Many options may be profitable for special applications, but the developable resource is too small for them to have a major role in satisfying national energy requirements.

8.2.7 *Environmental Charges*

Until recently, environmental impacts of energy production were ignored or charged to the public through state or local taxes. It has been proposed that these costs should be paid by the energy producers. The environmental impacts can be large, but the financial implications are difficult to quantify, particularly if the effects are cumulative. Compliance by existing nuclear installations with environmental regulations were estimated to require expenditures of $3 billion for 1990 and $4 billion for 1991 and following years until completion (*Energy Daily*, 1990). Environmental costs will be large also for oil shale and tar conversion systems and for coal conversion plants. Solar-based options are generally nonpolluting and are expected to involve minimal environmental costs.

8.2.8 *Vested Interests of Public, Industry, and Government Organizations*

Introduction of new energy sources is dependent on the support that can be developed from the existing energy producing and distributing organizations and their political support groups. The absence of a strong political base for solar energy options is a major impediment to development of this field.

8.3 ECONOMIC POTENTIAL OF EXISTING AND PROPOSED LIQUID FUEL SOURCES

The material in this section is based mainly on recent surveys by the Electric Power Research Institute (EPRI), the Pittsburgh Energy Technology Center of the U.S.

Department of Energy, and the National Research Council and on articles published in the Proceedings of the annual energy technology conferences. The systems described represent only a small fraction of the current and proposed systems but are believed to be representative of the range of options. For each option the sales prices of the direct energy product needed for profitable investment are estimated from the published estimates of capital cost, fuel costs, and operating and maintenance costs, by using the Mossman financial analysis procedure.

8.3.1 *Petroleum-Based Fuels*

Petroleum must be refined to make products suitable for use in transportation, in industrial or residential boilers, and in miscellaneous applications such as paint solvents and chemical feed stocks. The refining processes are designed to incorporate variations in distillation, thermal cracking, and catalysis to produce oil fractions that have boiling ranges suitable for motor vehicles (gasoline and diesel fuel) and other uses. The refining processes add to the cost so that a barrel of crude oil costing $18 ($0.43/gal) was associated with wholesale prices, before the Gulf War, of $0.66/gal for gasoline and $0.58/gal for diesel fuel (*DOE/EIA Monthly Energy Review,* 1989). Taxes added $0.40/gal to the price to the consumer (*DOE/ EIA Monthly Energy Review*, Table 9.4). We conclude that the wholesale gasoline price may be derived from the quoted oil price ($/gal) by multiplying by the factor $66/43 = 1.5$. Thus, for oil at $30/bbl (September 1990) the wholesale gasoline price was about $1 per gallon or $42 per barrel. In an unbiased market, an alternative fuel will be competitive with gasoline if its sales price is lower in vehicle miles per gallon than gasoline.

In assessing the competitive price for alternative fuels, appropriate credit should be given for environmental and national security benefits and allowance made for "disbenefits" associated with the introduction of a new fuel type. As an example, the "real cost" of imported oil to the national economy was estimated to be not the $17.50/bbl listed as the market price, but $80/bbl, when military costs to protect imported oil supplies, foreign aid to mid-East governments, and lost jobs in the United States were considered (Tonelson and Hurd, 1990). A report by the International Energy Agency (IEA) (1990) suggests that a tax of $50/mt on carbon produced by combustion would be desirable to deter fossil fuel production of greenhouse gases.

8.3.2 *Synthetic Fuels*

Programs to provide new sources of fuels for motor vehicles have included:

1. Development of technologies to produce fuels from hydrocarbon sources such as oil shale and tar sands that are present in large amounts in the United States.
2. Programs to produce liquid hydrocarbon fuels by coal liquefaction from the vast U.S. reserves of coal.
3. Development of synthetic alternative fuels such as methanol or ammonia as a replacement for gasoline. Natural gas reforming and coal gasification can

provide the feed stock, followed by purification and catalytic steps to produce a liquid fuel.

4. Processes based on hydrogen produced by water electrolysis followed by further reactions to produce a liquid fuel.

8.3.3 Fuels from Oil Shale

Two programs initiated with support by the now defunct Synfuels Corporation received major funding in the early 1980s to demonstrate technologies for recovering motor fuel from oil shale: an above-ground retorting method of Unocal Corporation and an underground gasification scheme of Occidental Oil Shale Co. The Occidental program was discontinued with the demise of the Synfuel Corporation, but Unocal completed development of a pilot plant producing 6000–7000 bbl/day of oil (mid-1990). Continued studies by Occidental appear to show that the Occidental approach will be more economical and have led to new interest in their approach. A demonstration program, planned for a plant requiring an investment of $225M, will produce 1200 bbl/day for motor fuel. The process will also generate 35 MW of electric power, which will add to the profitability (Maize, 1990). These data are used in the Mossman diagram shown in Fig. 8-1.

8.3.4 Liquid Fuels from Tar Sands

This approach seems to be worth developing, but cost information on U.S. tar sands fuel production facilities is not available (Woods, 1986).

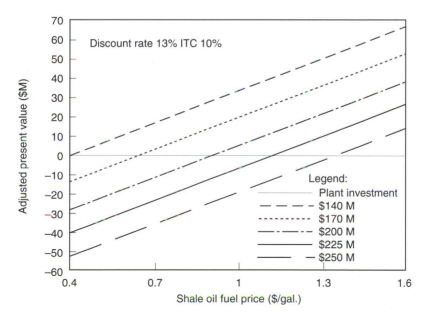

FIG. 8-1. Mossman profitability diagram for sale of 1200 bbl/d of motor fuel produced from oil shale.

8.3.5 *Liquid Hydrocarbons from Coal*

Approaches that have been supported include (McGurl, 1990):

1. Pyrolysis, which produces a liquid fuel (+ gas) and coke.
2. Catalytic addition of hydrogen to processed coal to produce a synthetic crude oil, with a hydrogen/carbon (H/C) ratio typical of liquid fuels. This is then refined to produce a vehicle fuel.
3. Coal gasification (reaction of coal with air or oxygen and steam to form CO + H_2) followed by catalytic combination of the products in successive steps to form liquid hydrocarbon fuels.

Research and development by industry with government participation during the past decade show that the product cost for the hydrogen addition processes and the coal gasification processes are nearly the same. A financial analysis that gives results typical of projected commercial processes in the 1990s can be based on the reported costs for the Shell Gasification/FT Slurry process, by Gray and Tomlinson (1989). Synthetic gasoline prices predicted by the Mossman analysis, as dependent on plant investment and coal prices, are presented in Figs. 8-2(a) and (b).

The pyrolysis approach is not considered to be competitive with the other processes if used alone but may be promising in combination with electric power generation using the residual coke.

8.3.6 *Alternative Liquid Fuels from Natural Gas Reforming or Coal Gasification*

Methanol and ammonia can be used directly as motor fuels (Section 7.3). Both are produced at low cost in large quantities (~ 20 million tons/year in the United States) by natural gas reforming. A detailed discussion of current processes and estimated costs for methanol production has been published by McGurl (1990). Estimated capital costs range from $237M for a U.S.-sited 2500-metric-tons-per-day (mt/d) methanol plant, to $558M for a 10,000-mt/d plant. Operation and maintenance costs are estimated to be 6% of fixed costs. These cost estimates are used in determining the profitability estimates shown in Figs. 8-3(a) and (b). In November 1990 the average cost of natural gas at the national level was $2.45 per million Btu ($2.45/mBtu) (DOE/EIA, 1991). Methanol prices in barge quantities at the Gulf coast were in the range $0.60-0.67/gal ($25-28/bbl, $9.8-10.9/mBtu) in April 1991 (*Chemical Marketing Reporter*, 1991).

Ammonia plant costs are similar to those for methanol. A 1000-mt/d plant based on natural gas reforming to supply hydrogen and air liquefaction to provide nitrogen, was estimated to cost $100M (Babbitt, 1981). The financial analysis for this process is shown in Figs. 8-4(a) and (b). Ammonia in tank-car quantities was priced at $0.33/gal ($13.9/bbl, $8.3/mBtu) in April 1991 (*Chemical Marketing Reporter*, 1991).

Coal gasification to produce fuel-grade methanol appears to be competitive with natural gas reforming at April 1991 prices (~ $0.65/gal). Large use of methanol or ammonia in transportation would cause the price of natural gas to increase and

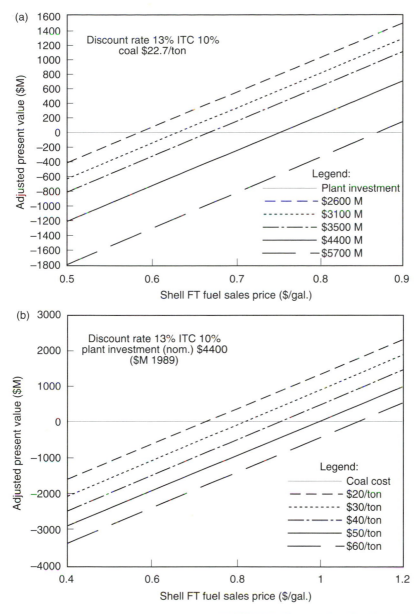

FIG. 8-2. Mossman profitability diagram for sale of 83,500 bbl/d of motor fuel produced by the Shell-FT coal conversion process. (a) Variable plant investment; (b) variable coal price.

would make the preparation of the fuels by coal gasification processes more competitive. The cost of methanol from coal gasification can be reduced to about $0.55/gal in a combined process that uses part of the gas for steam electric power generation (McGurl, 1990).

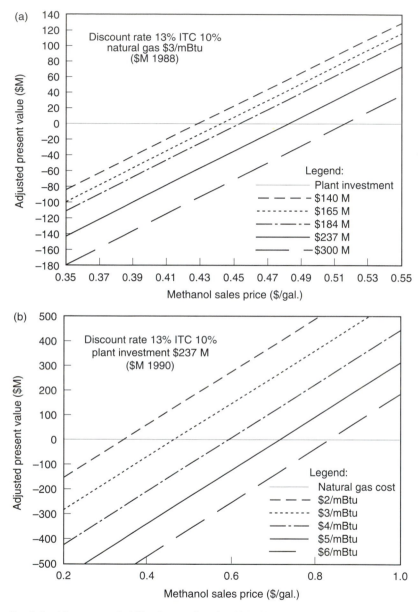

FIG. 8-3. Mossman profitability diagram for sale of 25,000 mt/d of methanol produced from natural gas. (a) Variable plant investment; (b) variable gas price.

8.4 ELECTRICITY FROM EXISTING AND PROPOSED POWER PLANTS

The sources of energy for electric power generation in the United States, which totaled 7.5 quads in November 1991, included coal, 58.4% of total power produc-

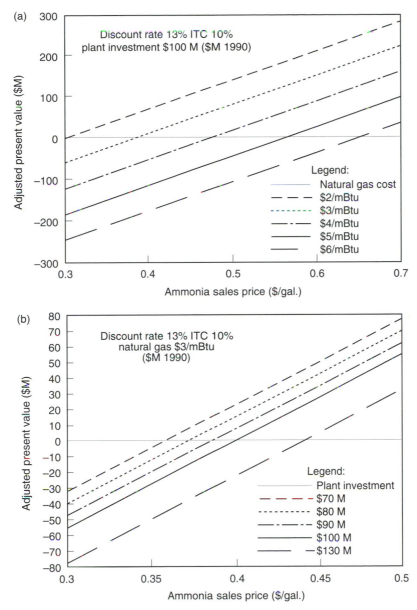

Fig. 8-4. Mossman profitability diagram for sale of 1000 mt/d of ammonia produced from natural gas. (a)Variable plant investment; (b) variable gas price.

tion; nuclear, 20.9%; natural gas, 8.3%; hydroelectric power, 8.3%; and petroleum products, 3.4% (*DOE/EIA Monthly Energy Review*, 1992). OTEC could be a significant, environmentally attractive, source of electric power by using OTEC-produced methanol or ammonia in fuel cells, as discussed in Chapter 7. Prices of the present and proposed sources of electric power with which it would have to compete are estimated in this section.

8.4.1 *Coal-Powered Steam Electric Plants with Flue Gas Desulfurization*

One kilogram of a typical coal with heating value of 14.8 MJ contains 0.88 kg of carbon and 0.018 kg of sulfur. Thus combustion of 1 mt of coal produces 3.2 mt of CO_2 and 96 kg of sulfur dioxide. Eighty to ninety percent of the sulfur dioxide can be removed by the flue gas desulfurization (FGD) process, with increases in the plant investment and operating costs. To produce the same heat output per kilogram with fuel oil would require only 0.33 kg of fuel oil and involve production of 2 mt of CO_2 per mt of fuel rather than 3.2 mt, with negligible sulfur dioxide production. These differences could lead the public utilities to favor the use of fuel oil or gas for electric power generation, but conversion to oil or gas greatly increases fuel costs and produces additional demands for oil imports. The costs of electric power generation by coal-powered steam plants with FGD (Yan and White, 1987) provide the data for the Mossman analysis shown in Figs. 8-5(a) and (b).

8.4.2 *Nuclear Steam Electric Power Plants (1000 MWe)*

Estimated costs of 1000-MWe nuclear power plants, based on estimated costs per kilowatt presented by Starr (1987), are as follows:

Plant investment	$1700–3000M
Fuel cost/year	$32M
Operating and maintenance cost/year	$42M
Delivery cost/year	$37M

The numbers do not include the cost of plant decommissioning, estimated as $1B for a 1000-MW plant (*Energy Daily*, 1990). Assuming a 30-year plant life and interest of 8%, the present value of the decommissioning cost is $100M. This is ignored in the present financial analysis because of the wide spread in the estimates of the plant investment. In addition, the costs of nuclear waste disposal are not included. No firm solution to this problem has been found, but programs seeking a solution were funded at $800M/y (1990). If prorated among the operating costs of the present nuclear plants, this would add approximately $8M/year to the operating costs of the typical plant.

With the data listed for plant investment, fuel cost, and operating and maintenance cost, excluding delivery cost, decommissioning, and waste disposal, the Mossman analysis gives the relationships shown in Fig. 8-6.

8.4.3 *Hydroelectric Power*

Hydroelectric power is the cheapest source of electricity and now provides 10% of U.S. electric power. No significant increase in output is foreseen, and alternative energy sources will not replace the current production.

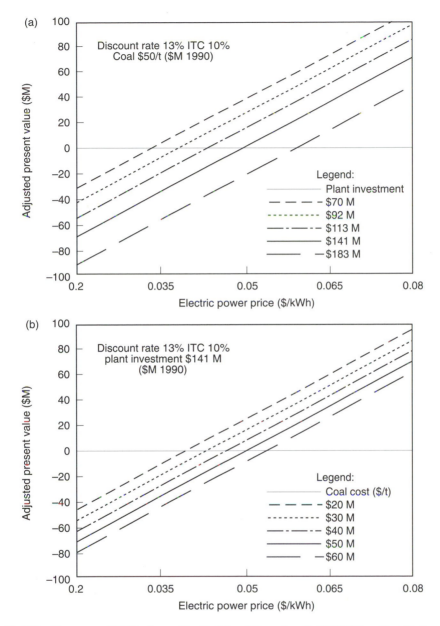

FIG. 8-5. Mossman profitability diagram for sale of electricity produced by 1000-MWe coal combustion power plants, with FGD. (a) Variable plant investment; (b) variable coal cost.

8.4.4   *Natural-Gas-Fueled Steam Electric Power*

Natural gas can substitute for coal in conventional coal-fired electric plants. The requirements and costs have been discussed by Yan and White (1987). The capital

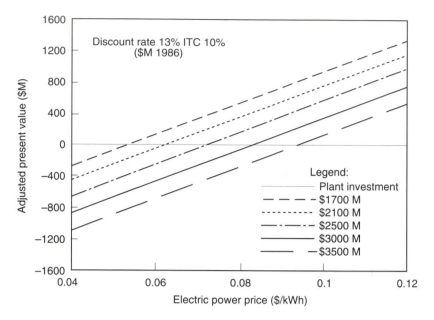

FIG. 8-6. Mossman profitability diagram for electric power produced in 1000-MWe nuclear power plants (variable plant investment).

cost is estimated at $780/kW (design capacity), and operating and maintenance at 2.6% of plant investment.

8.4.5 *Oil-Fueled Steam Electric Plants*

Fuel oil is a convenient source for steam-powered electric plants but has been more expensive than coal and therefore is used only in cases where the convenience outweighed the added cost. In the late 1970s construction of new oil plants in the United States was prohibited, as a means of reducing oil imports. The requirement for coal plants to incorporate flue gas desulfurization has reduced the cost advantage for coal, but it is unlikely that new oil plants will be built in the United States. The capital investments for the existing plants, excluding modifications to meet environmental requirements, have been amortized. Yan and White present a graph that indicates that electric power from these plants would cost in the range of $0.055–0.07/kW h for fuel oil prices of $20–30/bbl and capacity factors between 0.3 and 0.7 (Yan and White, 1987).

8.4.6 *Photovoltaic Electric Power Generation*

Photovoltaic cells convert solar energy directly into electricity, and could supply a significant fraction of electric power needs in the southwestern regions of the United States if system costs can be reduced to economically competitive levels. Energy conversion efficiencies of 10% or more of insolation are now being achieved. The annual 24-hour-average radiant energy at ground level in these regions is approximately 285 W/m^2 (Holl and Barron, 1989). Thus, if solar cell installations were designed that intercepted one-quarter of the incoming radiation, the electric power output of state-of-the-art photovoltaic cells would be 7 MW/km^2. This implies that the total electric power needs of Arizona (1000 MWe in 1988) could be supplied by photovoltaic installations occupying 140 km^2 (55 square miles), or 0.05% of the state area.

A recent study by the Jet Propulsion Laboratory of the California Institute of Technology predicts that a factory with a production of solar cells large enough in number to total 25 MWe/year of peak power capacity could provide 0.6 m × 1.2 m modules (2 × 4 ft) that would each produce 131 Wp at a cost of $197/m^2 and cost of $1120/kW of design capacity (Maycock, 1987). (Annual average photovoltaic power is one-quarter to one-fifth of peak power.) A photovoltaic power module would also require a support structure and power conversion equipment to enable it to be connected to a utility grid, which would furnish the backup power to maintain a desired load during daily cycles. Free-standing systems would contain batteries for this purpose. Installed costs for grid-connected photovoltaic systems are predicted by Maycock to be $5500/kW of design capacity in 1990 and to fall to $2500/kW by 1995. Experience with several megawatt-size installations has shown very high reliability and very low operation and maintenance costs (Jones, 1987; Risser and Stokes, 1987). Operating and maintenance costs were only $0.001/kW h at the 874-kW facility operated by the Sacramento Municipal Utility District. The Mossman analysis using these data is shown in Fig. 8-7.

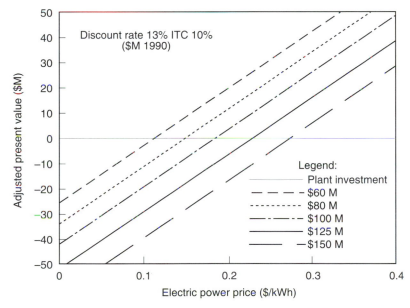

Fig. 8-7. Mossman profitability diagram for electricity produced in a 15-MWe photovoltaic power plant.

8.4.7 *Solar-Thermal Electric Systems*

Three types of solar-thermal electric concepts have received substantial development support. They are discussed in the following sections.

8.4.7.1   Central Receiver Systems

These systems employ an array of mirrors (heliostats) that focus solar energy on a central heat sink connected to a heat engine–electric-generator combination for electric power production. A system analysis based on the design of the 10-MWe installation at Barstow, California, led to an estimate of $362M for the first 100-MWe installation, dropping to $220M for the fiftieth installation (1980$). The efficiency of conversion of solar insolation on the heliostats to net electric power production was estimated to be 20% (Montague, 1981). An EPRI-sponsored review of the status of the technology was published by McDonnell–Douglas (Holl, 1989). Net efficiency was estimated at 8–17%. A cost range of $100–500M is estimated for installations of 100 MWe of design capacity.

8.4.7.2   Focused Dish Array

This system employs an array of parabolic mirrors with a small heat engine–electric generator at the focus of each mirror. It has demonstrated a 15% efficiency, and predicted efficiencies are in the range 16–28%. Development time of 10–20 years and cost of $100–200M are estimated for an installation of 100-MWe design capacity (Holl and Barron, 1989).

8.4.7.3   Cylindrical Mirror Assembly

This system has cylindrical one-axis parabolic mirrors that concentrate solar energy on an axial pipe at the focus that contains a heat-absorbing liquid. This is pumped to a central station where the hot liquid is used in a thermodynamic cycle to generate electric power (Roland and Kearney, 1986). Systems of this type designed by the Luz Engineering Corporation are in commercial production. The eighth unit with a design rating of 80 MWe (net) has recently been installed near Daggett, California, at a cost of approximately $200M. The power produced is supplied to the Southern California public utilities. Stored energy and supplemental natural gas are used to maintain the desired energy output when the solar power is inadequate to meet the demand. The ratio of annual power output to design power is 35%, of which 72% is supplied by solar-thermal energy. Energy costs are in the range of 8–12 cents/kW h (Jensen, 1990).

Mossman graphs for the three solar-thermal power systems are presented in Figs. 8-8(a), (b), and (c).

8.4.8 *Electric Power Generation by Wind Turbines*

Electric power generation by wind turbines has achieved commercial status and now has an installed capacity of over 1000 MWe in California alone. Average energy production is approximately 100 million kW h per month. Capital costs are

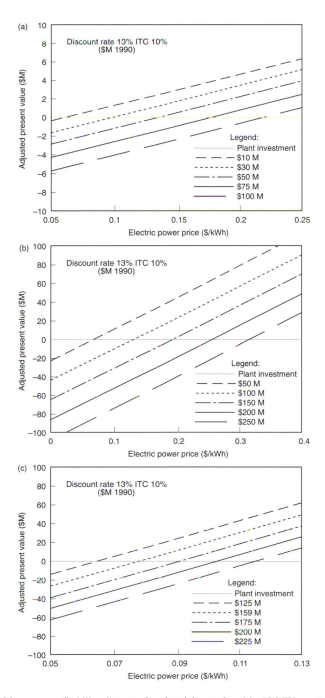

Fig. 8-8. (a) Mossman profitability diagram for electricity produced in 10-MWe central-receiver solar-thermal power plant. (b) Mossman profitability diagram for electricity produced in 100-MWe focused-dish solar-thermal power plant. (c) Mossman profitability diagram for electricity produced in 80-MWe cylindrical-mirror solar-thermal power plant.

in the range of $750–1000/kW of design capacity. Operating and maintenance costs are in the range of $0.008–0.012/kW h (Lynette, 1986). Wind turbines in Hawaii generated 90 MW h of electric power in December 1985 (Van Bibber and Andersen, 1986). These costs apply to grid-connected systems with power conditioning but without energy storage capabilities. Profitability estimates for wind power generation using the Mossman analysis are presented in Fig. 8-9.

8.4.9 *Fuel-Cell Electric Power Generation*

The use of fuel cells employing methanol or ammonia for electric power generation was discussed in Chapter 7. The application of this technology is independent of the source of the fuel. Whether methanol or ammonia produced by OTEC is used will depend on the cost relative to the costs of the fuels derived from natural gas. Mossman graphs showing the selling price of fuel-cell electric power derived from methanol or ammonia as feed stocks are presented in Figs. 8-10 and 8-11.

8.5   COMMENTS

The graphs presented in Chapters 7 and 8 provide information from which the competitive merits of alternative energy sources may be judged. The data for the alternative fuel systems presented in Chapter 7 are summarized in Table 8-1. The data for the representative alternative electric energy sources presented in Chapter 8 are summarized in Table 8-2. The use of the Mossman financial analysis for all of the systems allows an objective comparison to be made of the ability of each

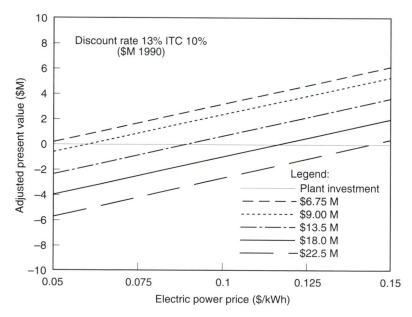

FIG. 8-9. Mossman profitability diagram for electricity produced in 9-MWe wind-turbine assembly in Hawaii.

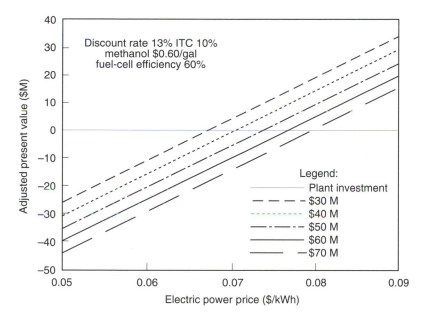

FIG. 8-10. Mossman profitability diagram for electricity produced in methanol fuel cells (variable plant investment).

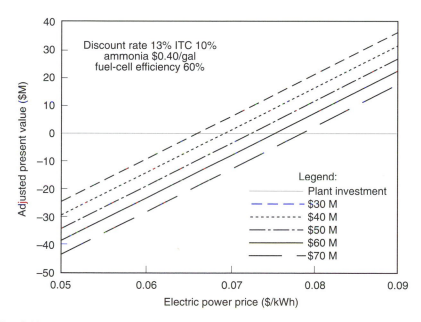

FIG. 8-11. Mossman profitability diagram for electricity produced in ammonia fuel cells (variable plant investment).

Table 8-1. Alternative motor fuels price comparison ($1990)[a] (ITC = 10%; APV = 0)

(1) Plant type	(2) PI ($M)	(3) Product	(4) Output (gal/y)	(5) Annual cost ($M/y)	(6) Sales price (APV = 0) ($/gal)	(7) Gasoline mpg equivalent ($/gal)	(8) Number of modules for 10^5 bbl/d gasoline equivalent	(9) Average plant investment ($B)	(10) Sales price (APV = 0) ($/gal)	(11) Gasoline equivalent price ($/gal)	(12) Investment for 10^5 bbl/d gasoline equivalent ($B)
OTEC CH$_3$OH (coal $50/t)	960	CH$_3$OH	1.99×10^8	51.1	0.70	1.27	13	0.72	0.620	1.12	9.3
OTEC NH$_3$	975	NH$_3$	1.70×10^8	25.6	0.74	1.98	24	0.70	0.540	1.44	16.9
Shale oil	225	HC fuel	1.47×10^7	7.5	1.51	1.51	83	0.14	0.662	0.66	12.0
Shell-FT coal liquefaction (coal $22.7/t)	4400	HC fuel	1.28×10^9	450	0.76	0.76	1	4.32	0.762	0.76	5.2
NG to CH$_3$OH (NG $3/mBtu)	237	CH$_3$OH	2.44×10^8	117	0.58	1.04	6.3	0.19	0.461	0.83	1.7
NG to NH$_3$ (NG $3/mBtu)	100	NH$_3$	1.53×10^8	44	0.37	1.02	26	0.074	0.337	0.93	1.9

[a] Does not include carbon tax or automobile modification costs.

Table 8-2. Alternative electricity sources price comparison (Does not include carbon tax) (ITC = 10%; APV = 0)

(1) Plant type	(2) Fuel	(3) Fuel cost ($/GJ)	(4) Plant investment ($M)	(5) Output (kWh/y)	(6) Annual cost ($M/y)	(7) Break-even sales price ($/kWh)	(8) Number of modules for 10^{10} kWh/y	(9) Sales price ($/kWh)	(10) Investment for 10^{10} kWh/y ($B)
Coal steam with FGD	Coal	2.2	141	6.1×10^8	20.1	0.055	16	0.047	1.7
Nuclear	Uranium	7.0	2350	5.3×10^9	84.0	0.069	2	0.063	4.5
Photovoltaics	Solar actinic energy	0.0	125	5.9×10^7	1.25	0.229	169	0.137	12.4
Solar-thermal electric									
Central receiver	Solar heat	0.0	70	3.1×10^7	2.30	0.288	326	0.164	12.5
Focused dish	Solar heat	0.0	200	8.8×10^7	12.0	0.258	114	0.160	13.9
Cylindrical mirror	Solar heat	0.0	225	2.5×10^8	6.8	0.117	46	0.071	6.3
Wind	Wind	0.0	18	1.6×10^7	0.16	0.118	633	0.064	5.8
Methanol fuel cell	CH_3OH	10.5	50	3.9×10^8	28.5	0.098	25	0.073	0.89[a]
		6.7	50	3.9×10^8	16.4	0.056	25	0.051	0.89[a]
Ammonia fuel cell	NH_3	12.7	50	3.9×10^8	29.9	0.090	25	0.085	0.89[a]
		8.4	50	3.9×10^8	20.1	0.065	25	0.060	0.89[a]

[a]Includes only the cost of the fuel-cell fuel plant.

system to compete with currently available sources of fuels and electric power, based on the quoted values of the plant investment, and operating and maintenance charges. The range of values of plant investment listed for each alternative in Figs. 7-1 to 7-6 and Figs. 8-1 to 8-11 enable one to judge how improvements or deficiencies in performance or cost, which might result from further research and development or manufacturing experience, would affect the profitability of a particular option.

The energy systems discussed in this and the preceding chapter are not of uniform size. Since construction costs are reduced as component size and numbers increase, it is desirable to estimate projected product prices based on the same total output for each system compared. A total system capable of supplying 100,000 barrels per day of fuel equivalent to gasoline in miles per gallon has been chosen arbitrarily for fuel-production cost estimates. This amount is roughly 1% of daily U.S. oil imports. Electrical systems designed to produce a total output of 10 billion kilowatt hours per year are compared. Total electrical power production in the United States in 1990 was approximately 2800 billion kilowatt hours.

For the estimates listed in columns 8 and 9 of Tables 8-1 and 8-2, a conservative value of 0.93 for the experience exponent has been selected for all of the options. This gives a basis for judging the importance of the scaling and experience factors in comparing the options discussed.

Starting with the plant investment for a particular option, shown in column 2 of Table 8-1, and plant output (module size) shown in column 4, column 8 lists the number of modules needed to give the desired system output. The number of doublings needed to reach this number gives the exponent of the experience factor that is used to calculate the lower plant investment achieved with larger production. The new sales price of product and equivalent gasoline price are then estimated as shown in columns 10 and 11. The last column lists the estimated investment for the total system of a particular type. In Table 8-2 the number of modules needed to generate the desired electric power output is shown in column 8, the sales price of electricity in column 9, and the total investment in column 10.

McGurl (1990) quotes a plant investment of $225M for a 2500-ton/d coal liquefaction plant and $585M for a 12,500-ton/d plant. Thus for this system the scaling factor is estimated to be 0.6 (that is, a fivefold increase in size increases the cost by a factor of $2.6 = 5^{0.6}$). U.S. Navy experience in construction of multiple ship orders has shown that an experience factor of 0.93, beginning with the second plant, is appropriate for ship construction (Fink, 1987).

In Tables 8-1 and 8-2 all of the estimates for the costs of multiple units are based on the use of an experience factor of 0.93. Although the use of these factors in actual construction estimates would be replaced by values based on engineering judgments for the particular option, they are worthwhile for the present comparison.

For the OTEC options it is assumed that a size of 200 MWe (nom) will be optimum for OTEC methanol plantships, and 364 MWe (nom) for ammonia plantships. Identical modules will be produced to give the required total annual output. For the land-based liquid-fuel systems it is assumed that plant size will be scaled up to 100,000 bbl/d gasoline equivalent by constructing identical plants to yield the required total output. Identical 1000-MWe nuclear plants are assumed for the calculation.

The effects on the national economy of including the fuel and electric power production options discussed here are presented in Chapter 9.

REFERENCES

Babbitt, J. F. (1981). *Closed cycle ocean thermal energy conversion pilot plant—Grazing ammonia plantship.* SOLARAMCO, Feb.

Chemical Marketing Reporter, 1991. "Chemical prices." *Chemical Marketing Reporter*, New York, N.Y. Verbal quotation.

DOE/EIA Monthly Energy Review, 1989. Washngton, D.C.: U.S. Department of Energy. Tables 9.1, 9.4, 9.6 (annual average values).

——, 1991. Washington, D.C.: U.S. Department of Energy.

——, 1992. Washington, D.C.: U.S. Department of Energy. Table 7.1.

Energy Daily, 1989. "DOE unveils 5 year clean-up plan." *The Energy Daily*, Washington, D.C.: The King Publishing Group. **17**, no. 167.

——, 1990. "DOE to get big budget hikes for cleaning up defense programs." *The Energy Daily*, Washington, D.C.: The King Publishing Group. **18**, no. 8.

Fink, L., 1987. U.S. Maritime Administration (personal communication to W. H. Avery).

Gray, D., and G. Tomlinson, 1989. *A review of the economic estimates of coal-derived transportation fuels documented in the National Research Council report.* MITRE Corp., SAND-89-7089.

Holl, R. J., 1989. *Molten salt solar electric experiment. Volume 1: Testing, operation and evaluation.* EPRI, GS-6577.

——, and D. R. Barron, 1989. *Status of solar thermal electric technology.* EPRI, GS-65.

International Energy Agency, 1990. *Energy and environment: policy overview.* Paris, France: International Energy Agency. **18**, no. 24.

Jensen, C. B., 1990. LUZ Corporation, Daggett, Cal. (personal communication to W. H. Avery).

Jones, G. J., 1987. "Photovoltaic system technology." *Proc. Energy Technology Conf. XIV*, Government Institutes, Inc., Rockville, Md., 899.

Lynette, R., 1986. "The California wind farm experience." *Proc. Energy Technology Conf. XIII*, Government Institutes, Inc., Rockville, Md., 485.

Maize, K., 1990. "OXY MIS oil shale project wins support in subcommittee." *Energy Daily* **18**, no. 149.

Maycock, P. D., 1987. "Photovoltaic technology: Installed system and module cost." *Proc. Energy Technology Conf. XIV*, Government Institutes, Inc., Rockville, Md., 893.

McGurl, G. V., 1990. *Assessment of costs and benefits of flexible and alternative fuel use in the U.S. transportation sector.* DOE Pittsburgh Research Center. Center Report.

Montague, J. E., 1981. "Economic viability of solar thermal central-receiver power plants." *Proc. Energy Technology Conf. VIII*, Government Institutes, Inc., Rockville, Md., 1288.

Risser, V. V., and K. W. Stokes, 1987. *Photovoltaic field test performance assessment.* EPRI, GS-6251.

Roland, J. R., and D. W. Kearney, 1986. "The world's largest solar electric power plant." *Proc. Energy Technology Conf. XIII*, Government Institutes, Inc., Rockville, Md., 449.

Starr, C., 1987. "Technology innovation and energy futures." *Proc. Energy Technology Conf. XIV*, Government Institutes, Inc., Rockville, Md., 21.

Tonelson, A., and A. K. Hurd, 1990. Reported in the *N.Y. Times*, September 4, 1990.

Woods, L. M., 1986. "Fuels from tar sands." *Proc. Energy Technology Conf. XIII*, Government Institutes, Inc., Rockville, Md., 1419.

Van Bibber, L. E., and T. S. Andersen, 1986. "Hawaii 9-MWe wind farm: fifteen 600 kW machines." *Proc. Energy Technology Conf. XIII*, Government Institutes, Inc., Rockville, Md., 493.

Yan, T. Y., and J. R. White, 1987. "Economics of electric power generation options and sulfur dioxide emission control." *Proc. Energy Technology Conf. XIV*, Government Institutes, Inc., Rockville, Md., 102.

9

ECONOMIC, ENVIRONMENTAL, AND SOCIAL ASPECTS OF OTEC IMPLEMENTATION

9.1 ECONOMIC AND SOCIAL BENEFITS OF OTEC COMMERCIAL-IZATION

The financial analyses presented in Chapters 7 and 8 indicate that commercial development of OTEC will have a significant impact on the economics of U.S. energy production and use. Two scenarios for commercial development are examined in this section:

1. Development of OTEC methanol capacity sufficient to replace all U.S. gasoline produced from imported oil.
2. Development of OTEC ammonia capacity sufficient to replace all gasoline used in U.S. transportation.

9.1.1 *OTEC Methanol Commercialization*

Commercialization of this option implies a project goal to produce methanol plantships with enough total methanol capacity to replace the gasoline used in the United States that is now produced from imported petroleum, 47 billion gallons of gasoline in 1990 (DOE/EIA, 1990). This would require a total of 427 200-MWe plantships, each producing 199 million gallons of methanol per year (1.8 gallons of methanol give the same automobile mileage as 1 gallon of gasoline.

9.1.1.1   Plant Investment

We assume financing based on an initial nominal plant investment of $960M (1990$) and an eighth plant investment of $664M. With repeated manufacture, the cost will be reduced to $438M for the 427th plantship, assuming that an experience exponent of 0.93 applies for all production of identical plantships after the first three. The average plant investment for the total production is then $507M.

9.1.1.2   Schedule

If financial support is maintained to complete the program, the year 2020 is a reasonable target date for achieving the full fuel production capacity. This implies construction of OTEC plantships at an average rate of 17 per year after commercial production is established. This rate could be accommodated in U.S. shipyards with

feasible modifications to satisfy specific OTEC requirements. The U.S. shipbuilding facilities are discussed in Section 4.1.

9.1.1.3   Investment Requirements and Cost Tradeoffs

In addition to the investments required for OTEC, methanol automobiles must be
in production, and distribution systems for methanol must be installed. The
associated costs must be included in the financial analysis. Offsetting these costs are
the savings resulting from:

1. Large improvements in the U.S. balance of trade through elimination of oil
    imports.
2. Tax receipts accruing from reinvigorated U.S. shipbuilding and associated
    manufacturing industries.
3. Economic benefits of stabilized world fuel prices.
4. Savings from reduced expenditures necessary in present refineries and motor
    vehicles to meet regional requirements to control air and water pollution.
5. Reduction in carbon emissions compared to gasoline (see footnote to Table
    9-1).

Additional benefits will be realized by using methanol fuel cells for electric
power generation. Adoption of this technology will eliminate the costs associated
with nuclear waste disposal, and the investments that will otherwise be needed to
replace nuclear power plants that become obsolete or unsafe to operate.

Savings will also result from replacing coal-fired power plants with low-
pollution methanol fuel cells, which will reduce environmental costs and can
compensate for lower fuel costs of coal-fired power generation.

A quantitative determination of the cost benefits is difficult to make and is
subject to argument, often spirited. Information is available, however, that can be
used to make approximate estimates of the potential future role of OTEC relative
to other energy alternatives (Freudmann, 1988; Commission on Engineering and
Technical Systems, 1990).

A balance sheet consistent with estimates reported by fuel and power producers, and by groups oriented toward preservation of the world environment, is
presented in Table 9-1. In this table the net impact on the U.S. economy of OTEC
methanol commercialization is estimated. The table includes the economic effects
of commercial investments in OTEC plantships, including an assessment of the
economic benefits of reduced oil imports and the social benefits of reduction of
carbon emissions to the atmosphere. For this calculation the net present value
(NPV) is judged to be a suitable financial measure of the overall economic and
social impacts. All equity financing is assumed, and no tax benefits or government
subsidies are included. A positive value of the NPV equates to an overall benefit to
the national economy. The estimates are based on 1990 dollars and assume zero
inflation. Details of the financing assumptions are given in the notes accompanying
the table.

The calculation of OTEC methanol selling price, based on the Mossman
analysis using nominal first plant cost, shows that the production of plantships will

Table 9-1. Economic impact of replacing U.S. gasoline requiring oil imports, with methanol produced on OTEC plantships (discount rate = 13%; construction interest = 12%; ITC = 0; NPV = 0)

Plant no.	Relative cost	Sum	nth CH₃OH plant PI[a] ($M)	CH₃OH total PI ($B)	Avg. PI ($M)	Total CH₃OH prod.[b] (Bgal/y)	CH₃OH sales price[c] ($/gal)	Equiv. gasoline price[d] ($/gal)	Gasoline market price[e] ($/gal)	Gasoline replaced (Bgal/y)	Gasoline "real price" with import impact[f,g] ($/gal)	CH₃OH equiv. price incl. auto costs[h] ($/gal)	Savings using CH₃OH to replace gasoline ($B/y)	Gasoline price incl. import and carbon tax[i] ($/gal)	CH₃OH price incl. auto cost and carbon tax[i] ($/gal)	Net saving to U.S. economy in replacing gaso. by CH₃OH[j] ($B/y)
1	1.000	1.0	960	0.96	960	0.199	1.109	1.996	0.786	0.111	1.358	2.056	−0.077	1.531	2.214	−0.075
2	0.800	1.8	768	1.73	864	0.398	1.024	1.843	0.786	0.221	1.358	1.903	−0.120	1.531	2.061	−0.117
3	0.800	2.6	768	2.50	832	0.597	0.996	1.793	0.786	0.332	1.358	1.852	−0.164	1.531	2.010	−0.159
4	0.744	3.3	714	3.21	803	0.796	0.970	1.746	0.786	0.442	1.358	1.806	−0.198	1.531	1.964	−0.191
8	0.692	6.3	664	6.07	758	1.592	0.930	1.674	0.786	0.884	1.358	1.734	−0.332	1.531	1.892	−0.319
16	0.643	11.9	618	11.38	711	3.184	0.889	1.600	0.786	1.769	1.358	1.660	−0.534	1.531	1.818	−0.507
32	0.598	22.2	575	21.27	665	6.368	0.848	1.526	0.786	3.538	1.358	1.586	−0.806	1.531	1.744	−0.753
64	0.557	41.3	534	39.65	620	12.736	0.809	1.456	0.786	7.076	1.358	1.516	−1.116	1.531	1.674	−1.010
128	0.518	76.9	497	73.84	577	25.472	0.771	1.388	0.786	14.151	1.358	1.447	−1.264	1.531	1.605	−1.051
256	0.481	143.2	462	137.45	537	50.944	0.735	1.323	0.786	28.302	1.358	1.383	−0.693	1.531	1.541	−0.269
427	0.456	225.5	438	216.47	507	84.973	0.709	1.276	0.786	47.207	1.358	1.336	1.053	1.531	1.494	1.761
512	0.448	266.4	430	255.74	500	101.888	0.703	1.265	0.786	56.604	1.358	1.325	1.874	1.531	1.483	2.723

[a] The nominal plant investment (PI) for the first plantship is $960M. The PI is calculated from the cost data listed in Table 7-2 assuming discount rate 13%, construction interest 12%. The PI for the second and third plants is 0.8 times the first plant cost. Thereafter, the plant investment decreases by the factor 0.93 for each doubling of the number produced.

[b] The 200-MWe OTEC CH₃OH plantship yield = 1.99×10^8 gal/y of fuel-grade CH₃OH.

[c] The sales price of methanol is calculated using the Mossman financial analysis with investment and income tax credit of 0%. There is no depreciation allowance. The price is determined by the requirement that the NPV $\geqq$ 0. Note that the lower sales prices for profitability quoted in Chapters 8 and 9 are based on the adjusted present value (APV) and assume that an investment tax credit (ITC) of 10% will be available.

[d] In automobile miles/gal, 1.8 gallons CH₃OH = 1.0 gallon gasoline. This is based on operating experience.

[e] The 1990 gasoline price = $0.786/gal (1990 avg. without taxes) (DOE/EIA, Oct. 1991, Table 9.6).

[f] The average U.S. gasoline supply in 1990 was 6.96×10^9 bbl/d = 106.6 billion gallons/year (Bgal/y). The crude oil fraction imported was 44.4% or the equivalent of 47.4 Bgal/y of imported gasoline (DOE/EIA, Oct. 1991, Tables 3.2a, 4). To supply methanol fuel that would give the same automobile mileage requires $47.2 \times 1.8/0.199 = 427$ OTEC CH₃OH plantships.

[g] The average import cost of crude oil (1990) was $21.78/bbl = $0.518/gal. The market price of gasoline was $0.786/gal (1990 avg) (DOE/EIA, Oct. 1991). The refinery markup = $(0.786 − 0.518)/gal = $0.270/gal. The "real cost" of imported oil to the U.S. economy = 2.1 × market price (National Defense Council Foundation, abstracted by Freudmann,1988). Therefore, the real cost of crude was $21.78/bbl × 2.1 = $45.7/bbl = $1.088/gal. Thus, the real cost of gasoline derived from imported oil was $1.088 + 0.270)/gal = $1.358/gal versus the market price of $0.786/gal.

[h] A study by the Commission on Engineering and Technical Systems (1990) arrived at an added cost of $2.5/bbl ($0.0595/gal) oil equivalent to adapt automobiles in production to use methanol as a fuel in place of gasoline.

[i] A tax of $70 per ton of carbon emitted was recommended by the United Nations Intergovernmental Panel on Climate Change (1984). This corresponds to $0.173/gal of gasoline used, and $0.158/gal of methanol for equivalent automobile mileage.

[j] It is assumed that by 2030 all U.S. gasoline will be priced at a value equivalent to the 1990 real cost of imported oil.

have a small negative value of the NPV until 256 are produced, after which the national economy is benefited.

The nearly neutral impact of OTEC commercialization on the national economy, shown in Table 9-1, suggests examination of the profitability of industrial production of OTEC plantships if a government commitment is made to support OTEC via a 10% investment tax credit. The financing rationale of Mossman that was explained in Chapter 7 is used. The results are presented in Table 9-2.

According to the analysis, methanol plantship production will be profitable beginning with the third plantship if the economic gains of reducing oil imports and the environmental benefits of lowered carbon emissions are quantified and included. When the total production of 427 plantships needed to replace imported oil is operational, the net profit to the investors using the Mossman financing plan will be $18B/y after taxes.

9.1.2 OTEC Ammonia Commercialization

Liquid ammonia is an automobile fuel that has a high octane number (111) and burns efficiently in automobile engines. It produces only water and nitrogen as products of combustion and can meet future U.S. transportation requirements for a technically feasible, nonpolluting automobile fuel that does not produce CO_2.

The physical properties of ammonia are nearly the same as those of liquid propane (bottled gas), and the requirements for automobile storage and handling will be similar. It is estimated that 2.67 gallons of liquid ammonia will give the same automobile mileage as a gallon of gasoline. A 368-MWe OTEC ammonia plantship will produce 170 million gallons of liquid ammonia fuel per year. Thus, 1681 plantships will supply enough fuel to replace the 107 billion gallons of gasoline used in the United States in 1990. The nominal plantship investment is $975M. The average investment per ammonia plantship for the total production would be $446M.

9.1.2.1   Schedule

A national goal to replace all U.S. gasoline and ammonia fuel, by the year 2050, could be established to comply with world desires to keep carbon emissions low enough to stabilize CO_2 buildup in the atmosphere. This would require an annual production of 30 ammonia plantships. A reasonable extension of U.S. shipbuilding capabilities would accomplish this.

9.1.2.2   Financing

The costs and benefits of a large-scale ammonia fuel program are similar to those outlined in the discussion of OTEC methanol financing. The principal differences are that replacement of gasoline with ammonia fuel would entirely eliminate carbon emissions, with a corresponding economic benefit. A larger automobile adaptation cost than that required for methanol fuel would be associated with storage, handling, and use of the liquefied ammonia fuel. Ammonia can also be used in fuel cells for electricity generation.

Table 9-2. Profit to investors replacing U.S. gasoline requiring oil imports, with methanol produced on OTEC plantships
(discount rate = 13%; construction interest = 12%; ITC = 10%; APV = 0)

Plant no.	Relative cost	Sum	nth CH$_3$OH plant PI[a] ($M)	CH$_3$OH total PI ($B)	Avg. PI ($M)	Total CH$_3$OH prod.[b] (Bgal/y)	CH$_3$OH sales price[c] ($/gal)	Equiv. gasoline price[d] ($/gal)	Gasoline market price[e] ($/gal)	Gasoline replaced[f] (Bgal/y)	Gasoline "real price" with import impact[g] ($/gal)	CH$_3$OH equiv. price incl. auto costs[h] ($/gal)	Savings using CH$_3$OH to replace gasoline ($B/y)	Gasoline price incl. import and cost and carbon tax[i] ($/gal)	CH$_3$OH price incl. auto cost and carbon tax[i] ($/gal)	Net profit to investors replacing gaso. by CH$_3$OH[j] ($B/y)
1	1.000	1.00	960	0.96	960	0.20	0.75	1.34	0.79	0.11	1.36	1.40	0.00	1.53	1.56	−0.003
2	0.800	1.80	768	1.73	864	0.40	0.70	1.25	0.79	0.22	1.36	1.31	0.01	1.53	1.47	0.013
3	0.800	2.60	768	2.50	832	0.60	0.68	1.23	0.79	0.33	1.36	1.29	0.02	1.53	1.44	0.029
4	0.744	3.34	714	3.21	803	0.80	0.67	1.20	0.79	0.44	1.36	1.26	0.04	1.53	1.42	0.051
8	0.692	6.32	664	6.07	758	1.6	0.64	1.15	0.79	0.88	1.36	1.21	0.13	1.53	1.37	0.143
16	0.64	11.9	618	11.4	711	3.2	0.62	1.12	0.79	1.77	1.36	1.18	0.32	1.53	1.33	0.35
32	0.60	22.2	575	21.3	665	6.4	0.60	1.07	0.79	3.54	1.36	1.13	0.79	1.53	1.29	0.85
64	0.56	41.3	534	39.6	620	13	0.58	1.04	0.79	7.08	1.36	1.09	1.86	1.53	1.25	1.97
128	0.52	76.9	497	73.8	577	25	0.55	1.00	0.79	14.15	1.36	1.05	4.29	1.53	1.21	4.50
256	0.48	143	462	137	537	51	0.53	0.96	0.79	28.30	1.36	1.02	9.60	1.53	1.18	10.0
427	0.46	225	438	216	507	85	0.52	0.93	0.79	47.21	1.36	0.99	17.3	1.53	1.15	18.0
512	0.45	266	430	256	500	102	0.51	0.93	0.79	56.60	1.36	0.98	21.1	1.53	1.14	22.0

[a] The nominal plant investment (PI) for the first plantship is $960M. The PI is calculated from the cost data listed in Table 7-2 assuming discount rate 13%, construction interest 12%. The PI for the second and third plants is 0.8 times the first plant cost. Thereafter, the plant investment decreases by the factor 0.93 for each doubling of the number produced.

[b] The 200-MWe OTEC CH$_3$OH plantship yield = 1.99 × 10^8 gal/y of fuel-grade CH$_3$OH.

[c] The sales price of methanol is calculated using the Mossman financial analysis with investment tax credit (ITC) of 10%. Note that the lower sales prices for profitability quoted in Chapters 8 and 9 are based on the adjusted present value (APV) and assume that an ITC of 10% will be available.

[d] In automobile miles/gal, 1.8 gallons CH$_3$OH = 1.0 gallon gasoline. This is based on operating experience.

[e] The 1990 gasoline price = $0.786/gal (1990 avg. without taxes) (DOE/EIA, Oct. 1991, Table 9.6).

[f] The average U.S. gasoline supply in 1990 was 6.96 × 10^6 bbl/d = 106.6 Bgal/y. The crude oil fraction imported was 44.4% or the equivalent of 47.4 Bgal/y of imported gasoline (DOE/EIA, Oct. 1991. Tables 3.2a, 4). To supply methanol fuel that would give the same automobile mileage requires 47.2 × 1.8/0.199 = 427 OTEC CH$_3$OH plantships.

[g] The average import cost of crude oil (1990) was $21.78/bbl = $0.518/gal. The market price of gasoline was $0.786/gal (1990 avg) (DOE/EIA, Oct. 1991). The refinery markup = $(0.786 − 0.518)/gal = $0.270/gal. The "real cost" of imported oil to the U.S. economy = 2.1 × market price (National Defense Council Foundation, abstracted by Freudmann,1988). Therefore, the real cost of crude was $21.78/bbl × 2.1 = $45.7/bbl = $1.088/gal. Thus, the real cost of gasoline derived from imported oil was $1.088 + 0.270/gal = $1.358/gal versus the market price of $0.786/gal.

[h] A study by the Commission on Engineering and Technical Systems (1990) arrived at an added cost of $2.5/bbl ($0.0595/gal) oil equivalent to adapt automobiles in production to use methanol as a fuel in place of gasoline.

[i] A tax of $70 per ton of carbon emitted was recommended by the United Nations Intergovernmental Panel on Climate Change (1984). This corresponds to $0.173/gal of gasoline used, and $0.158/gal of methanol for equivalent automobile mileage.

[j] It is assumed that by 2030 all U.S. gasoline will be priced at a value equivalent to the 1990 real cost of imported oil.

The financial evaluation of a program designed to meet a goal of supplying OTEC ammonia fuel to replace all gasoline used in U.S. automobiles is presented in Table 9-3. Details of the financial assumptions are given in the notes accompanying the table.

The overall economic impact of commercial production of OTEC NH_3 plantships shows a negative NPV. In this case the NPV rises to a net loss to the national economy of 34B/y when the production goal of 1680 plantships is reached.

An analysis of the financial returns to investors of carrying out the OTEC ammonia program with private funds, based on the Mossman rationale, is presented in Table 9-4. This shows that the program has a negative adjusted present value (APV) averaging $300M/y until the 500th plant is produced, after which the profitability rises to a value of $23B/y, when the production goal is reached.

9.1.2.3   Conclusions

Several significant conclusions about the feasibility of establishing an OTEC commercial industry are evident from the analyses presented. Their validity is supported by production and operational experience gained in the U.S. and foreign shipbuilding and chemical engineering industries.

1. Economic and social gains can be achieved at modest cost to the United States by large-scale substitution of OTEC methanol for gasoline made from imported oil, as shown in Table 9-1. The program will be financially attractive to investors with customary investment incentives, netting investors $18B/y when plantship production reaches 427 plantships, as shown in Table 9-2.
2. Large social gains resulting from the elimination of carbon emissions from U.S. automobiles can be achieved at reasonable cost by replacing gasoline with ammonia fuel (Table 9-3). An OTEC program that replaces all gasoline used in U.S. automobiles by ammonia, with customary investment incentives, will be financially attractive, netting investors a return of $23B per year when the total production is established (Table 9-4).
3. The cost estimates are sensitive to the values judged to be reasonable for discount rates and construction interest. For example, as shown in Tables 9-5 to 9-8, use of a discount rate of 11% and construction interest of 10% in the Mossman analysis yields a net economic value of $9B/y to the national economy for the 427 methanol plantship production, and profit to the investors of $22B/y. For ammonia plantship production the economic value (NPV) is slightly negative with the lower interest rates until the 500th plantship is produced and then rises to $11B/y when plantship number 1681 is delivered. The annual profits to investors (APV) is $38B at this time. For this case, the difference in savings to the national economy associated with the lower discount rate is $45B/y when full plantship production is achieved, and the net difference to investors increases to $38B/y.

The large difference in the financial benefits associated with the small differences in the assumed interest rates is important to note. The lower interest rates are typical of recent values (1990), but rates will change to higher or lower

Table 9-3. Economic impact of replacing U.S. gasoline with ammonia automobile fuel produced on OTEC plantships
(discount rate = 13%; ITC = 0; NPV = 0)

Plant no.	Relative cost	Sum	nth NH₃ plant investment[a] ($M)	NH₃ total PI ($B)	Avg. PI ($M)	Total NH₃ prod.[b] (Bgal/y)	NH₃ sales price[c] ($/gal)	Equiv. gasoline price[d] ($/gal)	Gasoline market price[e] ($/gal) (1990 avg)	Gasoline replaced[f] (Bgal/y)	Gasoline "real price" with import impact[g] ($/gal)	NH₃ equiv. price incl. auto costs[h] ($/gal)	Gasoline price incl. import and carbon tax[i] ($/gal)	Saving to U.S. economy using NH₃ to replace gasoline[j] ($B/y)
1	1.000	1.00	975	0.98	975	0.17	1.17	3.12	0.79	0.06	1.36	3.42	1.53	−0.12
2	0.800	1.80	780	1.76	878	0.34	1.06	2.84	0.79	0.13	1.36	3.14	1.53	−0.20
3	0.800	2.60	780	2.54	845	0.51	1.03	2.74	0.79	0.19	1.36	3.04	1.53	−0.29
4	0.744	3.34	725	3.26	815	0.68	0.99	2.65	0.79	0.25	1.36	2.95	1.53	−0.36
8	0.692	6.32	675	6.16	770	1.36	0.94	2.51	0.79	0.51	1.36	2.81	1.53	−0.65
16	0.643	11.9	627	11.6	722	2.72	0.89	2.37	0.79	1.02	1.36	2.67	1.53	−1.2
32	0.598	22.2	583	21.6	675	5.44	0.84	2.23	0.79	2.04	1.36	2.53	1.53	−2.0
64	0.557	41.3	543	40.3	629	10.9	0.78	2.09	0.79	4.07	1.36	2.39	1.53	−3.5
128	0.518	76.9	505	75.0	586	21.8	0.74	1.96	0.79	8.15	1.36	2.26	1.53	−6.0
256	0.481	143	469	140	545	43.5	0.69	1.84	0.79	16.3	1.36	2.14	1.53	−10.0
512	0.448	266	436	260	507	87.0	0.65	1.73	0.79	32.6	1.36	2.03	1.53	−16.2
1024	0.416	496	406	483	472	174	0.61	1.62	0.79	65.2	1.36	1.92	1.53	−25.6
1681	0.395	769	385	750	446	286	0.58	1.55	0.79	107	1.36	1.85	1.53	−33.7

[a] The nominal plant investment (PI) for the first plantship is $975M. The PI is calculated from the cost data listed in Table 7-2 assuming discount rate 13%, construction interest 12%. The PI for the second and third plants is 0.8 times the first plant cost. Thereafter, the plant investment decreases by the factor 0.93 for each doubling of the number produced.

[b] The 368-MWe OTEC NH₃ plantship yield of fuel-grade NH₃ is 1.70×10^9 gal/y.

[c] The sales price of ammonia is calculated using the Mossman financial analysis with investment and income tax credit of 0%. There is no depreciation allowance. The price is determined by the requirement that the NPV ≧ 0.

[d] In automobile miles/gal, 2.67 gallons NH₃ = 1.0 gallon gasoline, assuming the same benefit from high octane number that was found for CH₃OH.

[e] The 1990 gasoline price = $0.786/gal (1990 avg, without taxes) (DOE/EIA, Oct. 1991, Table 9.6).

[f] The average U.S. gasoline supply in 1990 was 6.96×10^6 bbl/d = 107 Bgal/y (DOE/EIA, Oct. 1991, Tables 3.2a, 4). To supply ammonia fuel that would give the same total automobile mileage requires (107×2.67)/0.17 = 1681 OTEC NH₃ plantships.

[g] The average import cost of crude oil (1990) was $21.78/bbl = $0.518/gal. The "real cost" of imported oil to the U.S. economy = $2.1 \times$ market price (National Defense Council Foundation, abstracted by Freudmann, 1988). Therefore, the real cost of imported oil was $45.7/bbl ($1.088/gal). The market price of gasoline was $0.786/gal (1990 avg) (DOE/EIA, Oct. 1991). Thus there was a refinery markup between oil cost and the sales price of $(0.786 − 0.518)/gal = $0.270/gal. The "real" price of imported oil was, then, $(0.278 + 1.088)/gal = $1.358/gal.

[h] A study by the Commission on Engineering and Technical Systems (1990) arrived at an added cost of $11.1/bbl ($0.254/gal oil equivalent) to adapt automobiles in production to use compressed natural gas as a fuel to replace gasoline. Ammonia fuel would have simpler storage and less volume but would need a modified fuel system. This suggests a cost increment of $0.30/gal (gasoline equivalent) above the gasoline market price as the ammonia fuel price for automobile use.

[i] A tax of $70 per ton of carbon emitted in energy production was recommended by the United Nations Intergovernmental Panel on Climate Change (1984). This corresponds to $0.173/gal of gasoline used in transportation versus zero dollars per gallon of ammonia for equivalent automobile mileage.

[j] It is assumed that by 2030 all U.S. gasoline will be priced at a value equivalent to the 1990 real cost of imported oil.

Table 9-4. Profit to investors replacing gasoline used in U.S. automobiles with ammonia vehicle fuel (discount rate = 13%; construction interest = 12%; ITC = 10%; APV = 0)

Plant no.	Relative cost	Sum	nth NH$_3$ plant investment[a] ($M)	NH$_3$ total PI[b] ($B)	Avg. PI ($M)	Total NH$_3$ prod.[b] (Bgal/y)	NH$_3$ sales price[c] ($/gal)	Equiv. gasoline price[d] ($/gal)	Gasoline market price[e] ($/gal) (1990 avg)	Gasoline replaced[f] (Bgal/y)	Gasoline "real price" with import impact[g] ($/gal)	NH$_3$ equiv. price incl. auto costs[h] ($/gal)	Gasoline real cost plus carbon tax[i] ($/gal)	Profit to investors using NH$_3$ to replace gasoline[j] ($B/y)
1	1.000	1.00	975	0.98	975	0.17	0.743	1.98	0.79	0.06	1.36	2.28	1.53	-0.05
2	0.800	1.80	780	1.76	878	0.34	0.673	1.98	0.79	0.13	1.36	2.28	1.53	-0.10
3	0.800	2.60	780	2.54	845	0.51	0.649	1.80	0.79	0.19	1.36	2.10	1.53	-0.11
4	0.744	3.34	725	3.26	815	0.68	0.628	1.73	0.79	0.25	1.36	2.03	1.53	-0.13
8	0.692	6.32	675	6.16	770	1.36	0.596	1.68	0.79	0.51	1.36	1.98	1.53	-0.23
16	0.643	11.9	627	11.6	722	2.72	0.561	1.59	0.79	1.02	1.36	1.89	1.53	-0.4
32	0.598	22.2	583	21.6	675	5.44	0.527	1.50	0.79	2.04	1.36	1.80	1.53	-0.5
64	0.557	41.3	543	40.3	629	10.9	0.494	1.41	0.79	4.07	1.36	1.71	1.53	-0.7
128	0.518	76.9	505	75.0	586	21.8	0.464	1.32	0.79	8.15	1.36	1.62	1.53	-0.7
256	0.481	143	469	140	545	43.5	0.434	1.24	0.79	16.3	1.36	1.54	1.53	-0.1
512	0.448	266	436	260	507	87.0	0.407	1.16	0.79	32.6	1.36	1.46	1.53	2.4
1024	0.416	496	406	483	472	174	0.382	1.09	0.79	65.2	1.36	1.39	1.53	9.4
1681	0.395	769	385	750	446	286	0.363	1.02	0.79	107	1.36	1.32	1.53	22.6

[a] The nominal plant investment (PI) for the first plantship is $975M. The PI is calculated from the cost data listed in Table 7-2 assuming discount rate 13%, construction interest 12%. The PI for the second and third plants is 0.8 times the first plant cost. Thereafter, the plant investment decreases by the factor 0.93 for each doubling of the number produced.

[b] The 368-MWe OTEC NH$_3$ plantship yield of fuel-grade NH$_3$ = 1.70×10^9 gal/y.

[c] The sales price of ammonia is calculated using the Mossman financial analysis with ITC of 10%. There is no depreciation allowance. The price is determined by the requirement that the APV $\equiv$ 0.

[d] In automobile miles/gal, 2.67 gallons NH$_3$ = 1.0 gallon gasoline, assuming the same benefit from high octane number that was found for CH$_3$OH.

[e] The 1990 gasoline price = $0.786/gal (1990 avg. without taxes) (DOE/EIA, Oct. 1991, Table 9.6).

[f] The average U.S. gasoline supply in 1990 was 6.96×10^6 bbl/d = 107 Bgal/y (DOE/EIA, Oct. 1991, Tables 3.2a, 4). To supply ammonia fuel that would give the same total automobile mileage requires ($107 \times 2.67)/0.17 = 1681$ OTEC NH$_3$ plantships.

[g] The average import cost of crude oil (1990) was $21.78/bbl = $0.518/gal. The "real cost" of imported oil to the U.S. economy = 2.1 × market price (National Defense Council Foundation, abstracted by Freudmann, 1988). Therefore, the real cost of imported oil was $45.7/bbl ($1.088/gal). The market price of gasoline was $0.786/gal (1990 avg) (DOE/EIA, Oct. 1991). Thus there was a refinery markup between oil cost and the sales price of $(0.786 − 0.518)/gal = $0.270/gal. The "real" price of imported oil was, then, $(0.278 + 1.088)/gal = $1.358/gal. The difference in ammonia production between the next-to-last and last rows is used to replace gasoline produced in the U.S. Therefore, the real price of gasoline for this increment is the market price.

[h] A study by the Commission on Engineering and Technical Systems (1990) arrives at an added cost of $11.11/bbl ($0.254/gal oil equivalent) to adapt automobiles in production to use compressed natural gas as a fuel to replace gasoline. Ammonia fuel would have simpler storage and less volume but would need a modified fuel system. This suggests a cost increment of $0.30/gal (gasoline equivalent) above the gasoline market price as the ammonia fuel price for automobile use.

[i] A tax of $70 per ton of carbon emitted in energy production was recommended by the United Nations Intergovernmental Panel on Climate Change (1984). This corresponds to $0.173/gal of gasoline used in transportation versus zero dollars per gallon of ammonia for equivalent automobile mileage.

[j] It is assumed that by 2030 all U.S. gasoline will be priced at a value equivalent to the 1990 real cost of imported oil.

Table 9-5. Economic impact of replacing U.S. gasoline requiring oil imports, with methanol produced on OTEC plantships (discount rate = 11%; construction interest = 10%; ITC = 0; NPV = 0)

Plant no.	Relative cost	Sum	nth CH_3OH plant PI[a] ($M)	CH_3OH total PI[b] ($B)	Avg. PI ($M)	Total CH_3OH prod.[b] (Bgal/y)	CH_3OH sales price[c] ($/gal)	Equiv. gasoline price[d] ($/gal)	Gasoline market price[e] ($/gal)	Gasoline replaced[f] (Bgal/y)	Gasoline "real price" with import impact[f] ($/gal)	CH_3OH equiv. price incl. auto costs[h] ($/gal)	Savings using CH_3OH to replace gasoline ($B/y)	Gasoline price incl. import and carbon tax[i] ($/gal)	CH_3OH price incl. auto cost and carbon tax[j] ($/gal)	Net saving to U.S. economy in replacing gaso. by CH_3OH[j] ($B/y)
1	1.000	1.0	918	0.96	960	0.20	0.95	1.70	0.79	0.11	1.36	1.76	-0.04	1.53	1.92	-0.04
2	0.800	1.8	826	1.73	864	0.40	0.88	1.58	0.79	0.22	1.36	1.64	-0.06	1.53	1.80	-0.06
3	0.800	2.6	796	2.50	832	0.60	0.86	1.54	0.79	0.33	1.36	1.60	-0.08	1.53	1.76	-0.07
4	0.744	3.3	767	3.21	803	0.80	0.83	1.50	0.79	0.44	1.36	1.56	-0.09	1.53	1.72	-0.08
8	0.692	6.3	725	6.07	758	1.59	0.80	1.44	0.79	0.88	1.36	1.50	-0.13	1.53	1.66	-0.12
16	0.643	11.9	680	11.4	711	3.18	0.77	1.38	0.79	1.77	1.36	1.44	-0.15	1.53	1.60	-0.12
32	0.598	22.2	635	21.3	665	6.37	0.74	1.32	0.79	3.54	1.36	1.38	-0.09	1.53	1.54	-0.03
64	0.557	41.3	592	39.6	620	12.7	0.70	1.27	0.79	7.08	1.36	1.32	0.23	1.53	1.48	0.34
128	0.518	76.9	552	73.8	577	25.5	0.67	1.21	0.79	14.2	1.36	1.27	1.23	1.53	1.43	1.44
256	0.481	143.2	513	137	537	50.9	0.64	1.16	0.79	28.3	1.36	1.22	3.94	1.53	1.38	4.37
427	0.456	225.5	485	216	507	85.0	0.62	1.12	0.79	47.2	1.36	1.18	8.36	1.53	1.34	9.07
512	0.448	266.4	478	256	500	102	0.62	1.11	0.79	56.6	1.36	1.17	10.53	1.53	1.33	11.38

a The nominal plant investment (PI) for the first plantship is $960M. The PI is calculated from the cost data listed in Table 7-2 assuming discount rate 11%, construction interest 10%. The PI for the second and third plants is 0.8 times the first plant cost. Thereafter, the plant investment decreases by the factor 0.93 for each doubling of the number produced.

b The 200-MWe OTEC CH_3OH plantship yield = 1.99×10^8 gal/y of fuel-grade CH_3OH.

c The sales price of methanol is calculated using the Mossman financial analysis with investment and income tax credits of 0%. There is no depreciation allowance. The price is determined by the requirement that the NPV ≧ 0. Note that the lower sales prices for profitability quoted in Chapters 8 and 9 are based on the adjusted present value (APV) and assume that an ITC of 10% will be available.

d In automobile miles/gal, 1.8 gallons CH_3OH = 1.0 gallon gasoline. This is based on operating experience.

e The 1990 gasoline price = $0.786/gal (1990 avg. without taxes) (DOE/EIA, Oct. 1991, Table 9.6).

f The average U.S. gasoline supply in 1990 was 6.96×10^6 bbl/d = 106.6 Bgal/y. The crude oil fraction imported was 44.4% or the equivalent of 47.4 Bgal/y of imported gasoline (DOE/EIA, Oct. 1991, Tables 3.2a, 4). To supply methanol fuel that would give the same automobile mileage requires $47.2 \times 1.8/0.199 = 427$ OTEC CH_3OH plantships.

g The average import cost of crude oil (1990) was $21.78/bbl = $0.518/gal. The market price of gasoline was $0.786/gal (1990 avg) (DOE/EIA, Oct. 1991). The refinery markup = $(0.786 − 0.518)/gal = $0.270/gal. The "real cost" of imported oil to the U.S. economy = 2.1 × market price (National Defense Council Foundation, abstracted by Freudmann,1988). Therefore, the real cost of crude was $21.78/bbl × 2.1 = $45.7/bbl = $1.088/gal. Thus, the real cost of gasoline derived from imported oil was $1.088 + 0.270)/gal = $1.358/gal versus the market price of $0.786/gal.

h A study by the Commission on Engineering and Technical Systems (1990) arrived at an added cost of $2.5/bbl ($0.0595/gal) oil equivalent to adapt automobiles in production to use methanol as a fuel in place of gasoline.

i A tax of $70 per ton of carbon emitted was recommended by the United Nations Intergovernmental Panel on Climate Change (1984). This corresponds to $0.173/gal of gasoline used, and $0.158/gal of methanol for equivalent automobile mileage.

j It is assumed that by 2030 all U.S. gasoline will be priced at a value equivalent to the 1990 real cost of imported oil.

Table 9-6. Profit to investors replacing U.S. gasoline requiring oil imports, with methanol produced on OTEC plantships (discount rate = 11%; construction interest = 10%; ITC = 10%; APV = 0)

Plant no.	Relative cost	nth CH₃OH plant PI[a] Sum	nth CH₃OH plant PI[a] ($M)	CH₃OH total PI ($B)	Avg. PI ($M)	Total CH₃OH prod.[b] (Bgal/y)	CH₃OH sales price[c] ($/gal)	Equiv. gasoline price[d] ($/gal)	Gasoline market price[e] ($/gal)	Gasoline replaced[f] (Bgal/y)	Gasoline "real price" with import impact[g] ($/gal)	CH₃OH price incl. auto costs[h] ($/gal)	Savings using CH₃OH to replace gasoline ($B/y)	Gasoline price incl. import and carbon tax[i] ($/gal)	CH₃OH price incl. auto cost and carbon tax[i] ($/gal)	Net profit to investors of replacing gaso. by CH₃OH[j] ($B/y)
1	1.000	1.0	918	0.92	918	0.20	0.69	1.24	0.79	0.11	1.36	1.30	0.01	1.53	1.45	0.01
2	0.800	1.8	826	1.65	826	0.40	0.64	1.15	0.79	0.22	1.36	1.21	0.03	1.53	1.37	0.04
3	0.800	2.6	796	2.39	796	0.60	0.63	1.13	0.79	0.33	1.36	1.19	0.06	1.53	1.34	0.06
4	0.744	3.3	767	3.07	767	0.80	0.61	1.10	0.79	0.44	1.36	1.16	0.09	1.53	1.32	0.09
8	0.692	6.3	725	5.80	725	1.59	0.59	1.06	0.79	0.88	1.36	1.12	0.21	1.53	1.28	0.22
16	0.643	11.9	680	10.88	680	3.18	0.57	1.02	0.79	1.77	1.36	1.08	0.49	1.53	1.24	0.51
32	0.598	22.2	635	20.33	635	6.37	0.55	0.98	0.79	3.54	1.36	1.04	1.11	1.53	1.20	1.16
64	0.557	41.3	592	37.91	592	12.7	0.53	0.95	0.79	7.08	1.36	1.01	2.49	1.53	1.16	2.59
128	0.518	76.9	552	70.61	552	25.5	0.51	0.91	0.79	14.2	1.36	0.97	5.49	1.53	1.13	5.70
256	0.481	143.2	513	131.43	513	50.9	0.49	0.87	0.79	28.3	1.36	0.93	11.99	1.53	1.09	12.42
427	0.456	225.5	485	207.00	485	85.0	0.47	0.85	0.79	47.2	1.36	0.91	21.11	1.53	1.07	21.81
512	0.448	266.4	478	244.56	478	102	0.47	0.84	0.79	56.6	1.36	0.90	25.72	1.53	1.06	26.56

[a] The nominal plant investment (PI) for the first plantship is $960M. The PI is calculated from the cost data listed in Table 7-2 assuming discount rate 11%, construction interest 10%. The PI for the second and third plants is 0.8 times the first plant cost. Thereafter, the plant investment decreases by the factor 0.93 for each doubling of the number produced.

[b] The 200-MWe OTEC CH₃OH plantship yield = 1.99×10^8 gal/y of fuel-grade CH₃OH.

[c] The sales price of methanol is calculated using the Mossman financial analysis with ITC of 10%. There is no depreciation allowance. The price is determined by the requirement that the NPV $\geqq$ 0. Note that the lower sales prices for profitability quoted in Chapters 8 and 9 are based on the adjusted present value (APV) and assume that an ITC of 10% will be available.

[d] In automobile miles/gal, 1.8 gallons CH₃OH = 1.0 gallon gasoline. This is based on operating experience.

[e] The 1990 gasoline price = $0.786/gal (1990 avg. without taxes) (DOE/EIA, Oct. 1991, Table 9.6).

[f] The average U.S. gasoline supply in 1990 was 6.96×10^6 bbl/d = 106.6 Bgal/y. The crude oil fraction imported was 44.4%, or the equivalent of 47.4 Bgal/y of imported gasoline (DOE/EIA, Oct. 1991, Tables 3.2a, 4). To supply methanol fuel that would give the same automobile mileage requires $47.2 \times 1.8/0.199 = 427$ OTEC CH₃OH plantships.

[g] The average import cost of crude oil (1990) was $21.78/bbl = $0.518/gal. The market price of gasoline was $0.786/gal (1990 avg) (DOE/EIA, Oct. 1991). The refinery markup = $(0.786 − 0.518)/gal = $0.270/gal. The "real cost" of imported oil to the U.S. economy = $2.1 \times$ market price (National Defense Council Foundation, abstracted by Freudmann, 1988). Therefore, the real cost of crude was $21.78/bbl × 2.1 = $45.7/bbl = $1.088/gal. Thus, the real cost of gasoline derived from imported oil was $1.088 + 0.270)/gal = $1.358/gal versus the market price of $0.786/gal.

[h] A study by the Commission on Engineering and Technical Systems (1990) arrived at an added cost of $2.5/bbl ($0.0595/gal) oil equivalent to adapt automobiles in production to use methanol as a fuel in place of gasoline.

[i] A tax of $70 per ton of carbon emitted was recommended by the United Nations Intergovernmental Panel on Climate Change (1984). This corresponds to $0.173/gal of gasoline used, and $0.158/gal of methanol for equivalent automobile mileage.

[j] It is assumed that by 2030 all U.S. gasoline will be priced at a value equivalent to the 1990 real cost of imported oil.

Table 9-7. Economic impact of replacing U.S. gasoline with ammonia automobile fuel produced on OTEC plantships (discount rate = 11%; construction interest = 10%; NPV = 0)

Plant no.	Relative cost	Sum	nth NH₃ plant investment[a] ($M)	NH₃ total PI[b] ($B)	Avg. PI ($M)	Total NH₃ prod.[b] (Bgal/y)	NH₃ sales price[c] ($/gal)	Equiv. gasoline price[d] ($/gal)	Gasoline market price[e] ($/gal) (1990 avg)	Gasoline replaced[f] (Bgal/y)	Gasoline "real price" with import impact[g] ($/gal)	NH₃ equiv. price incl. auto costs[h] ($/gal)	Gasoline real cost plus carbon tax[i] ($/gal)	Savings to U.S. economy using NH₃ to replace gasoline[j] ($B/y)
1	1.000	1.00	975	0.98	975	0.17	0.93	2.470	0.786	0.064	1.358	2.770	1.648	−0.07
2	0.800	1.80	780	1.76	878	0.34	0.84	2.243	0.786	0.127	1.358	2.543	1.648	−0.11
3	0.800	2.60	780	2.54	845	0.51	0.81	2.171	0.786	0.191	1.358	2.471	1.648	−0.16
4	0.744	3.34	725	3.26	815	0.68	0.79	2.099	0.786	0.255	1.358	2.399	1.648	−0.19
8	0.692	6.32	675	6.16	770	1.36	0.75	1.994	0.786	0.509	1.358	2.294	1.648	−0.33
16	0.643	11.86	627	11.56	722	2.72	0.71	1.885	0.786	1.019	1.358	2.185	1.648	−0.55
32	0.598	22.15	583	21.60	675	5.44	0.67	1.776	0.786	2.037	1.358	2.076	1.648	−0.87
64	0.557	41.30	543	40.27	629	10.88	0.63	1.669	0.786	4.075	1.358	1.969	1.648	−1.31
128	0.518	76.92	505	75.00	586	21.76	0.59	1.570	0.786	8.150	1.358	1.870	1.648	−1.81
256	0.481	143.17	469	139.59	545	43.52	0.55	1.477	0.786	16.300	1.358	1.777	1.648	−2.09
512	0.448	266.40	436	259.74	507	87.04	0.52	1.388	0.786	32.599	1.358	1.688	1.648	−1.32
1024	0.416	495.61	406	483.22	472	174.08	0.49	1.306	0.786	65.199	1.358	1.606	1.648	2.76
1681	0.395	769.13	385	749.91	446	285.77	0.47	1.247	0.786	107.030	1.358	1.547	1.648	10.82

[a] The nominal plant investment (PI) for the first plantship is $975M. The PI for the second and third plants is 0.8 times the first plant cost. Thereafter, the plant investment decreases by the factor 0.93 for each doubling of the number produced.

[b] The 368-MWe OTEC NH₃ plantship yield of fuel-grade NH₃ = 1.70×10^9 gal/y.

[c] The sales price of ammonia is calculated using the Mossman financial analysis with investment and income tax credit of 0%. The price is determined by the requirement that the NPV ≥ 0. Note that the financial analysis assumes 20% equity and 80% debt.

[d] In automobile miles/gal, 2.67 gallons NH₃ = 1.0 gallon gasoline, assuming the same benefit from high octane number that was found for CH_3OH.

[e] The 1990 gasoline price = $0.786/gal (1990 avg, without taxes) (DOE/EIA, Oct. 1991, Table 9.6).

[f] The average U.S. gasoline supply in 1990 was 6.96×10^6 bbl/d = 107 Bgal/y (DOE/EIA, Oct. 1991, Tables 3.2a, 4). To supply ammonia fuel that would give the same total automobile mileage requires (107 × 2.67)/0.17 = 1681 OTEC NH₃ plantships.

[g] The average import cost of crude oil (1990) was $21.78/bbl = $0.518/gal. The "real cost" of imported oil to the U.S. economy = 2.1 × market cost (National Defense Council Foundation, abstracted by Freudmann, 1988). Therefore, the real cost of imported oil was $45.7/bbl ($1.088/gal). The market price of gasoline was $0.786/gal (1990 avg) (DOE/EIA, Oct. 1991). Thus there was a refinery markup between oil cost and the sales price of $(0.786 − 0.518)/gal = $0.270/gal. The "real" price of imported oil was, then, $(0.278 + 1.088)/gal = $1.358/gal.

[h] A study by the Commission on Engineering and Technical Systems (1990) arrives at an added cost of $11.11/bbl ($0.254/gal oil equivalent) to adapt automobiles in production to use compressed natural gas as a fuel to replace gasoline. Ammonia fuel would have simpler storage and less volume but would need a modified fuel system. This suggests a cost increment of $0.30/gal (gasoline equivalent) above the market price as the ammonia fuel price for automobile use.

[i] A tax of $70 per ton of carbon emitted in energy production was recommended by the United Nations Intergovernmental Panel on Climate Change (1984). This corresponds to $0.173/gal of gasoline used in transportation versus zero dollars per gallon of ammonia for equivalent automobile mileage.

[j] It is assumed that by 2030 all U.S. gasoline will be priced at a value equivalent to the 1990 real cost of imported oil.

Table 9-8. Profit to investors replacing gasoline used in U.S. automobiles with ammonia vehicle fuel
(discount rate = 11%; construction interest = 10%; ITC = 10%; APV = 0)

Plant no.	Relative cost	Sum	nth NH$_3$ plant investment^a ($M)	NH$_3$ total PI^b ($B)	Avg. PI ($M)	Total NH$_3$ prod.^{b} (Bgal/y)	NH$_3$ sales price^c ($/gal)	Equiv. gasoline price^d ($/gal)	Gasoline market price^e ($/gal) (1990 avg)	Gasoline replaced^f (Bgal/y)	Gasoline "real price" with import impact^g ($/gal)	NH$_3$ equiv. price incl. auto costs^h ($/gal)	Gasoline real cost plus carbon tax^i ($/gal)	Profit to investors using NH$_3$ to replace gasoline^j ($B/y)
1	1.000	1.00	957	0.96	957	0.17	0.679	1.81	0.79	0.06	1.36	2.11	1.65	−0.03
2	0.800	1.80	878	1.72	861	0.34	0.627	1.81	0.79	0.13	1.36	2.11	1.65	−0.06
3	0.800	2.60	845	2.49	829	0.51	0.605	1.67	0.79	0.19	1.36	1.97	1.65	−0.06
4	0.744	3.34	815	3.20	800	0.68	0.585	1.62	0.79	0.25	1.36	1.92	1.65	−0.07
8	0.692	6.32	770	6.05	756	1.36	0.555	1.56	0.79	0.51	1.36	1.86	1.65	−0.11
16	0.643	11.9	722	11.3	709	2.72	0.524	1.48	0.79	1.02	1.36	1.78	1.65	−0.1
32	0.598	22.2	675	21.2	622	5.44	0.493	1.40	0.79	2.04	1.36	1.70	1.65	−0.1
64	0.557	41.3	629	39.5	618	10.9	0.463	1.32	0.79	4.07	1.36	1.62	1.65	0.1
128	0.518	76.9	586	73.6	575	21.8	0.434	1.24	0.79	8.15	1.36	1.54	1.65	0.9
256	0.481	143	545	137	535	43.5	0.407	1.16	0.79	16.3	1.36	1.46	1.65	3.1
512	0.448	266	507	255	498	87.0	0.382	1.09	0.79	32.6	1.36	1.39	1.65	8.5
1024	0.416	496	472	474	463	174	0.359	1.02	0.79	65.2	1.36	1.32	1.65	21.4
1681	0.395	769	446	736	438	286	0.342	0.96	0.79	107	1.36	1.26	1.65	37.7

^{a} The nominal plant investment (PI) for the first plantship is $975M. The PI is calculated from the cost data listed in Table 7-3 assuming discount rate 11%, construction interest 10%. The PI for the second and third plants is 0.8 times the first plant cost. Thereafter, the plant investment decreases by the factor 0.93 for each doubling of the number produced.

^{b} The 368-MWe OTEC NH$_3$ plantship yield of fuel-grade NH$_3$ = 1.70 × 10^8 gal/y.

^{c} The sales price of ammonia is calculated using the Mossman financial analysis with ITC of 0%. There is no depreciation allowance. The price is determined by the requirement that the NPV ≥ 0.

^{d} In automobile miles/gal, 2.67 gallons NH$_3$ = 1.0 gallon gasoline, assuming the same benefit from high octane number that was found for CH$_3$OH.

^{e} The 1990 gasoline price = $0.786/gal (1990 avg. without taxes) (DOE/EIA, Oct. 1991, Table 9.6).

^{f} The average U.S. gasoline supply in 1990 was 6.96 × 10^6 bbl/d = 107 Bgal/y (DOE/EIA, Oct. 1991, Tables 3.2a, 4). To supply ammonia fuel that would give the same total automobile mileage requires (107 × 2.67)/0.17 = 1681 OTEC NH$_3$ plantships.

^{g} The average import cost of crude oil (1990) was $21.78/bbl = $0.518/gal. The "real cost" of imported oil to the U.S. economy = 2.1 × market cost (National Defense Council Foundation, abstracted by Freudmann, 1988). Therefore, the real cost of imported oil was $45.7/bbl ($1.088/gal). The market price of gasoline was $0.786/gal (1990 avg) (DOE/EIA, Oct. 1991). Thus there was a refinery markup between oil cost and the sales price of $(0.786 − 0.518)/gal = $0.270/gal. The "real" price of imported oil was, then, $(0.278 + 1.088)/gal = $1.358/gal.

^{h} A study by the Commission on Engineering and Technical Systems (1990) arrives at an added cost of $11.11/bbl ($0.254/gal oil equivalent) to adapt automobiles in production to use compressed natural gas as a fuel to replace gasoline. Ammonia fuel would have simpler storage and less volume but would need a modified fuel system. This suggests a cost increment of $0.30/gal (gasoline equivalent) above the market price as the ammonia fuel price for automobile use.

^{i} A tax of $70 per ton of carbon emitted in energy production was recommended by the United Nations Intergovernmental Panel on Climate Change (1984). This corresponds to $0.173/gal of gasoline used in transportation versus zero dollars per gallon of equivalent automobile mileage.

^{j} It is assumed that by 2030 all U.S. gasoline will be priced at a value equivalent to the 1990 real cost of imported oil.

values with fluctuations in the world economy. The results show that national energy decisions concerning the role of OTEC, and of other capital-intensive energy sources, should not be based on narrow financial estimates but instead on broad recognition of the future benefits to the nation and the world of adopting renewable energy programs, which are inherently safe, will not degrade the environment, and will make a major contribution to the national economy and security.

9.2   ENVIRONMENTAL EFFECTS

Environmentalists have challenged the continued use of conventional sources of fuels and power in recent years because of their polluting nature. For human health and safety, it is essential to maintain and improve the environment. Increased energy use and a clean environment need not be mutually exclusive. OTEC offers a method of delivering energy in amounts that would significantly reduce U.S. use of fossil fuels and would be in full compliance with the environmental policy laid down by the U.S. Government.

Although OTEC plants do not consume irreplaceable natural resources and do not produce pollutants in their normal operating cycle, the operation of OTEC plants will have some interactions with the environment that should be quantified and evaluated. These potential effects include variation in local ocean temperatures and currents, fish attraction and production, local nutrient enhancement in surface layers, effects of biocontrol measures, accidents, and chemical pollution. Deployment of large numbers of OTEC plants could have subtle effects on climate. Quantitative information must be obtained to determine whether undesirable environmental effects could set an upper limit on the OTEC power production. Studies to date indicate that a substantial fraction of world fuel needs could be supplied by OTEC without significant impacts on ocean temperature currents, weather, or ecology.

9.2.1 *Variation in Local Temperatures and Currents*

A single OTEC plant will inject a large quantity of ocean water and discharge it at temperature about 3.5°C above or below the intake temperature. This will change the local temperature, salinity nutrient distribution, and mixed-layer depths and sea-surface temperatures, among others. The downstream behavior of water discharge from an OTEC plant or plantship is essential for assessing the effects of these disturbances.

As the plant discharge effluent enters the ocean, it will have a different density than the surrounding water (Fig. 9-1). The behavior of the discharge plume will be dominated by the local currents and by initial discharge momentum and buoyancy forces resulting from the initial density difference. As distance increases from the point of discharge, the discharge plume will be diluted by the ambient ocean water and will sink (or rise) to reach an equilibrium level within the water column, where the average density difference between the diluted plume and surrounding ambient water vanishes. The plume will lose velocity until the difference between the plume

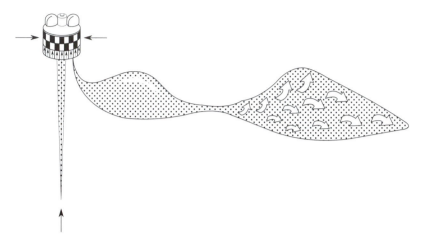

FIG. 9-1. Discharge flow field for an OTEC plant (Jirka et al., 1979).

velocity and the ambient current velocity is small. This initial region is the near-field regime.

When the discharge effluent from the plant has reached its equilibrium depth, it will have lost its jetlike characteristics and will have a velocity only slightly different from the ambient current. This region is the intermediate-field regime. The intrusion of the effluent into the stratified ocean results in the plume collapsing vertically due to residual buoyancy forces and spreading laterally due to gravity forces. Studies of the interaction of the spreading layer and the ambient current of a cylindrical plant designed for radial discharge of the exhaust streams showed that a plume will be produced in the far field that will extend up-current of the plant and grow in width down-current of the plant due to gravity spreading until gravity forces become small and turbulent diffusion takes over as the dominant mixing process. The mixing process in the intermediate field is greatly reduced compared to the near-field region. The magnitude of the ambient current dominates the behavior of the discharge plume in the intermediate field, although local ambient density stratification and initial near-field dilution will have some influence on the width and thickness of the resultant plume (Jirka et al., 1979).

Further downstream, buoyancy-driven motions become small and diffusion by means of ambient turbulence in the ocean becomes the dominant mixing and spreading mechanism. The region of passive turbulent diffusion is the far-field regime.

The brief description here provides a general model of the mixed-water discharge plume from a cylindrical OTEC plant, with radial inflow of surface water toward the intake of an OTEC plant and radial exhaust. The model is illustrated in Figs. 9-1 and 9-2. The external near-, intermediate-, and far-field flow regions of the OTEC plant discharge plume depend on many variables including the discharge structure design (discharge port location, discharge angle, discharge velocity, single or multiple discharge ports, etc.), plant size and intake and exhaust locations,

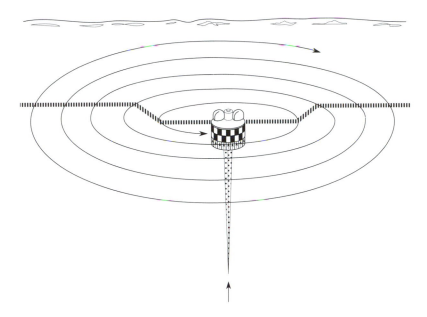

FIG. 9-2.   Surface water flow field toward an OTEC plant (Pritchard et al., 1980).

and discharge flow rate, ambient current velocity, ambient water temperature and density, and mixed layer depth among others. Several numerical and physical models developed by Jirka et al. (1979), Adams et al. (1979), Allender et al. (1978), Sundaram et al. (1978), Bathen (1977), Koh and Fan (1970), Brooks (1960), and others have been conducted to predict environmental impacts and losses in thermal resource due to the recirculation by the interaction of the OTEC plant intake and discharge flows with local ambient ocean currents.

The theoretical model by Jirka uses the two-dimensional continuity equation and potential flow equation that treats the discharge as a source and ambient current as a uniform flow. The combined flow pattern is illustrated in Fig. 9-3 with either a stepwise or linear density stratification. The first-order solution of the model is a rather tentative one because the model for plant interactions is extremely simple and two dimensional.

The experimental model by Adams et al. (1979) includes Q_i (evaporator intake flow rate, h_i (evaporator intake depth location below ocean surface), h_d (evaporator and condenser combined discharge depth below ocean surface), Q_o (combined discharge flow rate), α (discharge vertical angle), V_i, V_o, V_c, and b_o (plant characteristic sizes), u_o (ocean current velocity), ρ_o (discharge water density), ρ (ambient water density), ν (water kinematic viscosity), and H (depth below surface). Some of these parameters are shown in Fig. 9-4.

Dimensionless analysis shows that several dimensionless numbers characterize the flow. Three such numbers are the discharge Froude number, jet Reynolds number, and relative port size. Experiments have been done in many different cases.

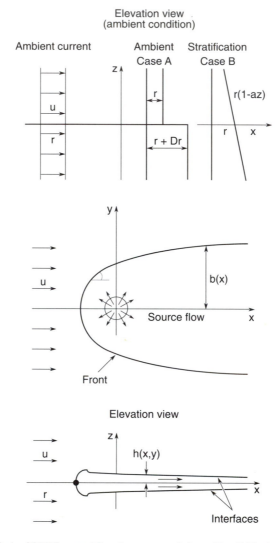

FIG. 9-3. Analysis of OTEC external flow from a moored plant with radial horizontal discharge (Jirka et al., 1979).

The results of these theoretical and experimental studies suggested that recirculation from a moored plant with radial discharge could be minimized by appropriate vertical separation of the intake and discharge. Discharge flow rates corresponding to a net capacity of 400 MWe, with mixed-layer depths in the range of 30–70 m and intake/discharge separations ranging from 30 to 90 m, have small recirculation and correspond to changes in intake temperatures of only 0.0–0.2°C (Ditmars and Paddock, 1979). Climatic alterations resulting from such small sea-surface temperature change over large ocean surface area (greater than 1000 km²,

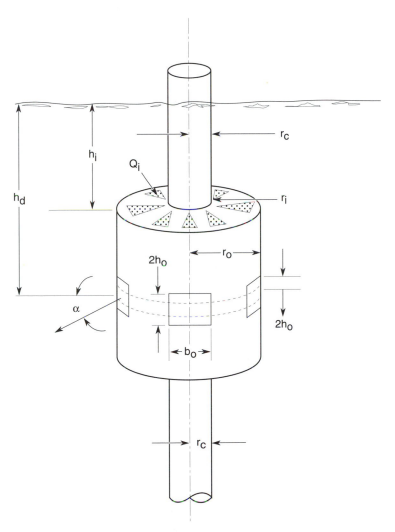

Fig. 9-4. OTEC flow schematic (Adams et al., 1979).

31-km radius) is estimated to be negligible or extremely localized. The potential effect may be more significant with a very large number of OTEC plants, which is discussed in the next section.

9.2.2 *Local Nutrient Enhancement in Surface Layers*

Deep-sea water is uniformly cold and rich in nutrients such as nitrate and phosphate that are necessary for plant growth. In tropical and subtropical areas, the temperature difference between the warm surface water and the cold deep water can be used for land-based ocean-thermal power generation or other cooling applications such as air conditioning, ice making, desalination, cooling of refineries, seawater temperature control, etc. Once the deep-sea water is brought up to the surface, the use of both the cold temperature and the nutrient content is likely to be more advantageous than the use of only one of them. Technical feasibility of artificial upwelling mariculture at Hawaii and St. Croix (Roels, 1975) indicated that the gross sales value of the potential mariculture yield from a given volume of deep-sea water may be several times that of the sales value of the power generated by OTEC from the same volume of deep-sea water. Utilizing both the nutrient content and the cold temperature of the deep-sea water may therefore make OTEC power generation from small installations (5-20 MW) economically attractive.

A comparison of nutrients of deep-sea and surface water at St. Croix (Roels, 1980) is shown in Table 9-9.

The artificial upwelling of the OTEC plant nutrients (nitrate, phosphate, and silicate) in the discharged water may locally enhance the growth of phytoplankton, which is the base of marine food chains. Thus the upwelling and nutrient redistribution may increase the local fishery production. During the OTEC-1 study (Sullivan et al., 1981), the nitrate concentration of the near field was estimated to be 15 μg atoms/liter and diluted by a factor of 10 by dispersion at the plume limits. Thus the overall effect of the local increase in the growth of phytoplankton was small.

However, deep-sea water can also be pumped into ponds on shore where planktonic algae are grown as food for filter-feeding shellfish (oysters, clams, and scallops) in a controlled food chain. The productivity of the controlled system may be much higher than that of the natural upwelling system because the deep water is not diluted with nutrient-poor surface water. The deep water is also free of pollutants, parasites, predators, and diseases harmful to shellfish. It appears that the nutrient content of deep water converted into animal protein could contribute greatly to the economic success of OTEC power generation (Gauthier, 1985; Daniel, 1990).

9.2.3 Possible Effects of Biocontrol Measures

Some concerns about OTEC have been expressed regarding biological impacts, such as impingement/entrainment, use of biocides, metallic discharges, etc.

Marine biota, particularly those with low mobility, may be harmed by impingement or entrainment in the pumping system by contact with the screens and walls of the pipe and heat exchanger system. Marine life mortality caused by impingement has been studied in relation to coastal nuclear power plants, and the

Table 9-9. Concentrations of nutrients in surface and 870-m-depth water(μg atoms/liter)

	$(NO_3 + NO_2)$	NO_2	NH_3	PO_4	SiO_4
Surface water	0.2	0.2	0.9	0.2	4.9
870-m-depth water	31.3	0.2	0.7	2.1	20.6

From Roels (1980).

effects during OTEC operation may be similar. For marine biota, impingement is expected to be confined predominantly to small fish, jellyfish, and pelagic invertebrates. For near-shore OTEC plants, crustaceans are likely to be impinged in the greatest numbers. The potential for ecologically or commercially significant losses is small.

Marine organisms small enough to pass through the screens and be entrained in the seawater flowing through the heat exchangers will be subjected to temperature and pressure changes in short time spans. In addition, marine life entrained in the deep ocean water pumped up to the surface is subjected to major changes in dissolved oxygen, turbidity, and light levels. Because of the very low level of marine life in the deep oceans, the effects should be negligible.

Chlorine has been tested as a biocide to prevent biofouling of the evaporator surfaces on the seawater side. Experiments in Hawaii have shown that addition of chlorine at a concentration of 0.050 ppm 1 h/d (0.02 ppm daily average) prevents biofouling. The U.S. EPA's standard for marine water quality allows an average chlorine concentration of 0.01 mg/liter (0.10 ppm). Thus it appears that the impact of OTEC biofouling control on the marine ecosystem will be minimal.

Protective hull coating materials released from ships stationed in harbors have been found to be toxic to resident organisms. The toxic substances released from the coatings can accumulate in the tissues of biofouling organisms, and be passed up the food chain. There is no evidence that such coatings would be necessary for commercial OTEC installations.

9.2.4 *Effects of Accidents*

Thousands of square meters of heat exchanger surface area in each OTEC plant will be exposed to constant physical and chemical stress; leaks may develop in the working fluid transport system. When the working fluid pressure exceeds the water pressure at a leakage point, the working fluid may seep into the sea, potentially causing environmental problems. On the other hand, if seawater intrudes into the working fluid, it will impede the OTEC operation, and a cleaning system will have to be devised.

Although ammonia is not the only working fluid candidate for OTEC plants, it seems most likely to be chosen because of its excellent thermal properties. The release of a small volume of ammonia into the environment would not endanger the local marine population; ammonia is a nutrient and would enhance marine growth. However, if ammonia were released into the surface water at a large rate, it could pose a serious health threat to the platform crew, the adjacent population, and the marine life.

Seawater flow through a baseline 40-MWe OTEC plant would be on the order of 2×10^7 m^3/d (Table 4-21). Ammonia leakage into the seawater flow would have to be smaller than 28×10^4 kg/d to remain under EPA's limit for ammonia concentration in marine water at 0.4 mg/liter. The EPA standard would not be exceeded during normal OTEC operation unless the OTEC plant were losing approximately 1% of its total ammonia inventory each day. This could only occur if there were a serious malfunction such as a major breakdown, a collision with an ocean-going vessel, a storm exceeding once in 100 years severity, military or political terrorism, or human error.

In addition to the large volume of working fluid (ammonia) necessary for OTEC plant operation, other materials used for operation or maintenance, such as diesel fuel for standby power generation, could be discharged accidentally. These would be of minor importance for the overall operation.

The working fluid leakage and incidental pollution problems can be controlled by modern technology. OTEC power would be safe and should be perceived to be safe by the public.

9.2.5 *Multiplant Impacts*

If OTEC assumes its full potential to provide world energy, eventually OTEC plants will be deployed to generate about two-tenths of a megawatt per square kilometer of ocean surface; that is 200-MWe plants will be sited about 30 km apart, either grazing on the tropical oceans or grouped near a large shore installation, such as in the Gulf of Mexico. The latter type of deployment might minimize undersea power transmission and shore-based power conditioning costs by allowing a single shore facility to draw power from many OTEC plants.

9.2.6 *Surface Temperature Versus Plant Spacing*

The concentrated deployment of OTEC plants raises additional questions concerning thermal resource utilization. The spacing between the plants must be selected such that the effluent from one plant does not degrade the resource of other nearby plants. Also, with large quantities of warm and cold water being removed from near the surface and from the deep sea and injected at some intermediate depth, the question of resource renewal arises.

The effect of concentrated OTEC operation on the thermal structure of the Gulf of Mexico was estimated by Martin and Roberts (1977), who used a one-dimensional unsteady heat conservation equation to predict the horizontal mean temperature. The surface heat fluxes, including solar radiation, back radiation, latent heat flux, and sensible heat flux, were parameterized in terms of the observed air–sea temperature difference and the predicted sea surface temperature. Advection of heat into the Gulf by the Yucatan current was treated as a heat source for the surface layer of the Gulf. A constant mean upwelling was calculated to balance the overall heat bulge. The operation of one thousand 200-MWe OTEC plants in the Gulf was parameterized by the addition to the model of a mean vertical velocity profile required to complete the circulation between the near-plant intake and discharge flows. The result was a slight surface cooling and a warming at depth. The sea surface temperature dropped about 0.3°C during the first 2 years and then remained fairly constant thereafter. However, the deep water in the region above the cold-water intake warmed continuously at the rate of about 0.3°C per year. For the operation of one hundred 200-MWe OTEC plants, the impact was correspondingly reduced. After 30 years, the model predicted a drop in sea surface temperature of

0.05°C and a warming in the water column above the cold-water intake of 0.8°C. The mean temperature profile in the Gulf of Mexico after 10 years of operation with one thousand 200-MW OTEC plants is shown in Fig. 9-5. These calculations are based on the utilization of 8.5 m³ of total seawater pumped per second per megawatt of power produced (8.5 m³/s is a high number; about 5 m³/s would be typical). The calculated rate of warming of the deep water resulted from the fact that the model (Martin and Roberts, 1977) does not allow for the removal of this heat input to the deep water from the Gulf by the currents.

The effluent from one OTEC plant could reduce the thermal resource available to other plants within a power park either by directly entering the warm-water intake of a plant down stream or by effectively thinning the mixed layer down stream. This thinning would increase the likelihood that cooler water near the bottom of the mixed layer would be drawn up into the intake of neighboring plants.

Jirka et al. (1977) have presented a preliminary analysis that can be used for estimating minimum plant spacing that will prevent the effluent from one plant from degrading the effective thermal resources available to neighboring plants. In his analysis, Jirka assumes both a mixed-discharge mode and a two-layer ambient ocean. After initial dilution due to jet entrainment the mixed-discharge effluent is assumed to form a layer of intermediate density between the upper and lower layers of the ambient ocean. The effluent then drifts with the ambient current and spreads laterally, due to buoyancy. Thus, the intermediate layer reduces the effective thickness of the mixed layer at neighboring plant sites and increases the potential for recirculation.

For a row of plants oriented orthogonal to the dominant current and an ambient mixed-layer depth of 700 m, a minimum lateral spacing of 4.8 km is predicted for

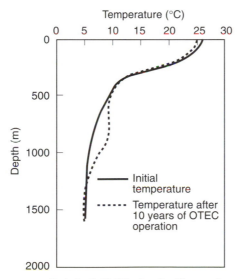

FIG. 9-5.   Mean temperature profile in the Gulf of Mexico after 10 years of operation of 1000 200-MWe OTEC plants (Martin and Roberts, 1977).

100-MW plants and 36 km for 400-MW plants. In the case of a rectangular grid of plants, the minimum spacing depends on the ambient mixed-layer depth, the magnitude of the ambient current, and the number of rows in the grid. In the case of a mixed-layer depth of 70 m, an ambient current of 0.1 m/s, and a configuration of five rows, a minimum spacing of 5.6 km is predicted for 100-MW plants and 160 km for 400-MW plants. Extremely simple and roughly approximate, Jirka's one-dimensional interaction model is surpassed by his later two-dimensional model. In the latter, Jirka (1978) deals directly with the problem of a source of buoyancy in a current. His modified results suggest that lateral spacing requirements may not be driven solely by thermal resource considerations. In addition, assessments of downstream environmental impacts may provide constraints on plant spacing.

9.2.7 Plant-to-Plant Interference

Even though a large spacing will probably not cause large-scale plant-to-plant interference, it is desirable to estimate the influence of the thermal effluent of the plant on the operation of a nearby plant. The mixed discharge from one plant sinks, entrains ambient fluid until buoyancy overbalances the inertia of the plume, rises slightly and begins to spread, governed mainly by gravitational and inertia forces, and leaves the vicinity of the plant. The ultimate height and width of the plume are controlled by the gravitational and rotational effects. The plant-to-plant interference depends on the intermediate hydrodynamic field of an OTEC plant. Both the interaction of buoyancy-driven currents and the ambient ocean current with the plant effluents must be taken into account. The intermediate field determines the area over which physical, chemical, and biological distribution from the plant can be felt (or measured) by its neighboring plants. Additional hydrodynamic drag and induced circulation are also generated by the plant and have effects on its neighboring plants. The intermediate field of an OTEC plant of approximately 100 MWe has a diameter of about 10 km.

Jirka et al. (1979) modeled the intermediate-field perturbations produced by OTEC plants. In their analysis, the OTEC effluent forms internal layers within the ambient stratified oceans. The motion within the layers is predominantly horizontal, as determined by the interaction of buoyant forces and of drag effects due to the ambient ocean current. While initial near-field (order 1 km) studies (Allender et al., 1978; Jirka et al., 1977) have demonstrated that moderately sized 100-MW single OTEC plants can be operated without significant recirculation, this may not be the case if several OTEC plants interact with each other. Several plants could alter the thermal and density profile at another site sufficiently to cause a decrease in thermal resource recovery. Thus, the plant-to-plant interference analysis must account for the intermediate-field or even far-field motions for multiple OTEC arrangements. Jirka considered that each plant is a source flow in a uniform ambient current as described in the previous section. He concluded that if another plant were located outside the density current zone of the first plant, the interference would be small in thermal resource utilization. The density current zone defines the region of biochemical activity, as related to nutrient upwelling and/or chemical releases from the plant. Other plants will both be subject to this influence and also add their incremental effect.

9.2.8 *Circulation*

A medium-size OTEC plant of 100-MW capacity and a temperature differential of 23.3°C (42°F) would require an intake of 300 to 500 m^3/s of warm surface water and an equal intake of cold deep water. The local withdrawal of the 300 m^3/s from the surface layer would induce a radial inflow toward this sink in a pattern dependent on surface currents. At a distance of 1 km from the intake, assuming that the flow is drawn from a layer 40 m thick, the in-flowing velocity would be about 0.12 cm/ s (0.05 in./s). However, a vortex flow pattern might develop as a result of the local sinking of surface water, as shown in Fig. 9-2. The induced circulation could possibly result from friction, mixing convective acceleration, Coriolis forces, as well as other terms. Pritchard et al. (1980) presents a surface water vortex flow pattern of near-field OTEC circulation for a cylindrical spar OTEC design. At any latitude other than the equator, Coriolis effects will cause the inflow to turn and approach the OTEC sink as a spiral. Pressure drop associated with this withdrawal of surface water would cause a slight depression in sea level above the plant and a rise in the isotherms in the upper part of the thermocline. The horizontal pressure field resulting from this rearrangement of the density field would support a geostrophically balanced, cyclonic flow characteristic of a cold core eddy. The first-order models (DOE, 1980) of an OTEC vortex suggest that a 100-MW plant drawing warm water from a mixed layer 100 m deep would produce an eddy with a radius of 10 to 20 km, having a maximum tangential velocity of about 15 cm/s at a distance of about 1.5 km from the sink. The thermocline below the sink would be raised about 15 m. Pritchard's simple two-layered model assumes that the bottom layer is infinitely deep and motionless, is constructed from a frictionless outer region, and has a solid body motion core. Similar models have been used to study the circulation flow in a hurricane. Such a distortion of the thermal field and the rise in the thermocline could, in fact, reduce the temperature of the surface water available as a heat source to the OTEC plant. However, calculations (Roels, 1980) suggest that the warm surface water temperature drop due to the circulation is insignificant for a 100-MW plant operating on a surface mixed layer 100 m deep. If the inter-plant space is sufficiently large, this temperature reduction is insignificant.

The interaction of the OTEC vortex with island coastal circulation should be specifically explored, since the first commercial OTEC plants may be located near islands. Thermal structures in coastal boundary layers are subject to natural transients related to wind-induced upwelling and downwelling. It is difficult to forecast the possible interaction without long-term observational data and numerical modeling. Currently, there is a lack of studies relating the possible structure and intensity of the expected OTEC vortex as a function of plant size and ambient oceanic conditions.

Similarly, circulation in the Gulf of Mexico also requires information about residence times and transport pathways for effluents. The impact of OTEC plant operations on the circulation in the Gulf of Mexico, particularly with regard to intrusions of the loop current into the regions of interest for large-scale OTEC deployment, have been studied by Thompson et al. (1978) and Blumberg and Mellor (1979). Thompson et al. (1978) developed a numerical model of a single ocean basin and reproduced observed aspects of its physical oceanography. In their analysis, each OTEC plant is inserted as a perturbing influence source on the basin. The model is time dependent and nonlinear and includes the Earth's rotation, wind force, loop current force, and sub-grid-scale turbulence, etc. The circulation characteristics of three sites in the Gulf are made in preliminary integrations of the numerical model. Near-surface and subsurface scalar discharges from each OTEC plant are traced. Circulation patterns similar to loop current structures have been noted in their preliminary integration. Their model indicates that discharges from OTEC plants and resulting circulation in the Gulf will be strongly time dependent.

Blumberg and Mellor (1979) developed a numerical model with the capability for greater spatial resolution in the vertical. It uses layers of variable thickness in the vertical. While the primary motivation for application of this model to the Gulf is assessment of environmental impacts, sensitivity studies regarding exchanges through the Yucatan Straits as well as simulated OTEC power park perturbations are relevant to resource renewal concerns.

9.2.9 Large-Scale Nutrient Enhancement

Natural physical aspects of the water flows in the oceans that produce the year-round vertical ocean temperature gradient also result in low surface biological production. The slow rate of vertical mixing between surface and deep waters allows the temperature gradient to persist, and limits the transfer of nutrients from the nutrient-rich deep to surface waters. Surface waters are thus nutrient-poor. A comparison of nutrients of deep-sea and surface ocean water was given in Table 9-9. OTEC plants may measurably alter the natural state of nutrient-poor marine ecosystems by artificially enriching surface water. For example, 100 commercial OTEC plants of 100-MW capacity each might have a seawater flow-through of about 5×10^9 m^3/day. Assuming an equal mixture of cold water from 1000 m depth and warm surface water from 10 m depth, nutrient-rich water (30 mg NO$_3$/m^3) would be artificially upwelled at a rate of 3×10^9 m^3/day. The induced nitrate flux of 80×10^9 mg NO$_3$/day, discharged at the surface over a 100-km^2 area of the upper 200 m, would add 0.5 μg NO$_3$/liter per day. This could be 20–25 times what is now found in the mixed layer of nutrient-poor near-surface ocean water.

A natural upwelling occurs off Peru. In this upwelling, the near-shore upwelled input of nitrate into a 20-m surface layer can be estimated by the Walsh equation (Walsh, 1975):

$$\delta NO_3 / \delta t = w\delta(NO_3) \, \delta z \,,$$

where w is the upwelling velocity (approximately 10 m/d), δNO_3 is the nitrate gradient between 15 and 25 m (about 5 μg NO$_3$/liter), and δz is 10 m. A nitrate rate of 5 μg NO_3/liter per day is therefore added to the surface water locally at the Peru coast. Thus, the nutrient increase from upwelling of an OTEC power park with 100-MW plants and area density of a 120-km^2 area per plant would be equal to that of an equivalent area of the ocean off Peru, which is the world's most productive coastal upwelling region.

The discharge and hence nutrient input of OTEC plants may be confined, however, to depths below the euphotic zone. For OTEC plants to be efficient, the temperature gradient between surface and deep water has to be maximally utilized. Therefore, the mixed discharge must not be recirculated into the plant surface intake, as a decrease of 1°C in the temperature difference would reduce the efficiency of the plant by 10%. For this reason, OTEC design has plants with combined discharge that will sink to an equilibrium depth below 150 m. Because the depth of the mixed layer in a typical tropical ocean is less than 100 m, nutrient enrichment of the euphotic zone, where plankton photosynthesis takes place, by OTEC discharge would be limited to vertical diffusion about 10^4 m^2/s in the immediate vicinity of the plant. The nutrient enhancement near the plant is certainly much less than estimated previously in the surface discharge scenario. At some distance from the OTEC plant, however, an uplifting of density surfaces and of the concentration gradient of plant nutrients may occur. Increased nutrient input to the euphotic zone may take place in this far-field zone, stimulating the growth of the plankton. It is not yet known if large-scale OTEC operations will result in a gain of surface plankton, but clearly plankton populations will be displaced in the vicinity of the plant. One could expect a trail of organisms in the wake of discharged upwelling water, as shown in Fig. 9-6 (Miller, 1977). In the order of their appearance one could expect the algae or phytoplankton, followed by the herbivores, and then the primary, secondary, and tertiary carnivores. Possibly this food-chain effect could be put to advantage by judicious planning and study leading eventually to harvesting. In any case, the "plowing" of the sea should prove to be beneficial to the natural environment whether the end products are harvested or not.

Lack of funding has prevented study of the ecological effects of grazing OTEC plantships. A quantitative picture was provided by Miller (1977) in the original studies of grazing OTEC plantships.

9.2.10 *Fish Production*

Net gains of plankton organisms may result some distance away from the OTEC plant as a result of increased nutrient input to euphotic zones that are associated with the shoaling of isopycnal and nutricline. Since plankton is important in the marine food chain, enhanced productivity due to redistribution of nutrients may improve fishing. Fish, which in general are attracted to offshore structures, are expected to increase their ambient concentration near OTEC plants. The world annual yield of marine fisheries is presently 70 million tons, with most fish caught on continental shelves. In fact, the open ocean (90% of the total ocean area) produces only about 0.7% of the fish because most of the nutrients in the surface water are extracted by plants and drift down to the ocean floor in the remains of plant or animal life. The

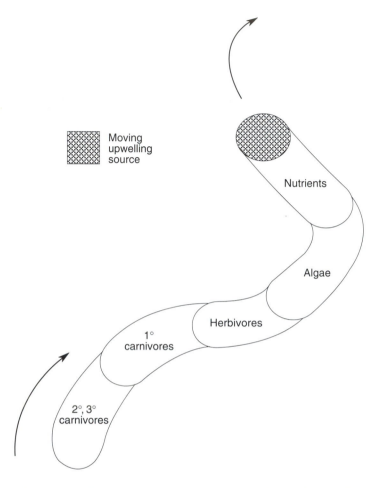

Fig. 9-6. Probable food-chain distribution in wake of moving OTEC ship (Miller, 1977).

water in the coastal zones is continually supplied with fresh nutrients in the runoff from the adjacent land and, hence, supports a high level of plant life activity and produces 54% of the fish. Only 0.1% of the ocean area lies in the upwelling regions, where nutrient-laden water is brought up from the ocean depths, yet these regions produce 44% of the fish. The reason for this spectacular difference can be seen in Table 9-9, which shows that the nitrate and phosphorus concentrations in deep seawater are about 150 and 5 times more, respectively, than their counterpart concentrations in surface water at a typical site (St. Croix in the Virgin Islands).

Proposals to produce artificial upwelling, including one using nuclear power, have concluded that the cost would be excessive. Roels (1980) studied the possibility of using a shore-based OTEC plant to supply nutrient-laden water to a mariculture system, with a series of experiments carried out at St. Croix in the U.S. Virgin Islands. At that site the ocean is 1000 m deep only 1.6 km offshore. Three polyethylene pipelines, 6.9 cm in diameter and 1830 m long, have brought approximately 250 liters/min of bottom water into 5-m^3 pools where diatoms from laboratory cultures are grown. The food-laden effluent flows through metered channels to pools where shellfish are raised. The resulting protein production rate was excellent; 78% of the inorganic nitrogen in the deep seawater was converted to phytoplankton-protein nitrogen, and 22% of that was converted to clam-meat-protein nitrogen. This compares with plant-protein/animal-protein conversion ratios of 31% for cows' milk production and 6.5% for feedlot beef production. The production of seafood is therefore more efficient than that of beef. Thus, shifts from beef to seafood, already underway in some societies for health reasons, could help to meet world needs for high-quality food.

Net gains of plankton organisms may result some distance away from the OTEC plant as a result of increased nutrient input to the euphotic zone associated with the shoaling of isopycnal and nutricline. Increased harvests of small oceanic fish, which feed on plankton, would result.

On the other hand, the daily intake of surface water is estimated to be about 3×10^7 m^3/d for a 100-MW OTEC plant, and entrainment on the heat exchangers may kill almost 100% of large zooplankton, fish eggs, and larvae. Acting as a predator on plankton, an OTEC plant would deprive many adult fish in its vicinity of their prey. It was estimated that 0.25 metric ton per day of oceanic fish could be lost at each OTEC plant, a small fraction of the potential increased production.

An OTEC park could convert its host area from an oligotrophic eutropic ocean area to a zone with production analogous to that in upwelling areas if an effort was purposefully made to do so. It is expected that this would result in wholesale changes of the species in the vicinity of the OTEC park, since experience in different regions of the oceans suggests that the size of the phytoplankton increases with increased nutrient input rate. These larger phytoplankton are associated with large crustacean zooplankton that provide food forage for harvestable fish. The size of the organisms at each level of the food web may be as important as total production in determining the ultimate yield of the food web to humans. However, building such a huge ocean fish farm in the open sea by artificial upwelling requires a large initial capital investment and poses many problems. The combination of mariculture with a coastal land-based or island-based small-scale OTEC is more feasible. Mariculture requires only a cold-water pipe and its associated pumps. The mariculture-power composite of a land-based OTEC plant could be a near-term commercial possibility. A composite OTEC plant that could provide a minimum of 13,000 gallons of deep ocean water per minute for mariculture and provide a net output power of 150–300 kW of electricity into a local electricity supply grid has been designed (Johnson, 1989) for a site close to the Natural Energy Laboratory of Hawaii at Keahole Point on the "Big Island" of Hawaii.

The current low price of oil has lessened OTEC activity. However, this slowdown has provided an opportunity for scientists and engineers to fine tune the mariculture components of OTEC fish production. Certainly, the greatest resource in the oceans is the cold ocean water itself. The deep ocean nutrient-rich and pollution-free water has attracted several commercial mariculture projects including raising fish, seaweed, abalone, and lobster in Hawaii. Daniel (1989) reported that a total of eight deep seawater systems have been installed and operating since 1987 at the Natural Energy Laboratory of Hawaii, Ocean Farms, Hawaii, and Hawaii Ocean Science and Technology Park. These pipes have provided the cold seawater to grow abalone, salmon, steelhead trout, flounder, and several commercially valuable seaweeds successfully.

9.3 GENERAL COMMENTS ON EFFECTS OF OTEC COMMERCIALIZATION

9.3.1 *Oil Imports and World Trade Balance*

The combustion of fossil fuels is the main source of thermal energy for the world. These fuels are a finite energy source that will eventually be depleted. The time of their depletion depends on extraction rates. Based on energy content, coal, oil, and natural gas are estimated to be about 90, 5, and 5%, respectively, of the total fossil fuel energy endowment of the Earth. Energy use according to fuel type in the United States since 1850, as indicated in Fig. 9-7, has had two transitions. The first, a transition from fuel wood to coal, occurred during the late nineteenth century. The second, a transition from coal to petroleum and natural gas, occurred early in the twentieth century. Energy use in the United States decreased after the energy crisis, but since 1987 has been increasing.

Only a few of the developed nations have substantial fossil fuel reserves and hence must rely upon imports, primarily of oil. Coal use tends to be small except in those few nations with large reserves. On the basis of energy content, the world trade in coal is insignificant compared to that of oil. For many of the developed nations, imported hydrocarbon fuels account for over half of the energy input.

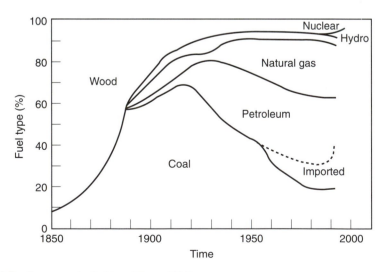

Fig. 9-7.   Energy use by fuel type (Krenz, 1984).

The world reserves of oil may be sufficient to maintain the current consumption rate for more than several decades. The political disposition of the oil-exporting nations, however, may limit the available supplies. In the United States (including Alaska), oil production has declined following a peak in 1979. U.S. oil production in that year reached 11 million barrels per day, but has declined in recent years to 9 million barrels per day in 1990. The U.S. dependence on oil net imports is given in Fig. 9-8. The energy trade balances of the United States have been negative in recent years and are the major contributing factor in the national annual trade deficit. In order to improve the U.S. economy, a reduction in energy imports is obviously needed.

Another energy transition may be on the horizon. Alternative energy sources such as OTEC and/or increased efficiency in the use of energy will be needed if the production of oil in the world continues to decline. Even if the world's oil production does not decline, additional alternative energy input should be provided to meet environmental requirements for the world's growing population and to improve human living standards.

High oil prices have affected developing countries more severely than the industrialized nations. Developing countries account for 15% of the world's energy consumption. Sixty percent of this energy is supplied by oil, and 30% of the oil is imported. In the case of Jamaica, the oil import cost increased from 17% of the country's budget in 1973 to 60% in 1983. Similarly, as a fraction of the cost of total imports, the oil bill increased from 10% to 34% in 1983. The Jamaican economy, and the economies of numerous other developing nations, have stagnated as a result of the high cost of oil imports.

Facing the large national deficit partly due to oil import cost, the United States has a different energy challenge today from that in the 1970s. We can predict how long easily recoverable oil reserves will last at present rates of consumption. The opportunity to develop energy alternatives for oil is greater than ever because of modern technology. OTEC is certainly one of the most promising alternative energy source candidates. As explained in previous chapters, this resource is more

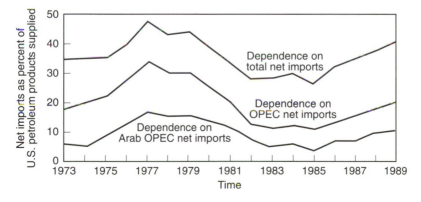

FIG. 9-8. U.S. dependence on foreign oil imports (DOE/EIA, 1989).

than enough to provide for current U.S. energy consumption and expected demands well into the future. The extra electricity generated by OTEC stored as energy-intensive products, particularly vehicle fuels to replace gasoline and diesel oil, can be exported to provide substantial future energy supplies for the world, as well as export merchandise values for the United States. As discussed in Chapters 7 and 8, the use of OTEC in combination with water electrolysis to produce liquid fuels that would reduce U.S. oil imports is economically attractive. The financial analyses show that costs of OTEC energy in the form of liquid delivered to U.S. users, if rapid commercial development was supported, could be low enough to compete with oil after the year 2000. With a production rate compatible with reasonable U.S. funding and shipbuilding capabilities, OTEC methanol plantships could be delivering enough motor vehicle fuel by 2025 to eliminate the need for imported fuel to supply gasoline in the United States. Alternatively, ammonia plantships could supply nonpolluting fuel (no carbon emissions) to eliminate all U.S. need for gasoline by mid-century. Large benefits would result for U.S. employment, industrial productivity, and balance of trade.

9.3.2 *Related Industry*

In addition to its use as a source of fuels, OTEC offers commercial opportunities for production and transport of electricity and nonfuel energy products to shore. The export of high-energy-consuming and energy-intensive products would stimulate economies of producing nations and reduce their reliance on other nations. The OTEC-related heavy industries would also provide jobs and promote the nation's technological capability. The range of products that could be produced by OTEC commercialization is depicted in Fig. 9-9. Several of the OTEC options are discussed in Chapter 4.

9.3.3 *Stability of Energy Cost*

Based on generally accepted estimates, more than half of the easily recoverable oil from the continental 48 states has been extracted, and hence the decline in production is likely to continue. On a global basis, production will begin to be limited by dwindling reserves within the lifetime of children now being born, if global consumption continues to increase at present rates. The U.S. reserves that remain are not only limited but also considerably more difficult to extract (Alaska and offshore, for example) than those utilized in the past. The production cost of these oil supplies is expected to be higher and higher. Forecasts of oil shortages combined with economic manipulations caused the price of oil to reach a peak of over $35 a barrel in 1981, five times the price in 1973 before the energy crisis.

The world oil outlook is brighter now (1992) than it has been for some time. The oil price declined from $35 per barrel in 1981 to $27 per barrel in 1985 to $15 per barrel in 1989, rising briefly to $30/bbl as the Gulf war approached, then dropping to $20/bbl in 1991. The impact of these price fluctuations on world economies has been enormous and imponderable.

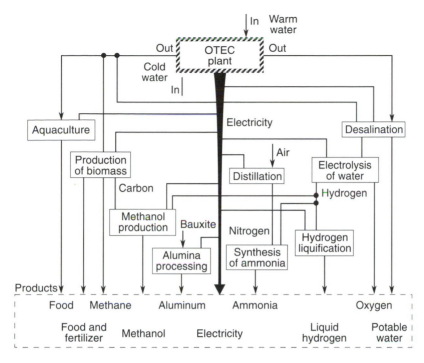

FIG. 9-9. Products that could be produced by OTEC commercialization.

The reductions of oil cost and consumption in recent years were made possible by impressive gains in energy efficiency around the world, the development of alternative energy sources, and the political situation in Mideast oil export countries. However, the current oil price is not stable. While the OPEC members are husbanding their oil reserves in an effort to maintain prices, non-OPEC producers such as the United States, the United Kingdom, and the Soviet Union are depleting their reserves at a rapid pace. The Middle East currently contains nearly 60% of the world's proven oil reserves. When non-OPEC petroleum production declines in the 1990s, Middle Eastern OPEC members may gain a major role in managing world oil supplies. World energy prices could again become subject to their control.

The drop of oil prices in recent years has undermined investments in energy alternatives. Declining energy prices have already led many countries to cut successful energy programs that would have provided substitutes for oil in the beginning of the next century. OTEC is a typical example. There has been little research and development activity in the past several years. OTEC is increasingly attractive because it can shield the economy from the effects of increasing oil import prices. Commercial development of OTEC can make it competitive with conventional fossil fuel and nuclear plants. As shown in Chapters 7 and 8, this objective can be achieved early in the next century if the latter sources take the steps needed to comply with environmental and safety requirements.

Since OTEC is a virtually unlimited renewable source of energy, large-scale OTEC energy production would stabilize the energy cost of the world.

9.3.4 *Mariculture and World Food Supplies*

The contemporary world consists mainly of food-deficient countries, with over 100 of them importing grains from the United States alone. Most of the national food deficits are the results of excessive population growth, land degradation, other factors, or combination of these. In recent years, grain deficits have worsened in Africa and the Middle East. Over the past decade, growth in food output has slowed markedly. The slowdown in food output dates from the first oil price increase. Between 1950 and 1973, grain production increased at 3.1% per year, outstripping population growth. During the years since, production has expanded at only 2.2% per year, dramatically narrowing the margin between the growth in food production and that of population. Per capita production has declined from 1.2 to 0.4. More than 500 million people now suffer from food and protein deficiency. The need for more and more food production, particularly high-quality protein, is great. Developed nations have achieved high agricultural productivity using petroleum-dependent machinery and fertilizers. Another factor limiting the expansion and agricultural production is the scarcity of arable land. Most of the arable land in the world is in use, and although some arid areas can be made suitable for cultivation through irrigation, fresh water needed to do so is scarce. Increasing the water supply would require large capital investments and vast amounts of energy.

Because of the limited supply and high costs of arable land, fresh water, and petroleum, it is essential that we explore alternative methods to provide the world's people with new sources of fresh water, energy, and food. The scarcity of arable land necessary to increase food production naturally points to the vast area of the oceans that cover approximately 70% of the earth's surface. Currently, the sea produces only about 5–10% of the protein consumed in the world. Throughout most of history, the human demand for seafood did not even remotely approach the sustainable yield of oceanic fisheries, but as world population has nearly doubled since mid-century, claims on many fisheries have become excessive. From 1950 to 1970, the world fish catch expanded at 5% per year, as fast as or even faster than the global economy growth. This changed abruptly after 1970, when the annual growth rate dropped to 1%. Overfishing, which had been the exception, became the rule, depleting the natural stocks. Fish catch per person, including fisheries from fish farming, has been down 15% since 1970. The biggest reductions in fish consumption have been in the third world countries such as the Philippines.

The promise of OTEC to become an important resource for mariculture and food production through use of the OTEC cold water has been discussed in Section 9.2. The potential protein production of the oceans exceeds that of land-based agricultural systems and is not dependent on petroleum for fertilizer and energy required by advanced agricultural technology. The projected protein yields of 27.8 tons per hectare per year exceed by far the protein yields of alfalfa at 0.71 ton per hectare per year. Alfalfa is land-based agriculture's highest protein-producing crop, with intensive use of fertilizers and energy-consuming mechanized systems. The shellfish meat protein conversion efficiency (from algae) ranges from 22% to 30%. In comparison, the plant protein to animal protein conversion efficiency is 31% for cows' milk and 6.5% for feedlot beef.

If 0.1% of the warm surface tropical ocean water were used per year to generate OTEC power, and if the ratio of warm surface water to cold deep seawater flow rate were 1 to 1, these plants would require 6.25×10^{12} m^3 of deep seawater per year. The potential mariculture yield from this deep seawater artificial upwelling flow would produce 750×10^6 tons of clam meat per year based on Roels's (1980) demonstration. This production should be compared to the total world meat production (beef, lamb, pork, etc.) for human consumption of 118×10^6 tons during 1975 and to 67×10^6 tons of fish produced in the same year. To achieve this production would require 8.3×10^6 hectares or 8300 km^2 surface area. (The tropical ocean area suitable for OTEC is about 60 million km^2.) The potential for increased food production (by the OTEC system) with high-quality protein is great.

To utilize the deep seawater in a mariculture operation, it would be necessary to contain the deep seawater in ponds or bays near shore for land-based OTEC plants or in some kind of floating structures, which would maintain it near the surface of the ocean, for floating OTEC plants in the open sea. Technologically, it appears that we are a long way from very large open-ocean mariculture farms, but the shore-based option is feasible in the near future. A comparison between power yield and the meat yield of a shore-based combined 1-MWe OTEC mariculture plant is given in Table 9-10. It shows that the combined utilization of the low temperatures and nutrients of deep seawater in the tropical ocean for power and mariculture could be economical and would contribute to solving shortages of food and energy now faced by humanity.

9.4   BENEFITS TO ISLAND ECONOMICS

A land-based OTEC plant would have to be located in a coastal area in the tropical ocean where a steep offshore slope brings deep ocean water near to shore. The shorter the cold-water pipe, the less the pipeline cost. Moreover, the longer the pipe the more the cold water will warm up in transit to the OTEC plant. An advantage of the land-based OTEC plants is that the power transmission from the plant to consumers uses existing technology. Steep offshore slopes are rare along the edges of the continents, which typically have wide continental shelves. However, many

Table 9-10.   Annual gross dollar value of production from a combined OTEC-mariculture system using the same deep-water flow rate of 3.76 m^3/s

System	Product	Unit price	Gross sales value
OTEC	1 MW	6¢/kWh	$ 473,000
Mariculture	Shellfish	$2.2/kg meat	$1,564,200
	1693 t	($1/lb meat)	

From Roels (1980).

oceanic islands rise steeply from the ocean floor and have steep offshore slopes. Therefore these islands in the tropical ocean are the most likely locations for economic land-based OTEC plant sites.

Tropical islands, which need electric baseload power and/or fresh-water supply, constitute excellent early markets for OTEC power plants. Worldwide there are 98 nations and territories with the prerequisites necessary to use land-based or near-shore OTEC technology. Of these, 53 candidates are islands in tropical areas (United Nations, 1984).

Many such islands share the following general characteristics: (1) heavy dependence on imported oil for electric power generation and shortage of natural precipitation for fresh-water supply, (2) high electricity costs due to small demand and high transportation costs, (3) small incremental power needs with baseload capacity addition in the 5- to 40-MWe range, (4) few domestic alternative energy resources, and (5) a developing economy with a growing population. Small-scale shelf-mounted OTEC plants are marginally economic, if the current state of the art, expected oil prices, and the environmental and social difficulties encountered by coal and nuclear power are considered.

It is estimated (Avery and Dugger, 1980; Gopalkrishman, 1984; Toms and Ford, 1984) that if OTEC supplied 10% of the potential tropical islands' OTEC market, 300 plants producing, in total, 57,700 MW of electric power (Bell, 1980; Sanchez, 1980; Pinckert, 1980) would be needed.

Coupling an OTEC plant to a mariculture operation of island-based plants is under active development in Hawaii and has been planned in Tahiti. Suitable sites with deep water close to shore are widely available at volcanic islands. Since the energy demands are in the range of 5-25 MW, a relatively high cost per kilowatt hour of OTEC power generation is implied. Roels (1980) estimated that the gross sales value of the seafood produced by a combined OTEC–mariculture system using a deep-water flow rate of 3.75 m^3/s would run about five times the value of the electricity produced by an OTEC 1-MWe plant with the same flow rate. A small land-based OTEC plant in Hawaii has been designed to supply deep nutrient-rich seawater to an aquaculture facility to produce a net electric power output of 300 kWe (Johnson, 1989).

The combination of aquaculture with a small-scale OTEC plant can significantly affect the economics of OTEC operation. Aquaculture can support the costs of installing a cold-water pipe and its associated seawater pumps. The OTEC plant can be viewed as just the warm-water pipe, heat exchangers, and power generation equipment. The composite plant appears to be commercially feasible. The design (Johnson, 1989) provides a deep ocean water rate of 13,000 gallons per minute for aquaculture and an electric power of 300 kW into a local electricity supply grid. In addition, the plant would produce a potable water rate of 2650 gallons per day.

OTEC processes that employ the flash evaporation of seawater are capable of producing fresh water as a byproduct of power production. In many island locations fresh water has a market value of $1-11 (1985 value) per 3875 liters (1000 gallons) (SERI, 1985). The price for island-based OTEC power could become competitive with fossil-produced power when water is also produced and sold at these prices.

Even when electrical power is not desired, OTEC appears attractive for the production of high-priced fresh water alone. Either open- or closed-cycle OTEC can be used.

It has been estimated (Johnson, 1989) that the cost of fresh water produced by OTEC at a favorable island site may be one-third that of fresh water produced by the standard multistage flash-evaporation process using fossil energy as a heat source. The technical feasibility of using ocean energy to desalinate seawater has been demonstrated successfully by an open-cycle heat and mass transfer test apparatus that produces 22 liters per second of potable water having a salinity of 86 ppm (Trenka, 1988).

Another marketable OTEC application in the islands involves hotel-resort developments. When cost analyses are performed factoring in not only fresh-water production but also space cooling and refrigeration via utilization of the cold seawater into the electrical energy product costs, the overall system economics improve dramatically.

In conclusion, the reduction in dependence on imported oil and the related benefits of OTEC could greatly improve the economic stability of many tropical islands. OTEC-produced metals (aluminum, magnesium, nickel, etc.) (Homma et al., 1979; Kajikawa and Hiramatsu, 1986) as well as other inorganic and organic chemical plants could be located on these islands. The deep seawater would enhance the local island fisheries and provide air conditioning and refrigeration. The potential industrial factories would create job opportunities and promote local economic growth.

9.5   REDUCTION IN ATMOSPHERIC POLLUTION

The most alarming effect of the conventional methods of mass power production may not be the depletion of fossil fuels but the large-scale damage to human health and the natural environment. Fossil fuel combustion, in addition to producing carbon dioxide, results in the emission of numerous undesirable and biologically harmful compounds including carbon monoxide, sulfur dioxide, nitrogen oxides, lead, hydrocarbons, and ozone. Not all pollutants are equally harmful. The air quality indices, which include the effect of a mixture of pollutants, is therefore based upon a weight average of the pollutants. A pollutant standard index is now used for reporting pollutant levels. In 1986, between 40 and 75 million Americans (Brown, 1986) were living in areas that failed to attain National Ambient Air Quality Standards for ozone, carbon monoxide, and particulates. Air pollution (the best known and most pervasive of the pollution is photochemical smog, the brown haze) causes health disorders, restricts visibility, erodes buildings, reduces crop yields, and is at least partly responsible for massive forest damage. Ozone, the most important component of smog, is the product of complex reactions between nitrogen oxides and hydrocarbons in the presence of sunlight. Nitrogen and sulfur oxides, together with unburnt hydrocarbons, are the principal components of acid rain. Acid precipitation is destroying fresh-water aquatic life and degrading marine life in coastal waters. It is also damaging forests throughout the world. The buildup of the "greenhouse gases"—carbon dioxide, nitrogen oxides, ozone, and fluorinated hydrocarbons—will probably produce the most serious long-term consequences of pollution. There is now virtual consensus among scientists that the increase in concentration of carbon dioxide in the world's atmosphere could have a serious impact on world climate if it continues to grow at present rates. The potential climate change could shift global precipitation patterns, disrupt crop-growing regions, raise sea levels, and threaten coastal cities worldwide with inundation. Some details of the greenhouse effect are given in the next section.

A substantial portion of air-polluting emissions is caused by energy-related fuel combustion. Estimated total emissions by the United States have increased steadily in the past 50 years. Although transportation accounts for only one-quarter of the total U.S. energy consumption, most of the carbon monoxide, and a major fraction of the hydrocarbon, and nitrogen oxide emissions are due to it.

The cost of air pollution is difficult to estimate. In addition to health costs, air pollution may result in damage to vegetation, corrosion and deterioration of materials, and a decline in property values. Controlling the emission of pollutants arising from energy consumption is a new technological challenge.

Pollution abatement efforts for motor vehicles have focused almost entirely on tail-pipe devices that seek to reduce exhaust emissions rather than on development of solutions that might prevent their formation in the first place. There is a strong social need for a new energy system to provide a practical and cost-effective method of supplying a nonpolluting substitute for petroleum-based fuels for transportation. Hydrogen produced by water electrolysis has been advocated as the ultimate fuel to meet the need, and a worldwide research and development effort is now in progress to find a practical method of using it in motor vehicles. Electric power for its electrolytic preparation can be supplied at reasonable cost by hydroelectric plants, but the resource is too limited to replace a significant fraction of motor fuel use. Two options would be able eventually to provide the required pollution-free electric power: large expansion of nuclear power plants or large production of OTEC plants. Only OTEC power production appears capable of satisfying public environmental requirements for safety and security of long-term operation. The logistic problems of using hydrogen as a motor vehicle fuel can be solved by combining it with ammonia to form a storable nonpolluting high-octane fuel that produces only water and nitrogen as combustion products (Avery, 1988). See the discussions in Chapters 4, 6, and 7.

9.5.1 Studies of Potential Impacts of OTEC Operation on CO_2 Emissions

Gas solubility in seawater decreases with increasing temperature. Also deep ocean water is enriched with inorganic carbon with respect to the surface waters. Operation of the OTEC condensers requires that large volumes of cold CO_2-rich water be brought to the surface. The decreased pressure and increased temperature will decrease the ability of the discharged water to retain CO_2 in solution. A net outgassing of CO_2 might occur.

At an OTEC facility, the CO_2 concentration in the cold-water effluent after discharge would approach equilibrium with CO_2 in the ambient water at the point of discharge, as a worst case in the mixed layer. The maximum CO_2 that could evolve due to OTEC operation is the difference between the CO_2 concentration in deep and surface waters. The CO_2 concentrations in surface water and at 700-m depth are approximately 2.0 and 2.4 moles CO_2/kg water, respectively. If a 100-MWe OTEC plant pumps 227 m^3/s (19.5×10^9 kg/d) of deep water to the surface, approximately 0.25×10^6 kg of CO_2 would be released each day if all excess CO_2 were outgassed (Quinby-Hunt et al., 1987). Sullivan et al. (1981) estimated 0.475×10^6 kg/d. These estimates assume that all excess CO_2 would be released. The Hawaii test program indicates that only a small fraction of the dissolved CO_2 is released.

For comparison, the amount of CO_2 released to the atmosphere by a fossil-fuel-fired power plant of the same electric power capacity is three to four times as much (Ditmars and Paddock, 1979). Release of CO_2 due to the operation of a small number of OTEC plants is not likely to cause significant effect on local or regional climate. If OTEC plants were deployed to supply approximately 10×10^6 MWe, at most 3×10^{10} kg/d of CO_2 would be released to the air. Release of CO_2 to the atmosphere by fossil-fuel plants of the same power would be approximately 10×10^{10} kg/d.

The effects of increases in CO_2 concentration in atmospheric air on global warming are uncertain but could cause a 2°C warming by the year 2040 and a 5°C warming by 2100 (Hileman, 1984) based on the past and current trend of fossil-fuel use. Such a temperature change would be accompanied by unpredictable changes in weather patterns and a rise in global average sea level, causing major shifts in agriculturally productive regions and disrupting established economic systems. With this threat to the world, one strategy, which has been examined to lessen the greenhouse effect, is a worldwide agreement on limitation of coal and oil combustion. This strategy (Hileman, 1984) would have a substantial effect on the temperature increase in the year 2100. The temperature increase would be reduced from 5 to 1°C with a total ban on fossil-fuel combustion and world power demand completely supplied by OTEC power. Increasing taxes on fossil fuels would have only a slight effect (Hileman, 1984) on CO_2 emissions. Energy demand of the world is ever increasing; so the CO_2 emission is increasing and will be increased with more conventional fossil-fuel combustion.

The speculation that energy production is now changing the Earth's climate and, with it, many of the natural systems on which humanity depends implies that food production, water supplies, forest products industries, and fisheries will all be at risk if global warming accompanies CO_2 buildup during the next several decades.

The United States urgently needs environmentally safe and inexpensive energy sources to stimulate the economy, maintain political independence, and carry on social programs. OTEC offers a near-term method of delivering substantial quantities of electricity and energy-intensive products. If large-scale OTEC power can be implemented, it will drastically cut foreign oil and gas imports; it will decrease or perhaps eliminate our deficit; it will help to stabilize the energy cost and fulfill the policy goal of energy independence; it will stimulate OTEC-related industries and generate thousands of additional jobs; it will also improve the air pollution and abate the greenhouse effect. Because of these environmental, economic, and socially attractive aspects, OTEC is one of the most promising solutions to the problems of supplying future needs for reliable, economic, and environmentally benign energy for the twentieth century and beyond.

REFERENCES

Adams, E. E., D. J. Fry, and D. H. Coxe, 1979. "Results of a near-field physical model study." *Proc. 6th OTEC Conf.*, Washington, D.C.

Allender, J. H., R. W. Blevins, G. L. Dugger, and E. J. Francis, 1978. "OTEC physical and climatic environmental impacts: An overview of modeling efforts and needs." *Proc. 5th Ocean Thermal Energy Conversion Conf.*, Miami Beach, Fla., Sept.

Avery, W. H., 1988. "A role for ammonia in the hydrogen economy." *Int. J. Hydrogen Energy* 13, 761.

——, R. W. Blevins, G. L. Dugger, and E. J. Francis, 1976. *Maritime and construction aspects of OTEC plantships*. Johns Hopkins Univ. Applied Physics Lab., SR76-1B.

——, and G. L. Dugger, 1980. "Contribution of OTEC to world energy needs." *Int. J. Ambient Energy*, 1(3), 177.

——, D. Richards, and G. L. Dugger, 1985. "Hydrogen generation by OTEC electrolysis, and economical energy transfer to world markets via ammonia and methanol." *Int. J. Hydrogen Energy* 10, 727.

Bathen, K. H., 1977. "A further evaluation of the oceanographic conditions found off Keahole Point, Hawaii, and the environmental impact of near-shore OTEC plants on subtropical Hawaiian waters." *Proc. 4th Annual Conf. on Ocean Thermal Energy Conversion,* New Orleans, La., July, IV-79.

Bell, R. E., 1980. "An island utility looks at OTEC." *Proc. 7th Ocean Energy Conf.*, Washington, D.C, June, II, 8.1.

Blumberg, A. F., and G. L. Mellor, 1979. "A whole basin model of the Gulf of Mexico." *Proc. 6th OTEC Conf.*, Washington, D.C., June II, 13.15.

Brooks, N. H., 1960. *Diffusion of sewage effluent in an ocean current*. California Inst. Technology. Research report, Pasadena, Cal.

Brown, L. R., 1986. *State of the World 1986*. New York: W. W. Norton.

——, 1989. *State of the World 1989*. New York: W. W. Norton.

Commission on Engineering and Technical Systems, 1990. *Fuels to drive our future*. National Academy Press, 0-309-04142-2, April.

Daniel, T., 1983. *Annual Report 1983*. Natural Energy Laboratory of Hawaii.

——, 1989. *New seawater delivery systems at the Natural Energy Laboratory of Hawaii*. ASME. Natural Energy Laboratory of Hawaii, Kailua-Kona, Hawaii, 323.

——, 1990. *Activities at the Natural Energy Laboratory of Hawaii/HOST*. Natural Laboratory of Hawaii Authority, Keahole Point, Hawaii.

Ditmars, J. D., and R. A. Paddock, 1979. "OTEC physical and climatic environmental impacts." *Proc. 6th OTEC Conf.*, Washington, D.C., 6a-3/1.

DOE, 1980. *DOE workshop on the potential environmental consequences of OTEC plants*, DOE Conf-800154, Upton, N.Y., Jan.

DOE/EIA, 1990. *Monthly energy review*. U.S. Department of Energy, Washington, D.C., Tables 3.1a,b, 3.4.

DOE/EIA, 1991. *Monthly energy review*. U.S. Department of Energy, Washington, D.C.

Energy Information Agency, 1989. *Monthly energy review*.

Francis, E. J., 1977. *Investment in commercial development of OTEC plantships*. Johns Hopkins Univ. Applied Physics Lab., Internal report, Dec.

Freudman, A., 1988. "Cost of foreign crude? '2.1 times the nominal price'." *Energy Daily*, **16** (184), 1.

Gauthier, M. M., 1985. *OTEC—The French program, a pilot electric power plant for French Polynesia*. IFREMER Report, March.

Gopalakrishnan, C., 1984. *The emerging marine economy of the Pacific*. Boston: Butterworth.

Hileman, B., 1984. "Recent reports on the greenhouse effect." *Environmental Science and Technology*, February, **18**(2), 45.

Homma, T., and H. Kamogawa, 1978. "Conceptual design and economic evaluation on OTEC power plants in Japan." *Proc. 5th Ocean Thermal Energy Conversion Conf.*, Miami Beach, Fla., Feb., **1**, V-91.

——, N. Kamogawa, and S. Nakasaki, 1979. "Design consideration on 100 commercial scale OTEC power plants and a 1 MWe class engineering test plant." *Proc. 6th OTEC Conf.*, Washington, D.C., June, **1**, 5.7.

Jirka, C. H., 1978. *Preliminary guideline for the siting of multiple OTEC plants*. Cornell Univ. Dept. of Environ. Eng., Ithaca, N.Y., June.

——, F. A. Johnson, D. J. Fry, and R. D. Harleman, 1977. *OTEC plant experimental and analytical study of mixing and recirculation*. Massachusetts Institute of Technology, R. M. Parsons Laboratory, MIT Report No. 231.

——, J. H. Jones, and P. E. Sargen, 1979. "Modeling of the intermediate field OTEC discharges." *Proc. 6th OTEC Conf.*, Washington, D.C., June, **1**, 13.14.

Johnson, F. A., 1989. "Small land based OTEC plant." *Proc. of the EEZ Resources: Technology Assessment Conf.*, Honolulu, Jan., 6.20.

Jones, Jr., M. S., K. Sathyanarayana, A. L. Markel, and J. E. Snyder III, 1979. "Integration issues of OTEC technology to the American aluminum industry." *Proc. 6th OTEC Conf.*, Washington, D.C., June, **1**, 10.9.

——, K. Sathyanarayana, V. Thiagarajan, J. E. Snyder, A. M. Sprouse, and D. Leshaw. "Aluminum industry applications for OTEC." *Proc. 7th Ocean Energy Conf.*, 8.5-1.

Koh, C. Y., and L. N. Fan, 1970. *Mathematical models for the prediction of temperature distributions resulting from the discharge of heated water into large bodies of water*. U.S. EPA Water Pollution Control Research Series 16130 DWO, Oct.

Kajikawa, T., and H. Hiramatsu, 1986. "Status and prospects of OTEC research and development in Japan." *Proc. 13th OTEC Congress of the World Energy Conf.*, Oct. (Exp. Mtg. #3), 1.

Krenz, J. H., 1984. *Energy Conversion and Utilization*. Second Ed. Boston: Allyn and Bacón.

Lavi, A., and G. H. Lavi, 1979. "OTEC: Social and environmental issues." *J. Energy* **4**, 833.

Lockheed Missiles and Space Co., Inc., 1975. *OTEC power plant technical and economical feasibility*. Lockheed Missiles and Space Co., April.

Martin, P. J., and G. O. Roberts, 1977. "An estimate of the impact of OTEC operations on the vertical distribution of heat in the Gulf of Mexico." *Proc. 4th Annual Conf. on Ocean Thermal Energy Conversion*, New Orleans, La., March, **IV**, 26.

Mayer, J., 1976. "The dimensions of human hunger" *Scientific American*, **253**, 40.

Miller, A. R., 1977. *Ranges and extremes of the natural environments related to the design criteria for OTEC*. Woods Hole Oceanographic Institution Report WH01-77-61, Oct.

Othmer, D. F., and O. A. Roels 1973. "Power, fresh water and food from cold, deep-sea water." *Science* **182**, 121.

Pinckert, W. F., 1980. "Overview of OTEC potential for electric power generation in Guam, Micronesia and the Western Pacific." *Proc. 7th Ocean Energy Conf.*, Washington, D.C., June, **II** c/3-1.

Pont, R. J., 1978. "Effect of changes in the relative costs of fuel and capital resources on OTEC's economic competitiveness." *Proc. 5th Ocean Thermal Energy Conversion Conf.*, Miami Beach, February, **II** 44—58.

Pritchard, D. W, J. Blanton, G. Canaday, J. Ditmars, T. Hopkins, G. Marz, G. L. Mellor, D. Pillsbury, and D. P. Wang, 1978. "Physical consequences of OTEC." Workshop on Potential Environmental Consequences of OTEC, Upton, N.Y., Jan.

——, 1980. "The physical consequence of OTEC." Potential Environmental Consequence of OTEC Plants Workshop, Upton, N.Y., Jan.

Quinby-Hunt, M. S., S. D. Sloan, and P. Wilde, 1987. "Potential environment impacts of closed-cycle OTEC." *Environ. Impact Access Rev.*, **7**, 169.

——, P. Wilde, and A. T. Dengler, 1986. "Potential environment impact of open-cycle OTEC." *Environ. Impact Access Rev.*, **6**, 7.

Roels, O. A., 1975. "The economic contribution of artificial up-welling mariculture to sea-thermal power generation." *Proc. 3rd Workshop on Ocean Thermal Energy Conversion*, Houston, Tex., May, 128.

——, 1980. "From the deep sea: food, energy, and fresh water." *J Mechanical Engineering* **102**, 37.

Rosen, S. J., 1979. "OTEC: A solution to the problem of energy versus the environment." *Proc. 6th OTEC Conf.*, Washington, D.C., 4D-5/1.

Sanchez, J. A., 1980. "The OTEC alternative for Puerto Rico versus coal-oil-nuclear." *Proc. 7th Ocean Energy Conf.*, Washington, D.C., June, **II**, 8.2.

SERI, 1985. *Thermoeconomic optimization of OC-OTEC electricity and water production plants.* Solar Energy Research Institute Report SERI/STR-251-2603 DE85121299, May.

Sullivan, S. M., M. D. Sands, J. R. Donat, P. Jetsen, M. Smooker, and J. F. Villa, 1981. *Environmental assessment OTEC pilot plants.* Lawrence Berkeley Lab., Research report LBL-12328.

Sundaram, T. R., S. K. Kapur, and A. M. Sinnarwalla, 1978. *Some further experimental results on the external flow field of an OTEC plant.* Hydronautics, Inc. Tech. report 7621-2, Laurel, Md.

Thompson, J. D., H. E. Hurlburt, and P. J. Martin, 1978. "Results from the Gulf of Mexico OTEC far-field numerical model." *Proc. 5th Ocean Thermal Energy Conversion Conf.*, Miami Beach, Fla., **1**, III-414, Feb.

Toms, A., and G. Ford, 1984. *OTEC for island applications.* Institute of Electrical Engineers, Energy Options No. 233, 220.

Trenka, A. R., 1988. *Research done by Creare R&D, Inc., under contract with SERI.* Solar Energy Research Institute, Oct.

——, L. J. Rogers, R. J. Hays, and L. A. Vega, 1989. "A status assessment of OTEC economics." *Resources Technology Assessment Conf.*, Honolulu, HI, Jan. **6**.32.

United Nations, 1984. *A guide to OTEC for developing countries.* United Nations, N.Y., United Nations report.

Walsh, J. J., 1975. "A spatial simulation model of the Peru upwelling ecosystem." *Deep Sea Research* **22**, 201.

Williams, R. H., 1975. "The greenhouse effect and ocean-based solar energy systems." Center for Environmental Studies, Princeton University, Oct. Working paper No. 21.

INDEX